Y0-BDP-830

Policy-Driven Mobile Ad hoc Network Management

THE WILEY BICENTENNIAL–KNOWLEDGE FOR GENERATIONS

$\mathcal{E}$ach generation has its unique needs and aspirations. When Charles Wiley first opened his small printing shop in lower Manhattan in 1807, it was a generation of boundless potential searching for an identity. And we were there, helping to define a new American literary tradition. Over half a century later, in the midst of the Second Industrial Revolution, it was a generation focused on building the future. Once again, we were there, supplying the critical scientific, technical, and engineering knowledge that helped frame the world. Throughout the 20th Century, and into the new millennium, nations began to reach out beyond their own borders and a new international community was born. Wiley was there, expanding its operations around the world to enable a global exchange of ideas, opinions, and know-how.

For 200 years, Wiley has been an integral part of each generation's journey, enabling the flow of information and understanding necessary to meet their needs and fulfill their aspirations. Today, bold new technologies are changing the way we live and learn. Wiley will be there, providing you the must-have knowledge you need to imagine new worlds, new possibilities, and new opportunities.

Generations come and go, but you can always count on Wiley to provide you the knowledge you need, when and where you need it!

WILLIAM J. PESCE
PRESIDENT AND CHIEF EXECUTIVE OFFICER

PETER BOOTH WILEY
CHAIRMAN OF THE BOARD

Policy-Driven Mobile Ad hoc Network Management

Ritu Chadha
Latha Kant
Telcordia Technologies

WILEY-INTERSCIENCE

A JOHN WILEY & SONS, INC., PUBLICATION

Published by John Wiley & Sons, Inc., Hoboken, New Jersey
Published simultaneously in Canada

For general information on our other products and services or for technical support, please contact our Customer Care Department within the United States at (800) 762-2974, outside the United States at (317) 572-3993 or fax (317) 572-4002.

Wiley also publishes its books in a variety of electronic formats. Some content that appears in print may not be available in electronic formats. For more information about Wiley products, visit our web site at www.wiley.com.

Wiley Bicentennial Logo: Richard J. Pacifico

Library of Congress Cataloging-in-Publication Data

Chadha, Ritu, 1964–
    Policy-driven mobile ad hoc network management / by Ritu Chadha, Latha Kant.
        p.   cm.
    ISBN 978-0-470-05537-3
1. Ad hoc networks (Computer networks)—Management.   2. Ad hoc networks (Computer networks)—Access control.   I. Kant, Latha, 1962–   II. Title.
    TK5105.77.C53 2008
    004.6—dc22

                                                                    2007026336

Printed in the United States of America

10 9 8 7 6 5 4 3 2 1

To my father, Indrajit Singh Chadha, who has been my inspiration throughout my life.
Thank you for being my mentor, supporter, cheerleader, and guide!

Ritu Chadha

To my Wonderful Family: my parents for having taught me perseverance and sincerity by example; my son and husband for their understanding, love, and support; and my brother for his confidence in me!

Latha Kant

CONTENTS

FOREWORD

This book provides a comprehensive overview of policy-based network management as applied to mobile ad hoc networks (MANETs), including technical foundations, lessons learned, standards initiatives, state-of-the-art approaches, and new research directions.

The challenge of network management is to synthesize coherent system-level behaviors from a large set of individual network elements. In policy-driven management approaches, a network operator's high-level statement of objectives is used to manage low-level control interfaces exposed by component software in the network elements. For traditional networks, policy-based approaches hold out the potential of automation leading to higher availability, superior performance, and tremendous cost savings. For MANETs, automation is a necessity and not just a desirable goal.

Mobile ad hoc networks are the linchpin of strategies for providing ubiquitous connectivity and computing in the absence of infrastructure. Defense applications for "network-centric operations" are a driving motivator, although closely related civilian applications are recently emerging in emergency response, monitoring and control, telematics, and home networking.

In the long run, simple and reliable multihop wireless transactions are likely to become an increasingly important fraction of all communications at the network edge. Even so, the technical obstacles to creating and operating adaptive, efficient, and assurable multihop radios are nearly overwhelming. Techniques and protocols that have long sufficed in infrastructure-based settings break down when applied to MANETs.

The first and most obvious casualty of network node mobility was traditional routing, and much of the early work on MANETs focused on new adaptive routing protocols. However, recently it has become clear that mobility in wireless networks has a deep impact on more than just the routing layer. The combined peculiarities of mobility and radio-frequency communication are now driving a deep reexamination of wider assumptions behind conventional networking.

For instance, researchers have recently demonstrated "cross-layer" designs that offer performance improvements by breaking down the separation of concerns implicit in conventional layered network models. Thus the conventional ISO-OSI 7-layer model may not be the most appropriate basis for a MANET software architecture. In addition, hop-by-hop mechanisms for congestion control appear to be better at avoiding hotspots and queue instability in wireless networks than end-to-end protocols. Thus the end-to-end designs that have been the basis for progress in the Internet may not be sufficient for MANETs. And finally, although the balance of evidence in favor of network coding has not yet tipped, it is possible that future wireless networks may even do altogether away with the traditional idea of packet routing and forwarding in favor of algebraically based information diffusion.

Researchers have much to do before the job is done, but unfortunately the need for MANETs is immediate and we do not have the luxury of waiting for theory to catch up with practice. And herein lies the importance of this book, which approaches the problem of MANETs from an entirely different angle: that of the network operator. Never mind that today's MANET designs are still evolving; there are, nevertheless, people tasked with the job of designing and operating such networks with minimal manual intervention. These are the people that will necessarily have an interest in policy-driven MANETs.

How does an operator view MANETs? Principally as a challenge in network management, with attention to subtopics such as fault management, configuration, accounting, performance, and security. Network operators have always faced the unenviable task of making up for shortfalls in design. The best example is fault tolerance, where operators are the last line of defense against software and hardware flaws, and good operating practices can go a long way to providing reliable solutions based on unreliable components. Policy-based network management is an emerging framework whereby operators can provide flexibility and adaptivity even to software systems that are not inherently adaptive; security to software that was not necessarily created with authorization and authentication in mind; and performance to software that leaves open many degrees of freedom.

Thus operators work from the top down to make the best of whatever solutions are available while at the same time researchers and software engineers work from the bottom up to provide inherently adaptive MANET solutions. Policy-based systems will define the boundary between these two activities: between territory that is theoretically well understood and territory that is not. This boundary is concretely defined by the network's sensing and configuration parameters, which are exposed by software developers precisely because there is no a priori understanding of how to tune these to the operating scenario at hand. Whenever a configuration parameter exists, it is because the software developer is passing a responsibility to the network operator.

A modern network has typically hundreds or thousands of parameters that need to be managed by the operator, and these parameters are rife with feature interactions. It should be no surprise that configuration is one of the largest of operating expenses for traditional telecommunication providers, that misconfiguration ("human error") is one of the most substantial sources of network downtime, and that misconfiguration is at the core of many security vulnerabilities. The challenge is even harder because these configuration parameters are an uncoordinated accident arising from the fact that every network is a system comprising many independent parts, often created by independent designers and manufacturers.

It is no secret that even traditional network operators find tasks daunting, even with the luxury of months or years to plan networks that will ultimately be highly static and predictable in nature. What then can be expected of mobile ad hoc networks that need to be deployed and operated in even more challenging and volatile conditions? MANETs exponentiate the difficulty of network management.

Manual management techniques that are barely tolerable in wired networks are entirely unworkable for MANETs, so we need an entirely new approach. The approach explored in this book is that of policy-driven management. Policy-based network management provides an architectural framework for bridging the gap between what software designers provide and how operators will use that software.

It is clear that at least three communities should find this book of practical and immediate interest. The first are those systems architects who must present the operator with a manageable overall system. The second is the community of operators, who must understand and use the high-level configuration capabilities of their policy-based MANET to meet the immediate needs of the users. And third, anyone developing component technologies for MANETs needs to understand the context in which their systems will be coordinated with other systems.

Moreover, this book should also be immensely interesting to those trying to decide what MANET research is worth undertaking and why. MANET protocol designers and theoreticians will be fascinated by this material because they will be able to see precisely what kinds of configuration parameters are not considered by operators and systems architects to be adequately handled by the fundamental MANET algorithms. Research might be undertaken to reclaim responsibility for parameters that are currently left in the hands of operators, and this book provides a roadmap of such concerns. Perhaps even more parameters associated with adaptation could lend themselves to direct and optimal algorithmic control, simplifying the remaining tasks for policy-based management. Conversely, wherever the policy enforcement algorithms are adequate, there may be little point in creating specialized protocols.

Beyond direct algorithmic control, there is the wider question of overall "design for manageability" that is really a software engineering problem in its own right, independent of any application in networks. What techniques, especially at the system level, will dramatically simplify the task of operators and therefore also simplify complexity of the policy-based framework? Within networking, current lines of research on "optimization decomposition" and "network utility maximization" hint at techniques for co-design of the exposed parameters and algorithms for automated adaptation. These techniques are interesting because they cross component boundaries and result in new, more manageable designs overall. There are doubtless many design patterns and techniques that component designers and system architects could use to simplify the remainder of policy issues. One productive perspective from which to study this book is to consider those opportunities.

Although the material in this book is of primary interest to those with a charter to create effective mobile ad hoc networks, the implications of successful MANET designs can be expected to have wider application. MANETs may be a niche, but it is often the case that niche solutions are the basis for disruptive innovation in much wider context. In this case, it is well known that network management is a substantial problem for all operators; any successful technique for automated management that is able to handle the particularly challenging case of MANETs has a substantial potential to be useful in less challenging instances as well. A solution to the hard problem may also address the "easy" problem.

At one point in the twentieth century, it seemed as though a large fraction of the population would need to serve as telephone switchboard operators if the growth in telecommunications continued. The invention of mechanical and digital switching changed the nature of the game. As anyone who manages even a small home network knows, we will similarly need to find new solutions to network management if the ideal of ubiquitous, reliable, secure computing and communications is to be realized. MANETs are at the leading edge of a revolution in adaptive networking protocols, and policy-based architectures are at the leading edge of a revolution in automated management.

For researchers, practitioners, entrepreneurs, and investors who are asking whether the effort to understand and conquer specialized MANET challenges is worthwhile, the answer must depend on whether MANETs are truly a specialized one-off niche, or simply the most pointed evidence that longstanding problems in network management must be solved in order to make overall progress in telecommunications. There is little doubt that the management issues which are taken to an extreme by MANETs are the same management issues that plague the operators of simpler networks. And there is little doubt that a future which truly does offer ubiquitous, secure, and reliable communications and computing will presuppose solutions to the self-configuration challenges brought into sharp relief by MANETs.

Washington, DC   J. CHRISTOPHER RAMMING

About the writer: Chris Ramming is presently a program manager in the Defense Advanced Research Projects Agency (DARPA) Strategic Technology Office. His primary charter is to vastly improve wireless network performance, self-management, and intrinsic security. Prior to joining DARPA, the bulk of Mr. Ramming's career was with AT&T/Bell Labs Research. He has a broad interest in telecommunications technology and has performed work on multimedia softswitches, WWW service creation, networking, provisioning, domain-specific programming languages, decision analysis, and large-scale software engineering. Mr. Ramming holds degrees in computer science from Yale College and the University of North Carolina, Chapel Hill, The views expressed in this foreword are his personal views only and do not necessarily reflect the official views of DARPA or the U.S. government.

PREFACE

Mobile ad hoc networks (MANETs) are rapidly gaining in importance, in both the commercial and the military arenas. In order to deploy and maintain these networks effectively, it is imperative that appropriate network management techniques and tools be used. While a significant amount of research has been dedicated to the development of *networking technologies* for MANETS over the past decade, not much attention has been paid so far to the unique *management needs* of these networks. This is mostly due to the fact that since MANET technologies are relatively new, the bulk of the research efforts in the area of MANETs were concentrated on solving fundamental problems in MANET networking, such as routing, mobility management, transmission schemes, and so on. However, now that technologies for implementing and deploying MANETs are becoming more mature, it is time to turn our attention to the effective *management of MANETs*.

MANETs differ fundamentally in both functionality and capability from their static wireline network counterparts due to a variety of reasons, including random node mobility, unpredictable network dynamics, fluctuating link quality, limited processing capabilities, power constraints, and so on. All of these characteristics give rise to a need for dynamic changes in both the functioning and management of the underlying network. Furthermore, unlike wireline networks, the dependencies or relationships between network elements are not fixed. Due to node mobility, network links are dynamic and of unpredictable quality, resulting in intermittent connectivity. Finally, an underlying problem is the scarcity of wireless network bandwidth. Unlike today's wireline networks, where bandwidth is plentiful and links are reliable, mobile ad hoc networks typically have very limited bandwidth and are relatively less reliable, due to environmental effects.

A great deal of work has been done in the area of network management under the auspices of standards bodies such as the ITU-T (International Telecommunication Union—Telecommunication Standardization Sector) and the IETF (Internet Engineering Task Force). The developed standards have resulted in valuable architecture definitions, abstractions, protocols, and process models for managing different types of networks. However, one aspect that has not been adequately addressed is the automation of network management and the integration of different management functions. The traditional network management functions include Fault, Configuration, Accounting, Performance, and Security (FCAPS) management. To see why these functions need to be performed in coordination with each other, consider the following illustrative examples. Statistics collected for fault and performance management can be processed and analyzed by various management applications, leading to diagnoses of network problems. In order to fix such problems, there may be a need for repairing or replacing faulty equipment (in the case of network faults); or it may be necessary to reengineer the network to add more capacity to deal with severe network congestion problems; and so on. Such requirements need to be addressed manually, due to the need to physically install or repair equipment in the field. However, many performance and some fault problems can be handled by network reconfiguration. For example, if a network link is severely congested, it may be possible to alleviate the problem by sharing the traffic load with another underutilized link, or by reducing the amount of management traffic, and so on. This can be accomplished simply by appropriately reconfiguring the network. Today, this is done manually by experienced network operators who examine outputs from fault and performance management systems and decide how to reconfigure the network appropriately. This is the fundamental problem with network management today: *There is too much human intervention required to run a network*. In order to reduce the cost of network operations, it is critical that human intervention be minimized by creating a feedback loop between fault/performance monitoring systems and configuration systems and by specifying policies that regulate how the system should be reconfigured in response to various network events.

Clearly, the above points are as valid for wireline networks as they are for MANETs; in other words, the lack of automation in network management is a problem for both wireline networks and MANETs. However, the critical point to note here is that *this lack of automation poses a much greater problem for MANETs than for wireline networks*. This is because *the characteristics of MANETs result in a requirement for much more frequent network reconfiguration and much more stringent monitoring* than needed in wireline networks. The characteristics of MANETs, such as the lack of fixed infrastructure, dynamic topology, power and processing constraints, intermittent connectivity, varying security requirements, scarce bandwidth, and the existence of high-loss, unreliable links, result in a need for much more network management control than is typically needed in a static wireline network, where network elements can be configured in a certain way and rarely need to be reconfigured. The heightened need for network management control stems from the fact that the network must be maintained in a functional state in spite of the dynamic nature of the network. Additionally, this increased network management control must be automated as far as possible, because otherwise the amount of manual intervention required to keep the network functioning optimally would result in prohibitive cost, making MANETs uneconomical and impractical to deploy.

This book discusses the management challenges associated with ad hoc networks, and it provides an in-depth description of how policy-based network management can be used for increasing automation in the management of mobile ad hoc networks. It describes the required components of a network management solution for such networks, using a policy-based management framework that integrates the traditional network management components (FCAPS, i.e., Fault, Configuration, Accounting, Performance, and Security management).

WHY IS THIS BOOK NEEDED?

The field of mobile ad hoc networking is gaining momentum in both the commercial and the military arenas. The IETF has several working groups devoted to various ad hoc networking issues, namely autoconf (ad hoc network auto-configuration) and MANET (mobile ad hoc networks), and several books devoted to mobile ad hoc networking have appeared recently (e.g., see Murthy and Manoj [2004], Aggelou [2004], Mohapatra and Krishnamurthy [2004], Toh [1997, 2002], Ilyas [2002], and Santi [2005]). However, relatively little work has been done in the area of network management that is specifically targeted at mobile ad hoc networks. Even in the area of traditional network management, there are very few textbooks that provide a comprehensive view of network management. Most of the books in the area of network management have a very narrow focus and primarily discuss SNMP [Case et al., 1990] and related MIBs (management information bases). There are no books currently on the market that address network management for mobile ad hoc networks. This can partly be explained by the fact that ad hoc networking is a relatively new field, and therefore the networking issues were the first ones that needed to be addressed by the networking community. However, now that the basics of ad hoc networking have been ironed out and standards for ad hoc networking are emerging, there is a void in the area of network management techniques for these networks. This book is the first of its kind in the marketplace and fills this void.

TARGET AUDIENCE

This book is targeted at professionals, researchers, and advanced graduate students in the field of IP network management who are interested in mobile ad hoc networks in particular. Not only does it open up key challenges in the areas of ad hoc networking and ad hoc network management, it also highlights important research aspects in this area that can be of potential interest to advanced researchers in the field of IP networking and network management, as well as to graduate students. Furthermore, there is a growing research community that is interested in mobile ad hoc networks. In particular, the U.S. military is spending billions of dollars on deploying mobile ad hoc networks for enhancing its warfighting capabilities, as well as hundreds of millions of dollars on building management systems for these networks. In fact, since a detailed understanding of the salient aspects of ad hoc networking has only recently been ironed out, there is now a critical need to turn the attention to managing the ad hoc networks that are being deployed.

To sum up, this book provides readers with an understanding of:

- Mobile ad hoc networking.
- Network management requirements for MANETs, with an emphasis on the differences between the management requirements for MANETs and those for static, wireline networks.
- The use of policies for managing mobile ad hoc networks to increase automation and to tie together management components via policies.
- Policy conflict detection and resolution.
- The aspects of mobile ad hoc networking that need to be configured and reconfigured at all layers of the protocol stack, including a discussion of interdomain policy management.
- Methodologies for providing service survivability in the face of both hard (deterministic) and soft (stochastic) failures in MANETs.
- The components of a quality of service (QoS) management solution for MANETs based on differentiated services (DiffServ).
- The intricacies of managing security in an ad hoc network environment.
- Important open research issues in the area of MANET management.

ABOUT THE AUTHORS

Ritu Chadha is Chief Scientist and Director of the Policy Management research group in Applied Research at Telcordia Technologies, where she has been working since 1992. She was the program manager for the U.S. Army CERDEC DRAMA (Dynamic Re-Addressing and Management for the Army) project, which focused on the design, prototyping, and field demonstration of a policy-based network management system for mobile ad hoc networks. She is also the Chief Engineer for Telcordia's Future Combat Systems (FCS) Network Management System project and is responsible for delivering Policy Management, FCAPS functionality, QoS, and Mobility Management for FCS. Dr. Chadha is a well-known expert in the area of policy management and has published over 50 refereed papers in journals and conferences. She was part of the IETF and DMTF teams that created the Policy Framework standards, and she co-authored IETF RFC 3460 [Moore, 2003], which describes extensions to the Policy Core Information Model. She spearheaded the development of Telcordia's policy-based network management effort and led a multi-million dollar, multi-year internal investment program within Telcordia to develop policy-based management technologies for Telcordia's customers. She received her Ph.D. in Computer Science from the University of North Carolina at Chapel Hill in 1991.

Latha Kant (Ph.D., Electrical and Computer Engineering) is Director of the Mobility Management research group in Applied Research at Telcordia Technologies, where she has been working since 1996. Her research interests and expertise span wireless ad hoc networking and network management, as well as performance modeling and analyses of MANETs. She has led and managed several research projects at Telcordia. Currently, she is the technical lead of the QoS team within Telcordia's Future Combat Systems (FCS) program, where she is leading a team of researchers to develop end-to-end QoS solutions for heterogeneous multilevel security networks. She has also led several projects on MANET performance modeling and analysis including Telcordia's internal research and development efforts on scalable mechanisms for MANET performance analysis and the CERDEC DRAMA subtask on scalable modeling and analysis of a dynamic policy-based network management system for MANETs. Dr. Kant is also currently leading a research effort on Network Science as part of the Army Research Laboratory (ARL) Communications & Networks (C&N) task on Survivable Wireless Mobile Networks (SWMN) with a focus on fundamental research in the area of MANETs. Prior to joining Telcordia Technologies, Dr. Kant worked with the communications systems division of the India Space Research Organization (ISRO), subsequent to completing her Bachelors of Engineering in Electrical and Electronics Engineering. As a research scientist with ISRO, she was part of the team that designed on-board and ground communications systems for satellite networks. Dr. Kant has published over 40 papers in journals and conferences, and she holds a patent on self-healing mechanisms in packet-switched networks.

RITU CHADHA
Telcordia Technologies
LATHA KANT
Telcordia Technologies

ACKNOWLEDGMENTS

As we journeyed through our book-writing adventure, there were several individuals who helped us in many ways. We begin by first acknowledging our friends and colleagues at Telcordia, who selflessly devoted long hours to reviewing various chapters of this book and providing us with their valuable comments, despite their numerous other pressing obligations. Thanks go out to (in alphabetical order): Jason Chiang, for providing useful observations on the policy chapters; Praveen Gopalakrishnan, for his helpful comments on performance management; Yitzchak Gottlieb, for his careful editing of the policy chapters; Flossie Hsu, for her many insights as an experienced systems engineer in fault and performance management; Petros Mouchtaris, for his review of security management that drew from his many years of experience in the field of security for ad hoc networks; Alex Poylisher, for his critical review, corrections, and valuable insights on policy management, stemming from his extensive work in this field; Gail Siegel, for giving us a useful perspective of this work from a different angle; and Michelle Wolberg, for her wonderful enthusiasm and willingness to review anything we asked her to and providing very insightful feedback based on her rich experience in the field of network management.

Next, many thanks are due to our colleagues at external institutions as well as at Telcordia who were kind enough to review our book proposal and provide useful feedback to shape the contents of this book. Again in alphabetical order, we would like to thank Professor J.J. Garcia-Luna-Aceves, Dr. S. Kandaswamy, Mr. Mohammed Afzal Khan, Dr. Bikash Sabata, Professor Adarsh Sethi, and Dr. Tsong-Ho Wu for their prompt and valuable feedback. Adarsh also reviewed and provided insightful comments on the fault management chapter, and Afzal was kind enough to review the Foreword and provide very timely and useful feedback.

Special thanks go out to Chris Ramming for his careful review of the book along with his constructive feedback and, most of all, for agreeing to write the Foreword for this book.

Last, but not least, we would like to acknowledge our families. Despite the geographical distance, our parents, Latha's brother Srikanth, and Ritu's sisters Kamal and Priti were a constant source of inspiration and encouragement. Our immediate families bore the brunt of the long evening and weekend hours spent working on this book. Ritu's daughter Trisha and Latha's son Shawn unfortunately had to hear the phrase "not now, Mom is busy" one too many times over the past year. Our husbands Arindam and Arun patiently put up with our frequent absences from social events and family obligations, in addition to long hours spent burning the midnight oil. We hope that the result of those long hours—this book—makes it all worthwhile.

RITU CHADHA
Telcordia Technologies
LATHA KANT
Telcordia Technologies

LIST OF FIGURES

1

INTRODUCTION

Mobile ad hoc wireless networks differ fundamentally in both functionality and capability from their static wireline network counterparts due to a variety of reasons, including random node mobility, unpredictable network dynamics, fluctuating link quality, limited processing capabilities, and power constraints, to name a few. All of these characteristics give rise to a need for "on the fly" changes with respect to both the functioning and the management of the underlying network. Furthermore, unlike wireline networks, the dependencies or relationships among network elements are not fixed. Due to node mobility, network links are dynamic and of unpredictable quality, resulting in intermittent connectivity. Finally, a pervasive problem is the scarcity of wireless network bandwidth. Whereas bandwidth is plentiful and links are reliable in wireline networks, mobile ad hoc networks typically have very limited bandwidth and are relatively less reliable, due to environmental effects. Thus traditional network management techniques used for wireline networks are inadequate for mobile ad hoc networks.

Policy-based network management is a new paradigm that holds the promise of enhanced network management automation. One of the most pressing problems in network management today is the lack of automation in configuring and reconfiguring networks. Current network management systems lack the ability to state long-term, network-wide configuration objectives and have them automatically realized in the network. A policy-based network management system, on the other hand, allows the network operator to enter these objectives as policies into the management system, and it ensures automatic enforcement of these policies so that no further manual action is required on the part of the network operator.

Another aspect that has not been adequately addressed by currently available commercial network management systems is the feedback loop between configuration and fault/performance management. Many performance and

Policy-Driven Mobile Ad hoc Network Management, by Ritu Chadha and Latha Kant
Copyright © 2008 John Wiley & Sons, Inc.

some fault problems can be handled by network reconfiguration. For example, if a network link is severely congested (a "soft" failure), it may be possible to alleviate the problem by sharing the traffic load with another underutilized link. This can be accomplished simply by appropriately reconfiguring the network. Today, this is done manually by experienced network operators who examine outputs from fault and performance management systems and decide how to reconfigure the network appropriately. This may be acceptable in static network environments, where network loads and topologies are predictable and where problems such as congestion rarely occur; however, in mobile ad hoc networks, due to the rapidly changing network environment, a much higher degree of automation is required. Thus, in order to be able to automate recovery from hard and soft failures, it is necessary to minimize human intervention by creating a feedback loop between fault/performance monitoring systems and configuration systems and by specifying policies that regulate how the system should be reconfigured in response to various network events. A management system using policy-based control can be used to complete the feedback loop between network monitoring and network reconfiguration.

Since a coordinated response is required to deal with a variety of events being generated both within and from outside the network, there is a need to be able to specify and store policies about the appropriate responses to events in the management system. The management system must be able to react automatically to network events by performing actions described in such policies. These policies can be created ahead of time by the network operator; once they are created and stored as part of the management system, the latter can automatically enforce these policies. This takes the human out of the loop and allows nearly fully automated, or "lights-out," operations. Note that such a system not only holds immense potential for significantly reducing network operations costs, but also is crucial in reducing response times and improving overall user satisfaction by automating recovery actions via a dynamic policy-based network management system.

Needless to say, the automation of network management operations is fraught with challenges. These challenges are especially daunting for mobile ad hoc networks, since they have a high degree of randomness. For example, since the underlying network is akin to a complex stochastic process, techniques to diagnose and correlate network problems and recommend judicious solutions are indeed challenging. This book discusses the management challenges associated with ad hoc networks and provides an in-depth description of how policy-based network management can be used for managing ad hoc networks. The next section describes the characteristics of different types of wireless networks, including cellular networks, wireless local area networks, and, finally, mobile ad hoc networks. Section 1.2 provides an overview of network management. Existing paradigms for network management are reviewed and examined in light of the management requirements for mobile ad hoc networks, also known as *MANETs* (an acronym for mobile ad hoc networks). The need for a new network management paradigm for MANETs is explained at the end of this section. A brief summary is provided in Section 1.3, along with a road map for the remainder of this book.

1.1   CHARACTERISTICS OF WIRELESS NETWORKS

Wireless networks are not new, and have been in use all over the world for decades. This section provides a brief overview of the types of wireless networks that are most familiar to users today: cellular networks and 802.11 networks, also known as wireless LANs (local area networks), or Wi-Fi (wireless fidelity). The purpose of this overview is not to give an in-depth tutorial about these networks, but rather to provide a foundation for differentiating such networks from mobile ad hoc networks, or MANETs. This is followed by an overview of mobile ad hoc networks, including their salient features.

The common characteristic of most wireless networks is that they support *mobility*. It is the extent to which they support mobility that varies. As will be discussed in this section, cellular networks support terminal (e.g., a cellular telephone handset) mobility and seamless handoff, or uninterrupted connectivity, for a user moving from one cell to another. Wireless LANs also support terminal (e.g., a laptop) mobility, but typically the mobility is much more limited than in the case of cellular networks. However, both of these types of networks require *fixed infrastructure*; in other words, components are required that are immobile and provide service to the mobile terminals. Mobile ad hoc networks, on the other hand, required *no fixed infrastructure*. It is this lack of fixed infrastructure that makes mobile ad hoc networks both (a) very useful in terms of their capabilities and (b) extremely challenging in terms of setup and maintenance.

1.1.1   Cellular Networks

1.1.1.1   Overview. Cellular networks (Figure 1.1), as the name indicates, are made up of a number of *cells*, where each cell is served by a fixed transmitter at a base station (also called a cellular tower). These transmitters provide radio coverage to the entire cell. Base stations are distributed in a manner that provides complete, and often overlapping, radio coverage to a given geographic area. Within this geographical area, transceivers are used by people to access the services provided by the network. The most commonly used service is mobile telephone service, also known as cellular telephone service. Cellular base stations connect to the public switched telephone network (popularly called the PSTN), thereby connecting mobile cellular telephone users with the rest of the world.

A carrier that operates a cellular network must purchase a license for a chunk, or block, of radio frequencies for its use. These frequencies can then be divided into smaller blocks for use by each of the base stations operated

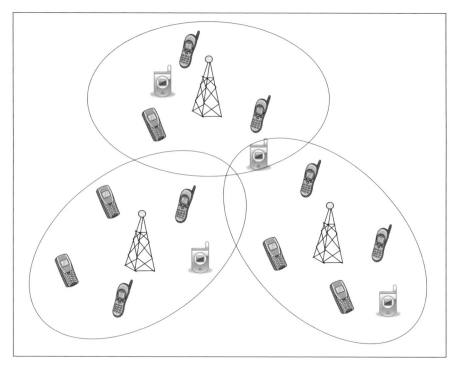

Figure 1.1. Cellular network.

by the carrier. Users within range of a base station share the frequency allocated to that base station.

Cellular networks use different underlying technologies to transmit signals. Frequency division multiple access (FDMA) uses different frequencies for adjacent cells, so that communications from one base station do not interfere with communications with nearby base stations; frequencies can be re-used in cells that are far enough apart that their signals do not interfere with each other. Code division multiple access (CDMA) uses a more complicated signaling mechanism that makes use of terminal-specific pseudonoise (PN) codes embedded within signals to differentiate communications belonging to one handset from those belonging to another. Base stations use broadcast mechanisms to set up channels for different handsets (or transceivers) to communicate with the base station.

1.1.1.2 Handling Mobility. In order to handle mobility, users (along with their handsets) need to be able to move from one cell to another without dropping calls. The mechanism for the handoff from one cell to another varies based on the technology used. In general, a channel is reserved for a user in a neighboring cell when the user approaches that cell, and the user's handset is then seamlessly transferred to the new channel, served by the base station for the new cell.

A critical service that must be provided by any mobile network is the ability to find a mobile user. For cellular networks, where telephony is the dominant service, users are "addressed" by their telephone numbers. Every telephone number is assigned to a fixed switching center, to which any incoming call is routed. When a given handset is turned on, it registers with the nearest base station (or the base station that has the strongest signal in that location). This informs the base station that the handset is in its cell. This information is conveyed by the base station to the switching center associated with the telephone number for that handset. Thus the switching center associated with a given telephone number always knows in which cell the corresponding handset is located—provided, of course, that the handset is switched on. When a call is placed to that telephone number, the corresponding handset is expected to ring and receive the call. This is accomplished as follows. Recall that every telephone number is assigned to a fixed switching center, to which any incoming call is routed. This switching center has knowledge of the cell in which this telephone number is currently registered, and it can therefore send a message to the base station for that cell to contact the corresponding handset and cause it to ring.

1.1.2 Wireless Local Area Networks

1.1.2.1 Overview. Wireless LANs (Figure 1.2), more popularly known as "Wi-Fi" (wireless fidelity), are in widespread use all over the world. They are used in homes, businesses, recreational areas (such as coffee shops), and airports—in other words, anywhere that a person with a mobile device such as a laptop computer may need Internet access.

Wireless LANs use 802.11 radio technology. 802.11 is a suite of standards developed by the IEEE for wireless access. As with cellular networks, wireless LANs depend on an *access point* with a transmitter to connect mobile users within range with other devices; typically, the access point connects to the Internet. 802.11 technology can also be used to form mobile ad hoc networks; however, this section focuses on the use of 802.11 to form wireless LANs.

Every access point periodically broadcasts its identifier, called a service set identifier (SSID), which can be heard by any client who has an 802.11-compatible device. The client can then choose to connect to the access point. This connection may be secured via encryption and access restricted to a certain set of users; or it may be open to all clients, depending on the configuration of the access point. For added security, some access points may not even broadcast their SSIDs; in such a case, a client wishing to connect to this access point must be preconfigured with the SSID of the access point.

The 802.11 networking standards include:

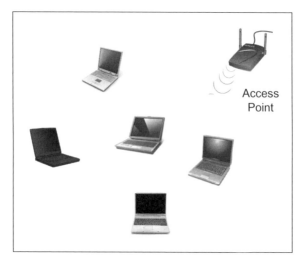

Figure 1.2. Wireless LAN.

- 802.11b: This was the first version to reach the marketplace, and it operates at 11 Mb/s.
- 802.11g: This version of 802.11 is substantially faster than 802.11b, and it transmits at up to 54 Mb/s.
- 802.11a: This version also transmits at up to 54 Mb/s; the main difference between 802.11g and 802.11a lies in the details of the physical layer, in particular the details that are associated with the symbols at the physical layer that correspond to the automatic gain control and frequency estimate for carrier signal.

One of the chief advantages of the 802.11 suite of standards is that standards-compliant implementations are universally interoperable—unlike cellular telephony where the proliferation of different standards renders interoperability among carriers difficult or impossible.

The major limitation of an 802.11 network is its limited range. Typical coverage ranges from 150 feet indoors to 300 feet outdoors, although these numbers vary greatly based on the frequency ranges used. It should be noted that longer ranges have been achieved in the neighborhood of 180 miles using special antennas. Thus, although some of the updated versions of 802.11 now include both "infrastructure" and "ad hoc" modes, range limitations restrict the usage of the 802.11-based networks for implementation of MANETs such as those envisioned for military deployments, or for use in disaster relief efforts, where the usage areas are typically spread over a large geographic region.

1.1.2.2 Handling Mobility. When a terminal (such as a laptop computer) connects to an access point, it typically obtains an IP address from the access point via DHCP (see Chapter 6, "Configuration Management," for more details about the functioning of DHCP). IP routing takes care of ensuring that any communications destined for that IP address are correctly routed to this laptop device. However, if a terminal moves from one access point to another and obtains a different IP address from the new access point, there is a need for a way to ensure that data sent to its original IP address continue to reach the mobile terminal. In order to handle such mobility of a terminal between access points, standards such as Mobile IP have been developed. Mobile IP (MIP) was defined by the IETF [Perkins, 2002; Johnson et al., 2004] and provides a simple method for locating terminals that change IP addresses. The concept of a *home network* is used, where a *home agent* is responsible for redirecting communications to the mobile terminal. The mobile terminal obtains an IP address, known as its *home address*, from its home network. When the terminal roams to a new network, it must register its new IP address with its home agent, which redirects messages destined to the roaming node's home address to the roaming node's latest location using a *foreign agent* as a proxy, as described below.

When a mobile terminal determines that it has moved to a foreign network, it obtains a *care-of address* on the foreign network. The care-of address can be determined either from a foreign agent's advertisements (a foreign agent care-of address) or by some external assignment mechanism such as DHCP (a co-located care-of address that is assigned to the mobile terminal itself). The mobile terminal operating away from home then registers its new care-of address with its home agent, possibly via a foreign agent. Datagrams sent to the mobile terminal's home address are intercepted by its home agent, tunneled by the home agent to the mobile terminal's care-of address, received at the tunnel endpoint (either at a foreign agent or at the mobile terminal itself), and finally delivered to the mobile terminal.

When away from home, Mobile IP uses protocol tunneling to hide a mobile terminal's home address from intervening routers between its home network and its current location. The tunnel terminates at the mobile terminal's care-of address. The care-of address must be an address to which datagrams can be delivered via conventional IP routing. At the care-of address, the original datagram is removed from the tunnel and delivered to the mobile terminal. In the reverse direction, datagrams sent by the mobile terminal are generally delivered to their destination using standard IP routing mechanisms, not necessarily passing through the home agent.

1.1.3   Mobile Ad hoc Networks

1.1.3.1   Overview. Mobile ad hoc networks (MANETs) are wireless networks without any fixed infrastructure. Thus, unlike cellular networks and wireless LANs, they do not rely on the existence of fixed towers or access points for relaying communications to the rest of the world. These types of networks are rapidly gaining importance due to the emerging need for rapid

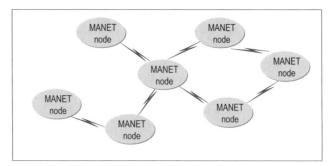

Figure 1.3. Mobile ad hoc network.

deployment of communications in areas without any preexisting infrastructure. The need for rapid deployment implies that there is no time to build the necessary infrastructure. These networks can be used for military operations, emergency rescue operations, disaster relief, temporary events such as concerts and rallies, and so on. MANETs can consist of a handful of nodes, or can contain hundreds or even thousands of nodes with differing capabilities and disparate communications requirements. Figure 1.3 shows a collection of wireless nodes networked to form a MANET.

The lack of fixed infrastructure results in the following requirement: Every node in a mobile ad hoc network must be capable of functioning as a router and must relay data for other nodes in the network, in addition to handling its own communications. Furthermore, functions such as routing and topology discovery must be handled in a distributed fashion, with no centralized infrastructure. Summarized below are a list of the key features that distinguish MANETs from their wireless network counterparts discussed in Sections 1.1.1 and 1.1.2.

- *No Fixed Infrastructure*: As mentioned above, there is no fixed infrastructure in a MANET. This means that routing must be performed by every node in a MANET; these nodes themselves may disappear or move out of range.
- *Dynamic Topology*: Although certain types of MANETs such as sensor networks are not always mobile, in general MANET nodes are mobile. This means that the topology of a MANET changes dynamically. This creates challenges for link scheduling, routing, organization of nodes into domains, and so on.
- *Power and Processing Constraints*: The nodes in a MANET may have limited battery power, CPU resources, storage capacity, and so on. These limitations may need to be taken into account when designing protocols and services for these environments. Note that limited power and processing constraints are also characteristics of cellular handsets and sometimes even of the terminals in wireless LANs; however, the impact of these

limitations is much greater in the case of MANETs. This is because MANET nodes must perform routing functions in addition to end host functions. The impact of the failure of one MANET node is therefore likely to extend to many other nodes in the MANET, if the node that failed was routing traffic for other MANET nodes. In contrast, the failure of a terminal in a cellular network or a wireless LAN only impacts the user of the failed terminal. The cellular towers and access points for cellular and Wi-Fi networks have no power and processing restrictions and can therefore continue to provide services to the remaining terminals in the network.

- *Intermittent Connectivity*: Due to mobility and limited battery power, nodes in a MANET may frequently get disconnected from the rest of the network. If a node moves out of range of all the other nodes, it will no longer be able to communicate with the other nodes. Similarly, if a node has limited power and wants to conserve energy, it may choose to periodically shut down, or "go to sleep," thus disconnecting itself from the network. This creates a challenge for the other nodes in the network, because they may have been relying on this node for routing purposes.

- *Varying Security Requirements*: MANETs use wireless communications and therefore will typically need to secure over-the-air communications from snooping. The extent of security will differ based on the application; for example, a military mission may require much more stringent security mechanisms than would an emergency disaster relief scenario.

- *Scarce Bandwidth*: Unlike today's wireline networks, where bandwidth is plentiful and links are reliable, mobile ad hoc networks are typically characterized by very limited over-the-air bandwidth. This results in a plethora of challenges that do not necessarily apply to the high-bandwidth Internet environment. Since bandwidth is so limited, reduction of bandwidth overhead is a critical consideration for all MANET protocols and management mechanisms. Again, cellular networks and wireless LANs also exhibit a scarcity of bandwidth, but the critical distinction here is that in cellular networks and wireless LANs, the bandwidth constraints are limited to the "access" portion of the data path (i.e., the path from the mobile handset to the cellular tower, or from the laptop or other 802.11 device to the access point). In MANETs, on the other hand, all the links connecting MANET nodes exhibit this characteristic, which translates to a scarcity of bandwidth for all links that traffic can traverse.

- *High-Loss, Unreliable Links*: Unlike the Internet, high loss and unreliability of links is the rule rather than the exception in MANETs. This challenges some fundamental assumptions that are embedded in existing Internet protocols. As an example, the TCP protocol is designed with the assumption that any network loss is caused by congestion; thus when packets start getting dropped, TCP reduces the size of its transmission window as a congestion control mechanism. However, this may be exactly the *wrong* thing to do in an environment where losses due to poor link quality are very high! Thus the nature of the MANET environment challenges some very fundamental assumptions made in existing Internet protocols.

While the last point above is an example of a challenge introduced in the data plane, the noisy and lossy nature of MANET links introduces a plethora of challenges in the management plane as well. For example, the root cause analysis (RCA) techniques in fault management systems designed for traditional wireline networks use a deterministic model, with the RCA algorithms requiring complete knowledge of the fault symptoms. However, due to the noisy nature of MANET links, it is unrealistic to assume that all of the symptoms and conditions are available to an RCA algorithm, since the fault reports and symptoms can get lost in transit. Furthermore, due to the inherent dynamic nature of MANETs, the RCA algorithms for MANETs should be probabilistic in nature—unlike their wireline counterparts, which are predominantly deterministic in nature.

Thus, as can be seen, MANETs are fundamentally different in terms of both functionality and capability from wireline and other wireless networks. These differences have spawned a significant amount of research in the field of mobile ad hoc networking. However, most of the research work to date has focused on the *networking* aspect of MANETs and not the *management* aspect of MANETs. In other words, a significant amount of MANET research has focused on getting functions such as link scheduling, IP routing, and so on, to work in an environment where nodes are mobile, where topology is highly dynamic, and where link quality fluctuates based on environmental factors. In particular, the problem of creating efficient IP routing protocols for MANETs has received a lot of attention. The IETF has chartered a working group (called "manet") in the routing area to develop routing protocols for MANETs. The principles used for routing in the Internet cannot be easily translated into similar principles for MANETs, due to the factors listed above. For example, in the Internet, the shortest path (with the smallest number of hops) is usually the best route between two points, but this is not necessarily true in a MANET. Other considerations, such as link quality, interference, mobility, fading, and so on, must also be taken into account to find optimal routes. The fact that routes may change rapidly and frequently, coupled with the fact that bandwidth is often scarce in MANETs, makes it necessary to consider whether reactive routing protocols are preferable to proactive routing protocols. A reactive routing protocol discovers routes between two nodes on demand, when the two nodes need to communicate, whereas a proactive routing protocol discovers routes in advance, so that routes are available when nodes need to communicate. Proactive routing protocols such as OSPF (Open Shortest Path First) have been used in the Internet with great success, due to the fact that the Internet topology is relatively stable and changes infrequently.

While the preference for proactive routing protocols can easily be justified in the case of the Internet and other wireline networks, the decision of whether to use a proactive or a reactive routing protocol for MANETs is not very clear-cut. While on one hand a strong case can be made for the use of reactive routing protocols in MANETs due to the fact the topology in a MANET is subject to changes (and hence one may argue that it is counterproductive to establish routes in advance), an equally strong case can be made for the use of proactive routing protocols in MANETs. A proactive routing protocol is a better choice when it is critical to have low route establishment latencies in order to service time-critical applications. The route establishment latencies for reactive routing protocols are typically higher than for proactive routing protocols, due to the fact that routes are essentially set up on demand when reactive routing is employed.

1.1.3.2 Handling Mobility. Handling mobility in MANETs is a challenging issue. In small MANETs where all nodes exchange routing information with each other, mobility is usually handled by routing updates; however, in larger networks, routing messages need to be aggregated for scalability. This means that nodes must be organized into routing domains, so that routes can be aggregated prior to being exchanged with other routing domains. Issues that arise include routing domain maintenance, dynamic assignment of IP addresses to nodes that move from one routing domain to another, and so on. These issues will be discussed in more detail in Chapter 6.

1.2 NETWORK MANAGEMENT

Network control and management plays a vital role in the well-being of any communications network. A vast amount of work has therefore been performed by the network management community in defining and standardizing network control and management functions. Work on standardizing network management began in the ITU-T (International Telecommunications Union-Telecommunication Standardization Sector; formerly known as the CCITT, the Comité Consultatif International Téléphonique et Télégraphique) with the development of the TMN (Telecommunications Management Network) standards. According to International Telecommunication Union [1996], "a TMN is conceptually a separate network that interfaces a telecommunications network at several different points." Although the TMN standards were developed for telecommunications networks rather than the Internet, certain concepts were defined in a very general manner and are applicable to management of IP networks too. An important point that separates the TMN concept from the Internet management concept is that the TMN refers to the use of a *separate network* for carrying management information. The IETF management paradigm, however, is based on the notion that the managed network is used to carry management traffic.

This remainder of this section is structured as follows. The next section describes TMN concepts that are valuable for any network management system. Note that the objective here is not to provide a comprehensive treatise of TMN, because most of the TMN standards are largely irrelevant for managing IP networks in general, and MANETs in particular. Instead, certain high-level concepts introduced by TMN are described, because they are relevant for managing IP networks too. Following this, an overview of the Internet management paradigm is provided, as standardized by the Internet Engineering Task Force (IETF). Section 1.2.3 describes the network management requirements for MANETs, which are sufficiently different from the management requirements for static wireline networks that they require a new management paradigm. This new management paradigm is introduced in Section 1.2.3.2 and is the central theme of the remainder of this book.

1.2.1   The TMN Standard

As mentioned earlier, the TMN standard was developed for telecommunications networks and has been successfully used in the management of circuit-switched wireline networks. More specifically, the operations support systems (OSSs) in circuit-switched wireline networks have been successfully structured around the TMN paradigm. As also mentioned earlier, while the TMN models are not necessarily relevant to managing IP networks in general (and MANETs in particular), they provide some fundamental management concepts that can be applied to the management of MANETs. In particular, the concept of FCAPS discussed below and the layered architecture discussed in Section 1.2.1.2 are two powerful concepts that can be applied to managing MANETs. The specific models and implementations of these principles and concepts for MANETs differ somewhat from their typical implementation in wireline networks, as will be pointed out in Chapters 6 through 9 of this book.

1.2.1.1   The TMN FCAPS Model. One of the concepts introduced by the TMN was the specification of five separate, distinct areas of network management:

- *Fault Management*: Management of network element failures, diagnosis, alarm management and correlation, root cause analysis, and so on.
- *Configuration Management*: Generation and storage of equipment configurations, configuration of network elements, configuration auditing and roll-back, inventory management, and so on.
- *Accounting Management*: Collection of usage data, which is often used to provide billing for services provided to network users.
- *Performance Management*: Monitoring of network element health and statistics via polling, thresholding, generation of threshold crossing alerts, performance report generation and analysis, trending, and so on.

- *Security Management:* Management of security-related services, authentication, encryption, intrusion detection, and so on.

These five areas of network management were collectively known as "FCAPS" (pronounced "ef-caps"), an acronym for Fault, Configuration, Accounting, Performance, and Security management. The FCAPS functions together help manage the underlying communications network by ensuring that

- Network faults are diagnosed and corrective actions performed.
- The network is and remains configured as intended in order to transport the applications and provide the services requested from the network.
- Proper tariffs are levied on the users of the network, especially when they cross administrative jurisdictions.
- The network is delivering services per the requested service level agreements (SLAs) and quality of service (QoS) assurances.
- The network is secure for all users.

Efficient FCAPS techniques are required so that the various components of the network work together harmoniously and consistently to deliver the services expected from the underlying network.

The network management and control functions defined and standardized by the TMN were broadly targeted for the static wireline networking environments. As noted in this chapter, however, MANETs differ significantly both in functionality and capability from their static wireline network counterparts. Due to the nature and significance of the differences, the current state-of-the-art management techniques designed for wireline networks do not fulfill all the management needs for MANETs. However, there is much to be learned and leveraged from the TMN network management standards for wireline networks. Therefore, rather than reinvent network management for MANETs from scratch, it is important to leverage existing work and enhance it appropriately, bearing in mind the unique characteristics of MANETs.

The TMN FCAPS classification is a concept that is useful for MANETs, and is used in this book to categorize the different management functions that must be performed for MANETs. The notable deviations from this model are:

- The TMN model does not mention Policy Management, which is a central theme of this book, and is an essential component of the suggested management scheme for MANETs.
- This book does not discuss Accounting Management, for several reasons. First, accounting practices and needs differ widely from company to company, and therefore it was decided that this was a topic best left to the network operator. Second, accounting is not a primary issue for users

such as the military or the government, who are expected to be major users of MANETs.

1.2.1.2 The TMN Layered Architecture. One of the most frequently referenced contributions of the TMN is the concept of a layered architecture for network management. The layered architecture is shown in Figure 1.4. It consists of the following five layers.

- *Business Management Layer:* The business management layer addresses functions that are related to management of the enterprise that owns the managed network. As such, it includes specifications of the business goals for the enterprise. It implements the functions that are related to these business goals.
- *Service Management Layer:* The service management layer manages services that are offered to users. At this layer, the network is not directly visible; however, the performance of the network directly impacts service quality, which is a primary concern at this layer. Functions at this layer include service creation and managing service quality.
- *Network Management Layer:* The network management layer manages the network, which includes all the network elements in the network. Network elements are visible as functional entities (e.g., routers, switches,

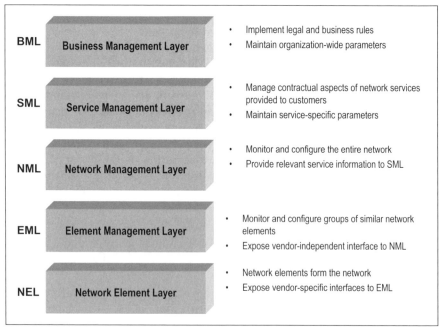

Figure 1.4. TMN layered architecture.

etc.) rather than vendor-specific equipment. The functions performed at this layer include (a) configuration of the network to provide services to users and (b) monitoring of the network to report service status to the service management layer.

- *Element Management Layer*: The element management layer deals with management of individual network elements. Typically, equipment manufacturers sell element management systems along with their network elements; these element management systems sometimes use proprietary protocols to communicate with the network elements for configuration and monitoring. The element management layer provides interfaces for the network management layer to send configuration data to the network elements, and retrieve monitoring data from the network elements.

- *Network Element Layer*: The network element layer contains all the network elements that form the network.

1.2.2   Internet Management Standards

As mentioned earlier, the TMN standards were developed for telecommunications networks and envisioned the use of a separate network for managing the telecommunications network. The Internet network management paradigm, by contrast, makes use of the same network for carrying data traffic as well as management traffic.

While the TMN standards developed a comprehensive specification of multiple telecommunications management architectures (including functional, physical, information, and logical layered architectures), the Internet management standards are largely focused on defining protocols for network management, the most notable of which is SNMP (Simple Network Management Protocol) [Case et al., 1990]. The SNMP management standards use a manager–agent model similar to that used by the TMN, as shown in Figure 1.5. The manager–agent model is based on the paradigm of having an *agent* resident on every managed network element. This agent populates information in a *Management Information Base* (MIB). A *manager*, which is part of the network management system, can poll the agent periodically to obtain the values of variables defined in the MIB (using the SNMP GET operation). It can also modify the values of variables in the MIB (using the SNMP SET operation). Also, the agent can asynchronously report problems (via SNMP traps) to the manager.

MIBs are defined using a standard language called SMI (Specification of Management Information) [Rose and McCloghrie, 1990]. The contents of the MIB are key to the management capabilities available for a given network element. A wide variety of MIBs have been defined by the IETF for managing different types of network elements and services running on those network elements. In addition, there has been a proliferation of proprietary MIBs, defined by different equipment vendors, that define management information for their own equipment. A variety of commercial as well as freeware tools

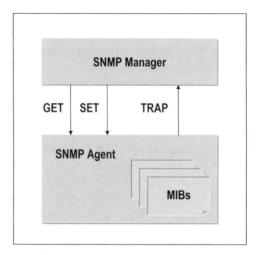

Figure 1.5. SNMP manager and agent.

are available that provide SNMP manager capabilities. These tools take MIB definitions as input, and they can collect MIB-specified information from all the network elements in their jurisdiction.

SNMP version 1 [Case et al., 1990] was relatively simple and lacked robust security mechanisms. The consequence of this was that SNMP quickly gained acceptance as a monitoring protocol, but not as a configuration protocol. It is easy to understand why; the trivial security mechanisms supported by SNMP version 1 meant that SNMP could not be used for network configuration, since the prospect of unwanted intruders reconfiguring a network was something that no network operator would want to risk. Fault and performance monitoring is fundamentally different from configuration because the former two activities (fault and performance monitoring) involve *read operations* on the concerned network elements, whereas the latter (configuration) involves *writing* to network elements. Even though it is certainly undesirable to have intruders monitor the status of an operator's network, the effects of such intrusion is generally nowhere as disastrous as an intruder breaking into a network and reconfiguring it at will. Thus network equipment vendors by and large implemented proprietary configuration interfaces to their equipment—for example, proprietary CLI (command-line interface), proprietary CORBA interfaces, and so on. With the standardization of robust security mechanisms in SNMPv3 [Case et al., 1999], however, SNMP was viewed as being suitable for use as a configuration protocol. The standardization of SNMPv3 included the definition of comprehensive security mechanisms that made it a viable protocol not only for monitoring, but also for configuration of network elements.

The SNMP management paradigm is built upon the paradigm of a centralized network manager that periodically polls all the network elements in the network and maintains a picture of the network health and performance based on the polled information. The network manager also receives SNMP traps from network elements that need to report problems asynchronously. When used for configuration, the centralized manager pushes out configuration information to all the network elements in the network. This paradigm is inherently unsuitable for MANETs, because, as will be discussed in the next section, any management paradigm for MANETs must be completely distributed and must minimize its use of over the air bandwidth. The centralized, polling-based management paradigm of SNMP cannot meet the management needs of MANETs.

1.2.3 Network Management in MANETs

Having provided an overview of existing network management models and standards, this section is devoted to providing a brief discussion of (a) network management requirements for MANETs (Section 1.2.3.1) and (b) why a new network management paradigm is required for MANETs (Section 1.2.3.2), thus providing the context for the remainder of this book.

1.2.3.1 Network Management Requirements for MANETs. As pointed out in the preceding sections, MANETs are highly dynamic in nature, since these networks are formed over wireless links. These wireless links are susceptible to failure because of mobility of nodes, or loss of connectivity due to volatility of wireless links. Strict requirements on security and reliability combined with the dynamic nature of the network provide a strong motivation for self-forming, self-configuring, and self-healing capabilities in the network. The conventional centralized network management paradigm collects network information at a designated management station, where a network operator analyzes the collected information and makes and disseminates management decisions across the network. The effectiveness of this paradigm depends on stable network connectivity and abundant network capacity, as provided by the Internet today. It also relies on relatively static network conditions that only require a modest amount of manual intervention. However, this paradigm does not suit MANETs because of network dynamicity and bandwidth instability introduced by mobile nodes. The characteristics of MANETs are sufficiently different from commercial wireline networks to have generated a great deal of work in alternate management paradigms. The task of managing MANETs involves frequent reconfiguration due to node mobility and consequently dynamically changing network topology. Network nodes often have limited battery power and storage capacity, and wireless radio link capacity and quality varies dynamically based on environmental conditions such as weather, terrain, foliage, and so on. These differences have resulted in a need for paradigms particularly suited for managing MANETs.

Section 1.1.3.1 described some of the salient characteristics of MANETs. In this section, the corresponding MANET-specific management requirements are derived based on those characteristics. Note that while there is still a need for the FCAPS management functions described in Section 1.2.1.1, MANETs have additional management needs that are outlined below.

- *Minimal Bandwidth Usage*: One of the most important requirements for a management system for MANETs is to minimize the use of over-the-air bandwidth. As was noted earlier, MANETs typically have very scarce bandwidth due to limitations in radio capacities and fluctuating link quality. Thus it is important to minimize any type of overhead so that capacity is available for all the applications that need to make use of the network.

- *Distributed Management*: An obvious corollary of the above requirement is that local management must be maximized, and therefore that management of a MANET must be distributed. This reduces the need for sending management information over the air. This is in direct contrast to a lot of networks, especially small and medium-sized enterprise networks, where all network management is centralized in one place.

- *Automated Reconfiguration*: Given the dynamic nature of MANETs, reconfiguration must occur much more frequently than in a static wireline network. Because of this, it becomes impractical to manually configure nodes, and therefore any management system must automate reconfiguration as much as possible.

- *Survivability*: The unreliable nature of wireless links leads to fundamental changes in the way information is disseminated. As was explained earlier, existing Internet protocols such as TCP do not perform well in high-loss environments that are typical of MANETs. Thus there is a need for survivable mechanisms for disseminating management information that do not depend on transport protocols such as TCP.

- *Self-Organization*: Even though management of MANETs must be distributed, there is still a need for sharing limited management information among network nodes for purposes such as fault correlation and viewing network status. MANET nodes therefore need to be organized in some kind of hierarchy for reporting management information. Such a hierarchy will necessarily be very dynamic due to the dynamic nature of the network. Thus MANET nodes must be able to organize themselves autonomously into a self-forming, self-maintaining management hierarchy.

- *Adaptability*: Given that MANETs have dynamically varying characteristics, any network management system must be able to adapt to the current network conditions. Since network conditions cannot be predicted ahead of time, management functionality must take into account existing network conditions and must adapt itself as needed. More specifically, the management system in a MANET should have the capability to adapt to the network dynamics and provide sustained Quality of Service (QoS) assurances, despite resource fluctuations. As an example, if the amount of available network capacity drops suddenly (e.g., due to weather deterioration), the management system may need to adapt by temporarily reducing the amount of information being exchanged among nodes, until network conditions improve.

1.2.3.2 Why Is a New Management Paradigm Needed for MANETs?

Although the TMN framework provides valuable abstractions for managing any type of network, one aspect that has not been addressed by the TMN standards or by existing network management systems is the feedback loop between configuration and fault/performance management [Kant, 2003]. The FCAPS functions of existing network management systems are typically implemented in a *stovepiped* fashion; that is, each of the FCAPS functions operate more or less in isolation. Statistics collected for fault and performance management are processed and analyzed by fault and performance management components, leading to diagnoses of network problems. In order to fix such problems, there may be a need for repairing or replacing faulty equipment (in the case of network faults); or it may be necessary to reengineer the network to add more capacity to deal with severe network congestion problems; and so on. Such requirements need to be addressed manually, due to the need to physically install or repair equipment in the field. However, many performance and some fault problems can be handled by network reconfiguration. For example, if a network link is severely congested, it may be possible to alleviate the problem by sharing the traffic load with another underutilized link or by reducing the amount of management traffic, and so on. This can be accomplished simply by appropriately reconfiguring the network. Today, this is done manually by experienced network operators who examine outputs from fault and performance management systems and decide how to appropriately reconfigure the network. This is the fundamental problem with network management today: There is too much human intervention required to run a network. In order to reduce the cost of network operations, it is critical that human intervention be minimized by creating a feedback loop between fault/ performance monitoring systems and configuration systems and by specifying policies that regulate how the system should be reconfigured in response to various network events.

Another aspect that is missing from current network management systems is the ability to state long-term, network-wide configuration objectives. Examples of such objectives include:

- If the network is congested, reduce the frequency of GPS information reporting for every network element.
- If link utilization exceeds 80%, reduce by half the bandwidth allocated for Best Effort traffic, and so on.

The observant reader will notice that the above points are as valid for wireline networks as they are for MANETs; in other words, the lack of automation in network management is a problem for both wireline networks as well as for MANETs. However, the critical point to note here is that *this lack of automation poses a much greater problem for MANETs than for wireline networks*. This is because *the characteristics of MANETs result in a requirement for much more frequent network reconfiguration and much more stringent monitoring* than needed in wireline networks. The characteristics of MANETs that were listed in Section 1.1.3—namely, the lack of fixed infrastructure, dynamic topology, power and processing constraints, intermittent connectivity, varying security requirements, scarce bandwidth, and the existence of high-loss, unreliable links—result in a need for much more network management control than is typically needed in a static wireline network, where network elements can be configured in a certain way and rarely need to be reconfigured. The heightened need for network management control stems from the fact that the network must be maintained in a functional state in spite of the dynamic nature of the network. Additionally, this increased network management control must be automated as far as possible, or else the amount of manual intervention required to keep the network functioning optimally would result in prohibitive cost, making MANETs uneconomical and impractical to deploy.

It should be noted that, due to the wireless nature of MANETs, a majority of failures are "soft" failures—that is, failures that are not caused by the malfunction of a network element (e.g., a router failure), but rather due to performance problems. An example of such a performance problem is excessive loss or excessive delay. Correlation engines that integrate fault and performance alarms and data are required to correctly diagnose the root cause of the problem. Once a root cause has been diagnosed, soft failures can often be corrected by dynamic reconfiguration of various aspects of the network. Some examples of dynamic reconfiguration are:

- Reorganization of routing domains to provide more efficient network organization and thereby reduce congestion and delay
- Reassignment of nodes to perform various gateway functions
- Dynamic reconfiguration of firewall functionality to contain network attacks, and so on

This implies that FCAPS operations *need to work together* and be closely coupled so that information from one or more of these functions can kick off the required processing and actions in the other functions *in an automated fashion*. Moreover, the timescales at which network dynamics change underscores the need for close, automated cooperation among the FCAPS functions—a notable deviation from the traditional stovepiped implementation of FCAPS operations for wireline networks. This paradigm is illustrated in Figure 1.6.

Figure 1.7 provides a high-level view of the close coupling that is required between the FCAPS functions. It shows network monitoring functions feeding

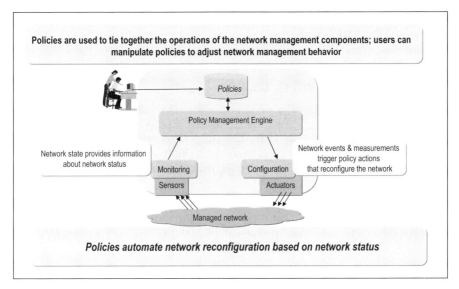

Figure 1.6. Policy-based managed paradigm for MANETs.

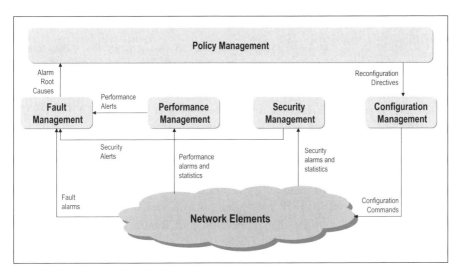

Figure 1.7. Coupling of FCAPS and policy management functions.

into a policy management engine, which, in turn, generates network configurations based on stored policies. This provides for automation of network configuration in response to current network status, which is one of the benefits of close coupling between the FCAPS functions. Thus, in contrast to the typical stovepiped implementation of the TMN FCAPS model, Figure 1.7 ties together the FCAPS functions via policies. MANET management is therefore accomplished via an *integrated* set of operations models that work collectively to manage the network.

1.3 SUMMARY AND ROAD MAP

The purpose of this section was to provide an overview of mobile ad hoc networks and their management requirements. The need for a new management paradigm for MANETs was discussed in light of the characteristics of MANETs. The concepts introduced in this chapter form the basis for the remainder of this book.

The next four chapters discuss various aspects of policy-based network management, with an emphasis on the use of policies for managing MANETs. This is followed by a detailed discussion of each of the FCAPS functions. The outline of this book is provided below:

- Chapter 2, "Policy Terminology," introduces terminology relating to policies that is used in the remainder of the book. Since the term "policy" is often used in many different contexts, this chapter lays out definitions that will enable the reader to understand the usage of the term "policy" in this book.
- Chapter 3, "Policy Languages and Frameworks," provides a comprehensive overview of relevant policy languages, models, and frameworks that have been developed by various standards bodies, academic groups, and industrial research groups. First, an overview of the relevant IETF standards is provided. This includes standards efforts such as COPS, RSVP, and DiffServ, which are related to the development and use of policies. The IETF Policy Framework, which was developed in the late 1990s, is also described. Following this, languages and frameworks such as Ponder and PECAN are described. The purpose of this chapter is to give the reader an understanding of (a) the state of the art in policy languages and frameworks and (b) their relevance for MANET management.
- Chapter 4, "Policy Conflict Detection and Resolution," introduces the difficult topic of policy conflict detection as well as resolution. Along with the power of policies comes the risk of conflicts among policies. This chapter provides a taxonomy of policy conflicts and then discusses various conflict resolution strategies for different types of conflicts that are likely to arise in policies created for the purpose of network management.
- Chapter 5, "Policy-Based Network Management," discusses the use of policies for managing networks and, more specifically, mobile ad hoc networks. This chapter takes a look at the shortcomings of current network management systems for managing MANETs, and it describes the use of policy-based network management for solving the management problems that are peculiar to MANETs. A reference architecture of MANETs is

provided in this chapter, which is referred to in the remainder of the book. Following this, the architecture of a policy-based network management system for MANETs is described. The chapter concludes with a collection of usage scenarios that describe simple use cases that showcase the use of policies for automating network management.

- Chapter 6, "Configuration Management," discusses the configuration of MANETs. Configuration management is the most complex of the FCAPS functions, since it involves the setup of the network. This chapter begins with an overview of the relevant network management standards, including COPS-PR and snmpconf. The chapter also discusses relevant network services, including DHCP and DNS. Following this, the configuration functions for a MANET are discussed for the lower layers of the protocol stack (layers 1 through 3). Interdomain policy management poses a special challenge when configuring a network that has to communicate with networks in other administrative domains; relevant aspects of interdomain policy management are discussed. The chapter concludes with a look at the configuration management architecture for a MANET, which includes a description of an innovative configuration optimization function that uses advanced reasoning techniques to reconfigure the network dynamically based on network state. This is an exciting area of exploration for MANET management, and Chapter 6 provides some insights into preliminary research in this area.

- Chapter 7, "Fault Management," addresses the complex topic of fault management for MANETs. The principal functions of fault management—namely, network monitoring and root cause analysis—are discussed. Root cause analysis, in particular, is a function that becomes much more complex in MANETs—as compared to wireline networks—due to the dynamic and stochastic nature of these networks. This chapter explains the complexities of analyzing faults in a MANET environment. Another very important aspect of fault management, self-healing, is also discussed. Given that MANETs are more prone to failures than wireline networks, it is imperative that MANETs provide the ability to automatically recover from different types of faults, whenever possible. The chapter concludes with a description of various fault scenarios, along with the corresponding policy-triggered self-healing actions that can be performed to recover from these faults.

- Chapter 8, "Performance Management," describes the functions required to manage the performance of applications in a MANET. Aside from the collection and aggregation of performance statistics, which is a requirement for any type of network, a performance management system for MANETs must also be able to provide quality of service guarantees for the applications running in the MANET. Unlike wireline networks that are typically characterized by an abundance of capacity, MANETs typically provide very limited bandwidth that needs to be carefully managed so that higher priority traffic receives better treatment than lower priority traffic when there is resource contention. This chapter describes the required functions and related algorithms that support management of quality of service in a MANET.

- Chapter 9, "Security Management," discusses security services and management functions for a MANET. The chapter begins with a discussion of policies for access control. Following this, the topic of key management is discussed, including a broad overview of related security services such as cryptography, confidentiality, message integrity, user authentication, and public key infrastructure (PKI). This is followed by an overview of communications security mechanisms and intrusion detection for MANETs. The need for securing different security domains and managing the flow of information across security domains is also discussed in this chapter.
- Chapter 10, "Concluding Remarks," provides a summary of the book, along with some final remarks about promising future research directions in this field.

2

POLICY TERMINOLOGY

The field of policy-based network management has been in existence for over a decade. The purpose of this introductory chapter is to provide some motivation for the use of policies for network management. This is followed by an overview of the terminology used in this field and a discussion of the different types of policies that can be defined for network management.

2.1 MOTIVATION

Why are policies needed? After all, management systems have been functioning for decades without explicitly using the concept of policies. This section philosophizes about why policies are really needed. The purpose of this section is not to define what the syntax of a policy is or how a policy should be implemented; rather, it attempts to provide an intuitive understanding of the purpose of policies as described in this book.

In the context of computer and communications networks, think of a policy as

a way to configure the behavior of software.

This statement might appear hard to understand at first. Not many people think of software as something that needs to be configured. To understand this a little better, as a parallel, think about how people deal with hardware— for example, a router. All routers support a relatively standard and well-understood set of functions, namely, IP routing using any of a number of routing protocols (such as RIP, OSPF, BGP, IS-IS, etc.); firewall access control functionality; security (e.g., IPSec); and so on. However, any router must be *configured* before it can be deployed. Configuration parameters include

Policy-Driven Mobile Ad hoc Network Management, by Ritu Chadha and Latha Kant
Copyright © 2008 John Wiley & Sons, Inc.

parameters such as IP addresses and subnet masks for router interfaces; choice of and parameters for one or more routing protocols; access control lists implementing firewall functionality; parameters for IPSec tunnels (source, destination, encryption protocols to be used, etc.); and so on.

One could ask, Why doesn't a router ship fully configured? Anyone who is familiar with router functionality would scoff at the question: It is obvious to them that there is a need to be able to configure a router so that it behaves as needed. The very idea of not being able to configure a router with IP addresses, routing protocols, and so on, is ludicrous; routers would be useless if they all came configured with the same IP addresses with no way to change them!

In exactly the same way, it is important to be able to configure software, and most software today is indeed configurable. For example, a web browser can be configured with a long list of user preferences; an e-mail client must be configured with information about the e-mail server, and so on.

So is this all that policy is? A way to configure software? If so, isn't it a trivial problem that has already been solved by most large applications today? Take the example of a web browser; its Graphical User Interface (GUI) provides an easy way to configure values for all kinds of options—for example, the address of the browser home page; number of days to keep pages in history; which programs should be used for e-mail, calendar, contact list, etc.; security and privacy settings; and so on. Another example is a Java program that reads the values of all configurable parameters from a properties file. One characteristic of each of these examples is that all of the configurable parameter values are defined statically. If these configurable parameters are called *policies*, then the behavior of the system is completely known at the time these policies are defined.

However, policies need to do more than this. In light of the above discussion, let us revise the statement above; a policy is

> *a mechanism to configure the behavior of software in ways that are not predictable at policy definition time.*

This statement indicates that policies specify more than just a collection of configuration parameters. They are intended to capture the intent of a user in terms of how the user wants the software to behave. However, users don't necessarily think in terms of configuration parameters; rather, they follow complex thought processes that include conditional statements and awareness of the environment. To continue with the example of a web browser, users don't necessarily want to configure all their web browser parameters in a static way; rather, they would like to be able to express the desired configuration as follows: "During working hours (9 am–5 pm), I want my home page to be set to my company's home page, but in the evenings and on weekends I want my home page to be my favorite web portal. However, I sometimes work from home, and on those days, I want my home page to be set to my home ISP's web portal." This description of how to configure a web browser is a set of policies. The software is configured via a set of rules rather than a set of static parameters. Here the behavior of the software is configured in a way that is not predictable at policy definition time because it is not possible to tell exactly how this person's web browser is configured at any point in time by looking at this set of rules, without knowledge of the variables that form an integral part of the rules. In this particular example, since the rules depend on a number of variables such as time of day, and whether the person is telecommuting, the exact configuration of the browser cannot be specified without knowledge of these variables. In order to know the configuration parameters that are in effect at any point in time, it is necessary to have access to additional information, such as time of day, day of the week, and the user's telecommuting status. These inputs, along with the specified rules, completely specify the web browser's configuration state.

Now let us make the analogy more specific to network management. Network management systems are applications, or software, whose behavior needs to be configured in a manner analogous to the web browser example in the last few paragraphs. As an example, a network management application must collect device status from all managed devices. Collection of device status is an operation that must be parameterized with the following values: at a minimum, the list of devices to be monitored (e.g., a list of IP addresses and the corresponding SNMP community strings) and the monitoring interval (e.g., monitor every device every minute). In most existing network management systems, these configuration parameters could be specified statically. As an alternative, policies could be specified that provide a much more flexible specification of how to configure the network management software for monitoring. The following example describes the kind of functionality that could be supported via a very simple sample set of policies outlined below.

- Under normal network conditions, monitor every device at 2-minute intervals.
- If the network is congested, monitor every device at 5-minute intervals.
- If the network is critically congested, halt all device monitoring.
- The network is congested if average number of packet drops exceeds 50%.
- The network is critically congested if average number of packet drops exceeds 70%.

This set of policies is very simple, but the intent is to illustrate the power of policies for controlling the behavior of a network management application. Provided that the above policies can be implemented, the network management system will be able to automatically change its behavior (by using different "configuration" parameters) based on events occurring in the network.

As another example, consider a scenario where Quality of Service (QoS) is being provided to different classes of traffic, as supported by the DiffServ (Differentiated Services) paradigm, for a military mission. QoS configuration parameters must be defined to specify the percentage of bandwidth that will be allocated to different traffic classes. This could be done via static configuration parameters. However, QoS configuration parameters must be adjusted based on projected traffic profiles. Assume that the military mission has several phases and that each phase has different network traffic QoS requirements. For example, the first phase could be largely reconnaissance-oriented, where the bulk of the traffic is generated by sensors that collect and report information about the enemy targets. The second phase could be one where the forces move into position; here command and control messages might make up the bulk of the traffic. Finally, the third phase could involve active combat, and networked fire messages would require increased bandwidth. Policies could be put in place that would automatically reconfigure the network with the appropriate QoS configuration parameters every time the mission phase changed; these QoS configuration parameters would adjust the amount of bandwidth allocated to different classes of traffic based on projected mission traffic requirements. Since the changing of a mission phase is not an observable network event, some additional input would have to be provided to the network management system to trigger the configuration changes for a new mission phase; either the configuration changes could be activated at some predetermined times, or an external trigger indicating a mission phase change could be sent to the network management system.

As a final example, let us look at the authorization functions of a network management system. One of the functions of the security management component of a network management system is to control access to resources such as network devices, network services, and so on. Access to such resources should be controlled based on a variety of parameters, such as user information, the resources that they are trying to access, the operations that they are trying to perform on these resources, and so on. Again, a security management component could be statically configured with a list of authorized users who can perform specified operations on specific resources. However, there may be a need for much greater flexibility in defining access control information. For example, in a military mission, during normal combat mode, all users may be granted access to video-conferencing services; however, during crisis situations (e.g., when there is a critical shortage of network bandwidth), video-conferencing could be limited to officers of a certain rank and above. Policies can be used to configure the access control permissions granted by the security management component to achieve this flexibility.

Having looked in some detail at the flexibility that is desired in the configuration of a network management system, the next question is, How is this kind of functionality achieved today? The answer is simple: Human operators manually adjust the "configuration" of the network management system based

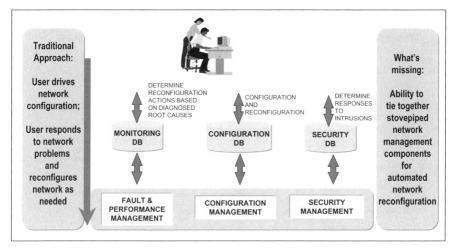

Figure 2.1. Current state of the art in IP management.

on observed network conditions. An experienced network operator can determine that a network is highly congested (e.g., based on performance statistics being collected by the network management system), and he or she will know how to adjust the configuration of the system by manipulating the values of the appropriate parameters. This paradigm is illustrated in Figure 2.1. The important thing to note about this illustration is the following: The human operator is part of the decision-making process ("human in the loop") about how to adjust the behavior of the network management system in response to network conditions.

In contrast to this, the next figure (Figure 2.2) shows a policy-based network management system in which a human operator is no longer part of the decision-making process about how to adapt to network changes. There is still a need for human participation in network management, but the participation required is at the level of defining the appropriate policies for the network management system to do its job and to handle exception conditions that cannot or should not be handled automatically. Instead of having human beings implement the policies held in their heads (as was being done in Figure 2.1), here is a piece of software that implements the policies that the human operator expresses in some machine-understandable format.

2.2 INTRODUCTION TO POLICY TERMINOLOGY

The term "policy" has been widely used with a broad range of meanings in very diverse fields, ranging from law to networking. There is no single definition of policy that is widely accepted. Two reasons why it is very difficult to provide a single definition of a policy are (a) the variety of applications for

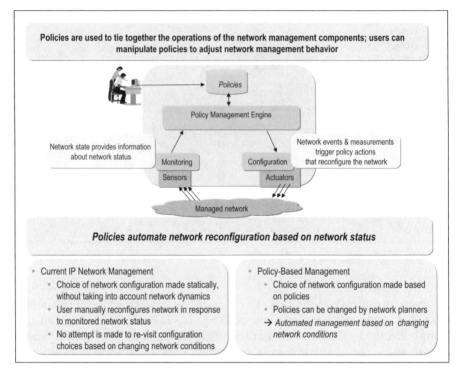

Figure 2.2. Vision for policy-based IP management.

policy-based management and (b) the plethora of policy tools available today. According to Saperia [2002], a policy implies "... a predetermined action pattern that is repeated by an entity whenever certain system conditions appear." The IETF defines a policy as a definite goal, course, or method of action to guide and determine present and future decisions [Westerinen et al., 2001], but also as a set of rules to administer, manage, and control access to network resources [Moore et al., 2001].

In order to provide some background to the reader, this section provides an overview of some of the terminology used when discussing policies for network management. Policy-based management has been discussed in the literature for close to two decades, but it really began to gain widespread attention in the late 1990s, when the IETF (Internet Engineering Task Force) formed the Policy Framework Working Group. The goals of this working group were to define an architecture and an information model for policy-based management of QoS in IP networks. At the same time, the Distributed Management Task Force (DMTF) was working on developing a wide range of information models for applications including network and policy management. The Policy Working Group within the DMTF (also known as the Service

Level Agreement, or SLA Working group) joined hands with the IETF Policy Working Group in order to push for the standardization of a policy information model within the IETF. Although the DMTF is also a standards body, it has a much smaller following than the IETF. The inclusion of the Policy work within the charter of an IETF working group was expected to result in a much wider audience for the policy standardization work than could be achieved within the DMTF alone. The information models developed within the IETF and the DMTF were to be kept closely aligned by having the same group of people lead the development of the IETF and DMTF standards.

The IETF began by developing a description of the policy-based management architecture, shown in Figure 2.3. The following are the components depicted in the figure:

- *Policy Server or Policy Decision Point (PDP)*: RFC 3198 [Westerinen et al., 2001] defines the policy decision point as "a logical entity that makes policy decisions for itself or for other network elements that request such decisions." Here "network element" refers to network equipment such as a switch or a router. It also describes two perspectives of a policy decision, namely:
  - A "process" perspective that deals with the evaluation of a policy rule's conditions
  - A "result" perspective that deals with the actions for enforcement, when the conditions of a policy rule are true

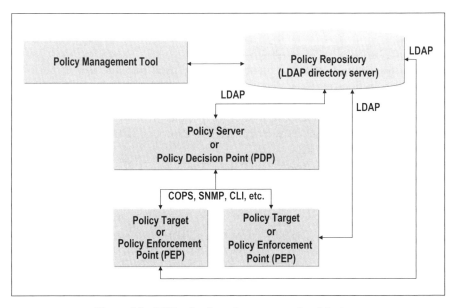

Figure 2.3. IETF policy-based management architecture.

In other words, the PDP is the "brains" of the policy system. It is the engine that evaluates policy rules and ensures that the appropriate actions are triggered as needed. More details about the functioning of a policy decision point will require a better understanding of what policy rules look like, so this discussion will be postponed until then.

- *Policy Target or Policy Enforcement Point (PEP)*: RFC 3198 [Westerinen et al., 2001] defines the policy enforcement point as "a logical entity that enforces policy decisions," where policy enforcement is defined as "the execution of a policy decision." In other words, the PEP is the entity that performs the action(s) mandated by a policy rule. A simple way to think about a PDP and a PEP is the following: The PDP determines what actions need to be taken, based on stored policy rules; and the PEPs perform these actions. As an example, the PDP may decide that a certain configuration action must be performed; the PEP contains the code to actually perform that configuration action. In addition, there may be multiple PEPs controlled by a single PDP, so the relationship between a PDP and a PEP is one-to-many.

- *Policy Repository*: According to RFC 3198, a policy repository can be defined from three perspectives:

  - A specific data store that holds policy rules, their conditions and actions, and related policy data (e.g. a database or directory).

  - A logical container representing the administrative scope and naming of policy rules, their conditions and actions, and related policy data. A "QoS policy" domain would be an example of such a container.

  - In Moore et al. [2001], a more restrictive definition than the prior one exists. It defines an object named "PolicyRepository," which is a model abstraction representing an administratively defined, logical container for reusable policy elements.

  The first definition is the most commonly used one and is the one that will be used in this book. The IETF Policy standards do not mandate any particular storage technology for storing policy rules. However, in order to provide a concrete data model for policies, a sample implementation using a Lightweight Directory Access Protocol (LDAP) schema was defined by the IETF.

- *Policy Management Tool*: The Policy Management Tool is a user interface that allows administrators and network operators to create and manipulate policies. It can range in complexity from a very simple graphical user interface (GUI) that allows access to policies stored in the Policy Repository, to a more complex tool that provides a higher-level interface to the user and translates the user's input into policies of the appropriate format that can be stored in the Policy Repository. The Policy Management Tool can provide utilities such policy conflict resolution. Policy conflicts are defined in RFC 3198 as occurring when the actions of two policy rules that are to be performed simultaneously contradict each other. Some form of policy conflict resolution must then be implemented in order to detect conflicts and provide appropriate resolution. Policy conflict resolution is an active field of research and will be discussed in more detail in Chapter 4.

The following interfaces are shown in Figure 2.3:

- *PDP–PEP Interface*: The interface between the PDP and the PEP allows the PDP to control the behavior of PEPs based on policies stored in the Policy Repository. The IETF Policy standards do not mandate the use of any particular protocol for this interface. Protocols that are mentioned as potentially suitable for PDP–PEP communication are COPS (Common Open Policy Service), SNMP, or proprietary protocols.
- *PDP–Policy Repository Interface*: The PDP interacts with the Policy Repository to retrieve policies. Since the Policy Repository itself is not restricted to any particular implementation, the interface between the PDP and the Policy Repository—which depends on this implementation—is therefore not dictated by the IETF standards either. If an LDAP directory is used to store policies, then the protocol used by the PDP to talk to the Policy Repository will obviously be LDAP; if the Policy Repository is implemented using a relational database, then the PDP uses SQL queries to retrieve policies; and so on.
- *PEP–Policy Repository Interface*: PEPs typically do not interact directly with the Policy Repository, since they usually receive instruction from the PDP about actions that they must enforce. However, in some cases it may be appropriate for PEPs to retrieve policies directly from the Policy Repository. This can be the case when policies are defined at a low enough level so that network elements can directly interpret them and implement them. In such a case, PEPs use the interface supported by the Policy Repository to talk to it; as mentioned above, this interface depends on the technology used to implement the Policy Repository.
- *Policy Management Tool–Policy Repository Interface*: As in the case of the PDP, the Policy Management tool uses whatever interface is supported by the Policy Repository, depending on the technology used to implement the Policy Repository.

2.3 COMMON USAGE OF THE WORD "POLICY" IN NETWORK MANAGEMENT

In the context of networking and network management, the term "policy" is often used in a wide variety of ways. This section examines the various ways that people use the term "policy," and it finally arrives at a categorization of the different types of policies for the purpose of network management.

2.3.1 Policies as Rules Dictating Behavior

The first and most obvious usage of policy is to define rules for network management in a manner analogous to the examples given earlier in this section. The rules are prescriptive in nature; that is, they dictate what management actions should be taken in different circumstances and also dictate how a network should be managed. These rules are specified in a procedural way, and they typically indicate specific management actions that must be performed under certain conditions. The actions usually involve configuring or reconfiguring network elements or network services. Some examples of policies that are rules that dictate behavior are:

- Upon receipt of failure root cause X, if automated reconfiguration is permitted, perform corrective action Y.
- At 8 pm every day, download new virus signature files.

The first of the above policies specifies that upon receipt of a certain event (root cause X), a condition should be checked (is automated reconfiguration permitted?); and if the condition is true, a certain action should be performed (action Y). In the second policy, a specific time is given (8 pm) at which a specific action must be performed (download new virus signature files). Both policies specify exactly what needs to be done, and how, since the explicit action(s) to be performed is (are) called out in each rule.

2.3.2 Policies as Rules Granting or Denying Permission

The next usage of policies is again in the form of rules, but this time the rules are declarative rather than procedural and are used to describe whether permission should be granted or denied to entities that are requesting permission to do something. This permission could be for access to a resource such as a database, for the right to request QoS, or for the ability to access a management application. Some examples of such policies are:

- User A is authorized to receive gold QoS for up to 1 Mb/s of network traffic.
- Only managers can access the salary database.

What should be noted about each of these policies is that they both specify rules about the functioning of a network (i.e., they specify *what* should be done), but do not specify *how* to achieve the desired result.

2.3.3 Policies as Constraints or Parameters

Policies are often used to specify constraints or parameters that guide the functioning of a network. These constraints are specified in a declarative way;

they make statements about certain relationships that must always hold true in a network. In certain situations, one can think about this type of policy as *invariants* about the network. Such policies provide an expression of high-level goals and objectives to be achieved, such as commander's intent (in military contexts), administrative guidelines, and so on. Such constraints comes from a variety of sources, including published organizational directives, operational guidance, administration manuals, interactive operator input, and guidance from higher headquarters. This policy may not be written in a form that is directly usable by the network management system, and it may need to be translated by system operators into a usable structure. The constraints specified in these policies basically dictate *parameter values for software components within the network management system*. These constraints or parameters are called policies because they dictate rules, or "policy" about the operation of various software components. As an example, Multi-level precedence and preemption (MLPP) policies provide information about the relative importance of different types of traffic, so that network planning and management components can use this information to determine how to allocate scarce resources to multiple contending traffic requests. Such MLPP policies are really configuration parameters that could be used by a network planning component to determine whether to grant mission requests, as well as by a QoS Management component to determine whether to grant flow admission requests and whether to preempt certain traffic flows in favor of others.

Some examples of policies that are constraints are:

- Network status should be monitored every 60 seconds.
- At most 20% of EF bandwidth can be granted to a certain traffic class C.

What should be noted about each of these policies is that they both specify a constraint on the functioning of a network management software component. The first policy sets the value of a parameter within the monitoring component and indicates how frequently the monitoring component should collect network status from network elements. The second policy is a constraint on the component that handles allocation of expedited forwarding (EF) bandwidth to users. The latter component could be a QoS management component that handles admission control. The behavior of each of these components (monitoring and QoS management) can easily be altered by changing the above policies to specify different values for monitoring frequency and for percentage of EF bandwidth to be granted to a certain class of users.

2.3.4 Policies as Configurations

Configuration policies specify configuration parameters for network elements, services, and protocols. This type of policy is probably the most misunderstood one in existence today. As such, these policies are really just configuration parameters (and not policies); however, commonly used industry terminology uses the term policy for a variety of configuration parameters, most notably parameters related to QoS, routing, firewalls, and so on. For this reason, rather than ignore such policies, they are classified as configuration policies. The reason that configuration policies are misunderstood is that they are often confused with policies that are specified as rules. As an example, DiffServ classifier configuration parameters in a router are commonly referred to as a QoS classifier policy. The function of this configuration is to mark IP traffic with the appropriate DiffServ Code Point (DSCP), based on information such as source/destination IP addresses and ports. Since the configuration information could be expressed in the form of an event–condition–action triple, where the event is the arrival of an IP packet, the condition is some check on the source/destination IP addresses and ports, and the action is to mark the packet with a DSCP, a QoS classifier policy is very often classified as an event–condition–action rule, when in fact it is really just a configuration policy. The way to distinguish between event–condition–action rules and configuration policies is that in order to implement an event–condition–action rule, a policy engine is required that listens for events, evaluates conditions, and invokes actions. This policy engine functions at the *management plane*, and policy decision-making happens outside of the network device. Configuration policies, on the other hand, can be implemented simply by conveying configuration directives to a network device (such as a router). The network device is then responsible for implementation of the policy *at the data plane.*

The following are examples of configuration policies:

- Twenty percent of the bandwidth must always be allocated to best effort traffic.
- All IP packets going from subnet 192.168.1.0/24 to subnet 192.168.2.0/24 should be marked with DSCP 26 (i.e., AF3).

The first policy above is used to configure routers to reserve a certain percentage of bandwidth for traffic that is marked as best effort. Such policies are typically configured on all network routers. The second policy is used to configure the ingress router for subnet 192.168.1.0/24 (i.e., the router via which traffic from subnet 192.168.1.0/24 enters the network to go to subnet 192.168.2.0/24) to mark packets with the specified DSCP.

2.4 DEFINITIONS: POLICY TYPES

The previous section described at a high level the various usages of the term *policy* in the area of network management. Next, the different types of policies that are required to be able to capture all of these usages need to be defined. This is one of the key steps in this chapter, since every policy in the remainder of this book will be defined in terms of one of the three types of policy being defined here. Another reason that it is critical to define the types of policies that will be used in a system is to be able to choose appropriate policy tools for the chosen application. Different policy tools support different types of policies; and as was seen in the previous section, there are many different ways in which the term "policy" is commonly used in the industry today. This section arms the reader with the definitions of policy that can then be used to analyze policy tools in order to determine which ones will support the types of policy required in any system. The scope will be restricted to policies used for network management purposes. Network management policies can be classified into the following three major categories:

Event–Condition–Action Policies. Event–condition–action (ECA) policies describe rules dictating behavior of a management system, as described in Section 2.3.1. They describe management actions that should be taken in response to network or external events, provided that certain conditions hold. An example of such a policy is one where an event such as a network intrusion or a performance problem triggers an action that reconfigures the network to mitigate the problem. Another example deals with time-based reconfiguration, such as weekly installation of new virus signatures; here the triggering event is a weekly timer.

Access Control Policies. Access control policies describe rules that specify whether to grant or deny permission to entities to perform various actions, as described in Section 2.3.2. They prescribe which operations are permitted by which users on which resources. These policies specify targets, which include resources (the entity being accessed), subjects (the entity accessing the resource), and actions (the operations that the subject performs on the target); conditions, which describe conditions under which the access is allowed or disallowed; and effect, which is either "allow" or "deny."

Configuration Policies. Configuration policies specify configuration parameters for network elements, services, protocols, and software components, as described in Sections 2.3.3 and 2.3.4. They can be divided into two categories:

- *Administrative Policy*: Administrative policy is an expression of high-level goals and objectives to be achieved, such as commander's intent, administrative guidelines, and so on. In military parlance, such policies are often called operational policies. Note that this type of policy was categorized as *constraints* or *parameters* in the previous section. Such policies contain configuration information for software components, as they place constraints on the operation of these software components. An example of such a policy is an MLPP policy that describes the priority of different types of network traffic.

- *Configuration Policies for Network Elements, Protocols, and Services*: Such policies contain configuration information for network elements or services/protocols such as routing and QoS. Examples of these include QoS policies that specify bandwidth allocations for different classes of service; firewall configuration policies that specify the types of traffic that should be permitted or not permitted to traverse a firewall; BGP routing policies that describe what routing updates should be redistributed into another routing protocol; and so on.

The next subsections delve into more detail about each of these types of policies.

2.4.1 High-Level View of an ECA Policy

This section provides a high-level look at ECA policies—that is, those that have a rule format (as opposed to constraints).

2.4.1.1 *Definition of an ECA Policy.* At a very high level, an ECA policy contains the following:

- *Events*: Events are used to trigger policies. Events represent asynchronous occurrences of interest in the relevant domain. For example, in a network management application, events of interest include network faults, performance threshold crossings, network intrusions, and so on. Events are optional components of policies.
- *Conditions*: Conditions are guards that are used to check whether certain expressions are true before the specified actions are performed. For example, conditions that might be checked include checking whether a monitored variable exceeds a specified value; whether a network element is of a certain type; and so on. Conditions are optional components of policies; if no condition is specified, it defaults to "true."
- *Actions*: Actions are different types of network management operations that are used to configure and monitor network elements or services. Actions may also be used to trigger other management components. For example, an action triggered by network congestion could invoke a traffic engineering component that can determine how to alleviate the congestion.

Policy rules are enforced by a PDP as follows. First consider the case of policy rules that have only one event in their event section. Whenever an event occurs, the PDP checks to see whether any policies contain this event in their event section. If so, then for every such policy rule, the PDP checks whether all the conditions in the condition section evaluate to "true." If they do, then the action or actions of this policy rule are performed. The collection of all of these activities performed by the ECA PDP is called *policy enforcement*. When the event(s) in the event section of a policy occur, that policy is said to be *triggered*. If a policy rule has no event, it is triggered immediately upon activation.

If there is more than one event in the event section of a policy rule, things become more complicated. To understand this, suppose a policy rule has events A and B in its event section. The question then arises: Must events A and B occur simultaneously? Since events will probably be processed sequentially by the PDP, the definition of simultaneous events becomes problematic. It is therefore useful to introduce the notion of an *epoch*, where an epoch is a configurable interval of time; any events occurring within the same epoch are treated as occurring simultaneously. Thus if an epoch is 5 seconds long and if events A and B occur within 5 seconds of each other, then the PDP will assume that the events for the corresponding policy rule have both occurred, will evaluate the policy rule conditions, and will perform the specified actions if the conditions are true.

The above notion of a policy is referred to as the event–condition–action (ECA) type of policy.

2.4.1.2 *Examples of ECA Policies.* This section describes some examples of how to define useful ECA policies in the context of network management.

Example 1. Consider a network management system where an intrusion detection component monitors the network for any intrusions or attacks, and generates alerts when such attacks are detected. This intrusion detection system can diagnose the attack to various degrees: At one extreme, it could diagnose the problem precisely enough so that an automated corrective action could be triggered to solve the problem, and at the other extreme it might simply produce a warning with a description of the anomalies detected, but with no diagnosis that could lead to a possible solution.

As an example, suppose that a denial of service on a certain port on a machine is detected; a possible solution could be to deny all access to that port. The corresponding policy would be:

Event: Denial of service attack on port X on machine Y
Condition: None
Action: Deny access to port X on machine Y.

In the above policy, X and Y would be parameters in the policy event and action. When the intrusion detection system detects the attack, it generates an event indicating a denial of service attack, with specific values for X and Y. The above policy gets triggered and the corresponding action is invoked, which configures the system to deny access to the specified port on the specified machine.

As a second example, the intrusion detection system might observe excessive traffic on a given subnet, but be unable to suggest a cause for this heightened activity level. A policy such as the following could be used:

Event: Problem: X

Condition: None

Action: Alert operator about X.

In the above policy, X is a parameter in the policy event and the action. When the intrusion detection system detects the above problem, it generates an event indicating the problem, replacing X in the above event by a description of the problem. For example, it could generate the following event:

Problem: *Excessive traffic on subnet 192.168.1.0*

The action of the above policy would be triggered, resulting in an alert to the operator which would contain the following descriptive text: "Excessive traffic on subnet 192.168.1.0."

Example 2. In this example, consider a situation where the network configuration must be changed at specific times of day. This scenario could arise, for example, if application requirements are very different during the day and night. Take the case where a network is configured to support DiffServ QoS, and certain percentages of bandwidth are reserved for different types of traffic. The percentages of traffic dedicated to specific types of traffic depend on the expected traffic mix in the network. If application requirements exhibit a marked difference based on time of day, then it makes sense to reconfigure the bandwidth allocations based on time of day. A realistic example of this is that during the night, certain files may need to be backed up over the network, and therefore there will be large volumes of FTP traffic at night; and during the day, more bandwidth must be dedicated to real-time voice and video applications, which are not that prevalent during night hours.

Policies can be put in place to reconfigure the network based on time of day. A special kind of event, a timed event, can be generated by a simple application at the desired times specified in a policy. The following policies can trigger configuration changes at specific times of day:

Policy 1

Event: Time is 08:00 UTC.

Condition: None

Action: Set DiffServ configuration on all network routers to allocate 50% of the bandwidth for EF (expedited forwarding) traffic, 30% for AF (assured forwarding) traffic, and 20% for BE (best effort) traffic.

Policy 2

Event: Time is 20:00 UTC.

Condition: None

Action: Set DiffServ configuration on all network routers to allocate 20% of the bandwidth for EF (expedited forwarding) traffic, 60% for AF (assured forwarding) traffic, and 20% for BE (best effort) traffic.

Policy 1 above will modify DiffServ configurations at 8 am, and Policy 2 will modify these configurations at 8 pm. The two policies together will automatically keep the network correctly configured to obtain the desired QoS.

Example 3. This example shows how policies can be used to perform data filtering in a monitoring component of a management system. All network management systems must periodically monitor various network elements that are within their management domains. In a distributed management system, an instance of the network management system monitors local network elements and sometimes needs to communicate monitored information to another instance of the network management system that is running on a different node in the network. In an ad hoc network, it is especially important to limit these over-the-air communications to the bare minimum required for network operations, because over-the-air bandwidth is scarce. The obvious way to do this is to filter the information being sent over the air based on criteria such as threshold crossings. For example, battery power is an important quantity that needs to be monitored on ad hoc network nodes; but this quantity doesn't necessarily need to be reported over the air to another manager in the network unless there is a problem that the other network manager needs to know about. The following policy can be used to specify such a filter:

Event: 60-second timer

Condition: Battery power <10%

Action: Report battery power to higher-level manager.

This policy specifies an event that is a 60-second timer. It triggers every 60 seconds. The policy condition to be checked is whether the battery power has dropped below a certain level (here 10%). If this condition is true, then the value of the battery power is reported to a higher-level manager so that it can take appropriate action. Thus, the battery power is monitored every 60 seconds, but it isn't sent over the air until there is a problem (the battery power is dangerously low).

2.4.2 High-Level View of an Access Control Policy

As discussed earlier, access control policies are those that are used by a PDP to provide yes/no decisions in response to PEP requests. The most common example of such policies is authorization policies, which control access to resources.

It should be noted that there are cases when policies about access control can be directly enforced by a device. The most common example of this is access control lists that are implemented by firewalls or routers in order to restrict the types of traffic that are allowed to pass through these devices in a network. Such policies are classified as configuration policies and not as access control policies, since no PDP is required to make decisions about the policies. Note that in this case, the access control actions are performed at the data plane (by a firewall or router) and not by a PDP at the management plane.

2.4.2.1 Definition of an Access Control Policy. A general form of an access control policy can be represented by the following:

- *Subject*: The entity that is requesting a decision.
- *Request*: Parameters describing the request for which a decision is needed.
- *Target*: Object that the subject is requesting permission to manipulate.
- *Conditions*: Optionally, conditions can be specified that restrict the circumstances under which a decision applies.
- *Permit/Deny*: Indicates whether the request should be permitted or denied.

The semantics of these fields are as follows: The PDP responds affirmatively or negatively (based on the permit/deny field) to the request from the subject, provided that the specified conditions (if any) are satisfied. As an example, consider a policy that states that:

- Subject: Alice
- Request: Permission to configure router X
- Target: Router X
- Conditions: Router X is not in standby mode.
- Permit/Deny: Permit

This policy states that Alice should be granted permission to configure router X, provided that the router is not in standby mode.

Note that it is usually not possible to have an exhaustive list of policies that covers all possible subjects and targets. Thus it is important for any policy decision mechanism that implements access control policies to specify the default outcome for requests that are not handled by any policy. The default could be to deny all requests that do not have explicit "permit" policies defined; such a default ensures that only requests that have explicit policies allowing access to the target succeed, and all others are denied. On the other hand, the default could be to allow all requests that do not have explicit "deny" policies.

2.4.2.2 *Examples of Access Control Policies.* This section provides some examples of access control policies.

Example 1. Assume that a user signals a request for QoS using RSVP. The ingress router for this request has a local policy decision point that must make the following policy decision for this request: Is this user entitled to the requested level of QoS? This decision is made based on stored policies that indicate the level of QoS to which users/applications are entitled. A sample policy has the following information:

- Subject: [user 1, user 2, user 3, user 4]
- Request: Guaranteed service; 1 Mb/s
- Target: High-resolution video
- Condition: Time is between 08:00 and 17:00 UTC.
- Permit/Deny: Permit

This policy states that users 1, 2, 3, and 4 are entitled to receive guaranteed service of 1 Mb/s for high-resolution video traffic flows between 8:00 am and 5:00 pm.

Example 2. In this example, consider a scenario where there is a protected resource in the network, such as a server containing confidential data. In such a case, policies can be defined that list which users are allowed access to the resource. A sample policy could be:

- Subject: user 1
- Request: Read/write
- Target: ConfDB
- Condition: None
- Permit/Deny: Permit

TABLE 2.1 Sample Policies for Content Filtering

	Policy 1	Policy 2
Subject:	Bob	Bob
Request:	Web access	Web access
Target:	URL categories 5, 8, 9, 11	URL categories 5, 8, 9, 11, 16, 18
Conditions:	09:00–18:00	18:00–9:00
Permit/Deny:	Permit	Permit

The above policy states that user 1 is allowed to read and write information in the database named ConfDB.

Example 3. This example makes use of policies for content filtering. Two policies are specified in Table 2.1; these two policies state that Bob is permitted to access URLs that fall into categories 5, 8, 9, and 11 (and no others) between 9 am and 6 pm. If he tries to access any other URL during this time period, the attempt is denied. Between 6 pm and 9 am, he can access URLs belonging to categories 5, 8, 9, 11, 16, and 18. If he tries to access any other URL during this time period, the attempt is denied.

2.4.3 High-Level View of a Configuration Policy

As discussed earlier, there are two types of configuration policies: (a) administrative policies and (b) configuration policies for network elements, protocols, and services. These policies have the common characteristic that they specify configuration parameters for some entity. Below, a description of what these policies look like is provided, along with some examples.

2.4.3.1 Definition of a Configuration Policy. In the case of configuration policies for network elements, protocols, and services, the entity being configured is either a network device or a network service. Such entities typically have well-defined, standard or proprietary configuration interfaces. As an example, a Cisco router has a command-line interface that accepts Internetwork Operating System (IOS) commands as configuration commands. As another example, a DHCP service running on a Linux router has a well-defined interface for configuring it, via configuration files placed in a well-known location.

In the case of administrative policies, the entity being configured is a software component that is usually part of the network management software. Such entities typically do *not* have well-defined configuration interfaces. Furthermore, every time a new administrative policy is defined, it may require software changes to support the semantics of the new policy. As an example, consider a policy that states that no more than 5 Mb/s of bandwidth should be allocated to any one customer. The management component that is responsible for allocating bandwidth to customers could be the configuration management component. In light of the need to have a configurable upper bound on the amount of bandwidth allocated to each user, there must be a way to configure the appropriate component with this parameter.

Given that configuration policies simply specify configuration parameters, they are defined by a configuration action. The action specifies the following:

- *Configuration Target*: This is the entity being configured, e.g. the router, service, or component being configured. The configuration target is sometimes implicit or not included as part of the configuration policy.
- *Configuration Parameters*: Specifies one or more configuration parameters for the entity being configured.

Recall that there are cases when policies about access control can be directly enforced by a device. The most common example of this are access control lists that are implemented by firewalls or routers in order to restrict the types of traffic that are allowed to pass through these devices in a network. Such policies are classified as configuration policies and not as access control policies, since no PDP is required to make decisions about the policies. Here the configuration target is the device (e.g., a firewall), and the configuration parameters specify the characteristics of the traffic that is being blocked or permitted to flow through the device. A sample set of parameters is:

Source IP address: 192.168.0.0/16
Destination IP address: 10.0.0.0/8
Source port: *
Destination port: 21
Operation: Permit

The above parameters specify a five-tuple consisting of source IP address, destination IP address, source port, destination port, and permission (permit or deny). The parameters listed here state that any traffic originating in subnet 192.168.0.0 from any port and destined for subnet 10.0.0.0, port 21 (FTP server port) should be permitted.

2.4.3.2 *Examples of Configuration Policies.* This section provides examples of configuration policies.

Example 1. As mentioned earlier, QoS configuration parameters are often referred to as policies. An example of such a policy is a rule for traffic conditioning. In the DiffServ paradigm, traffic must be marked with a DiffServ Code point (DSCP) at the ingress to the network, so that routers throughout the network can provide it with differentiated treatment based on this marking. In addition to marking traffic, it is important to police DiffServ traffic to ensure that there is a bound on the amount of traffic in each traffic class. The following policy states how traffic should be marked and policed:

- Rate-limit traffic originating in subnet 192.168.1.0/24 and destined to subnet 192.168.2.0/24 so that it does not exceed 1 Mb/s with a burst size of 100 Kb/s and an excess burst size of 200 Kb/s. Traffic conforming to the rate limit should be marked with a DSCP of 26, and traffic exceeding the excess burst size should be dropped.

The above configuration policy states that at most 1 Mb/s of traffic will be allowed from subnet 192.168.1.0/24 to subnet 192.168.2.0/24. A burst amount of 100 Kb/s is allowed for this traffic, and any traffic within the rate limit will be marked with a DSCP of 26. This policy, which performs both policing and marking, will ensure that an upper bound is maintained on the amount of traffic that is marked with a DSCP of 26 between the above subnets; any excess traffic will be dropped. Another reasonable option would be to mark excess traffic as best effort. Such a policy can be configured directly into routers using the interface supported by the device.

Example 2. This example describes configuration policies for BGP routing. A router sends and receives routing information to and from its peer routers via a routing protocol; and route filtering allows it to select a subset of this information for use in its routing table or for advertising to other routers. BGP policies can be configured to control the processing of prefixes and route filtering from route advertisements. The purpose of route filtering is to control which paths are used by a router to forward incoming packets. As an example of the use of route filtering, a router may have a number of stub networks that connect to the network core via a router. Since each of the stub networks have only one path to the rest of the world, the router can send just one default route to each of these stub networks. The following is a sample routing policy:

- Advertise only a default route on the interface connecting a stub network to this router.

Example 3. The two preceding examples showed policies for configuration of QoS and routing parameters, respectively. This example describes the configuration of network management software for performance management. The following is a (nonexhaustive) list of parameters that can be configured within the performance management component of a network management system:

- *Polling Intervals for Different Variables*: These are used to specify how frequently performance management polls different sets of variables, such as traffic statistics, QoS statistics, and so on.

- *Reporting Intervals for Different Variables*: These are used to specify how frequently to report the values of different sets of variables from distributed managers to a central or higher-level manager.
- *Aggregation Parameters*: These parameters specify how monitored data should be aggregated before being reported. Aggregation methods include computing averages over the last few monitored values; computing maxima and minima; computing sums of different monitored values, and so on.
- *Threshold Parameters*: These parameters specify values of thresholds for different monitored variables. Thresholds are used to trigger various types of actions, such as alarm generation.
- *Impairment and Congestion Declaration Criteria*: These criteria are used to determine whether network elements are impaired, and whether the network is congested, by specifying meaningful thresholds that can be changed dynamically.
- *Trending/Forecasting Algorithms*: These parameters allow different algorithms to be activated at different times, based on specified criteria.
- *Raw Data Filtering*: Filtering parameters such as upper and lower bounds can be specified as configurable parameters to filter the data that are used in report generation, trending, and so on.

The above parameters can be changed and communicated to the performance management component via a predefined interface. Changes in these parameters affect the functioning of the performance management component as described.

2.5 POLICY DECISION PARADIGMS: OUTSOURCED VERSUS PROVISIONED

The previous section categorized policies based on their content—that is, whether they were specified as ECA (event–condition–action) policies versus access control policies versus configuration policies. Another way to categorize policies is based on the paradigm used for communication between the PDP and the PEPs. Policies whose policy decisions are explicitly requested from the PDP by the PEPs are called *outsourced* policies, whereas policies whose policy decisions are pushed by the PDP to the PEPs in an unsolicited fashion are called *provisioned* policies.

The above terminology originated from the policy-related work that began in the IETF Resource Allocation Protocol (RAP) working group in the late 1990s. This working group was originally chartered to establish a scalable policy control model for RSVP (Resource ReSerVation Protocol) (see Section 3.1.1 in Chapter 3). As defined by the Integrated Services working group, RSVP [Braden et al., 1997] is a signaling protocol that allows data flows to describe their desired QoS to the network. RSVP carries the service requests to every router along the path from the source to the destination of the traffic, and these routers each must determine whether or not they have sufficient resources to provide the requested level of QoS to each flow. RSVP message formats contain a place-holder for policy data elements, which may contain information relevant to each network element's decision to grant or deny a reservation request.

Rather than require each network element to make a decision in isolation about whether or not to grant QoS to a traffic flow, the RAP working group defined a protocol, the Common Open Policy Service (COPS) for use among RSVP-capable network nodes and *policy servers*, that would allow the network nodes to ask a policy server for a decision about whether to grant QoS resources to a flow or not. In this architecture, a policy server is an entity that can make policy decisions on behalf of network elements (see Figure 2.4). When a flow is being set up, RSVP signaling is used to request QoS. Network elements, which are the PEPs in this scenario, then ask the policy server for a decision about whether to admit the flow or not. This is the act of *outsourcing* a policy decision. Note that in this paradigm, the policy server responds to requests from PEPs in a synchronous fashion. The protocol used to communicate between PEPs and the policy server is COPS [Durham et al., 2000]. Chapter 3 describes the COPS–RSVP policy-based decision-making framework in greater detail.

Although the original charter of the RAP working group was to define the COPS protocol, the emergence of the Differentiated Services QoS standards [Braden et al., 1997] led to a need for a different paradigm for PDP–PEP communications. Although RSVP guarantees that resources are reserved for each

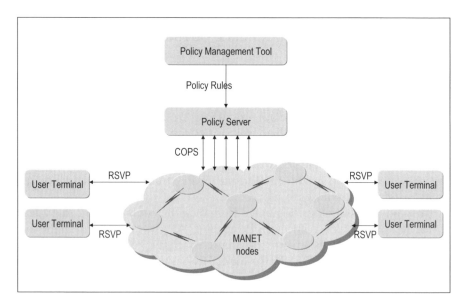

Figure 2.4. Usage of COPS with RSVP.

admitted flow, the drawback of using RSVP for QoS is that it lacks scalability, since every router in the network must keep state about every flow traversing that router. The amount of state kept in each router grows linearly as the number of flows traversing that router. The Differentiated Services approach, or DiffServ, aggregates traffic into a small number of classes and provides differentiated services to each class of traffic (see Section 3.1.2 in Chapter 3). Traffic is *conditioned* at the network edge (i.e., classified into one of several classes of traffic) so that the network routers do not need to be aware of every flow; rather, they are configured with mechanisms to treat the different classes of service in a manner that provides the treatment most appropriate for that class of traffic. Routers in a network need to be configured in a consistent manner to ensure that IP packets belonging to the same DiffServ class are treated in a similar fashion throughout the network. Thus there arose a need for a protocol to enable *provisioning*—that is, asynchronously pushing configuration information to network routers. Here, rather than have network elements request policy decisions and have a policy server respond to requests synchronously (as in the case of RSVP described above), configuration had to be pushed asynchronously to the network elements. This need led to the development of a new protocol, COPS-PR (COPS for PRovisioning) [Chan et al., 2001]. This protocol is used by policy servers that need to push configuration information to network elements. The pushing of configuration information to network elements was referred to as provisioning of policies (note that in this case what is being pushed is configuration information), and the configuration policies that were pushed to network elements were referred to as *provisioned policies*. Figure 2.5 illustrates the COPS-PR paradigm.

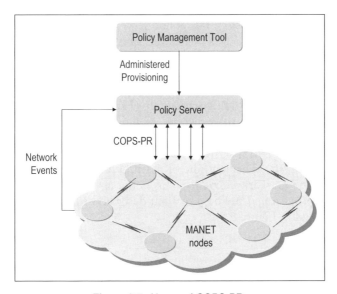

Figure 2.5. Usage of COPS-PR.

2.6 ANOTHER LOOK AT PROVISIONED AND OUTSOURCED POLICIES

The categorization of policies into ECA (event–condition–action) policies versus AC (access control) policies and into outsourced policies versus provisioned policies is not completely independent. At first, it may appear that the two categorizations are orthogonal, since the first one (policies as ECA policies versus AC policies) is concerned with the content of the policies themselves, whereas the second categorization (outsourced versus provisioned policies) deals with the communication paradigm between the PDP (or policy server) and the PEPs (synchronous versus asynchronous). However, it turns out that policies expressed as AC policies are implemented in a policy management system using the outsourcing policy paradigm, whereas policies expressed as ECA policies are implemented using the provisioned policy paradigm. Let us look at each of these in turn.

Policies expressed as AC policies are implemented in a policy management system by creating a process whereby a policy server is asked for permission to perform a certain action whenever there is a protected resource. This is akin to the concept of outsourced policy: Here the potential performer of the action is the PEP, and it asks the policy server for permission to perform an action (e.g., admit a traffic flow, in the case of the RSVP example in an earlier section). The policy server looks up its policies, which are typically constraints on which flows should be admitted, and responds with a decision. Another example of this paradigm is the use of policies for authorization. Such policies limit the users of resources based on a set of constraints (or policies) that describe which users can use which resources. An access control system provides a mechanism whereby users who wish to access a resource must first ask a policy server for permission; the policy server looks up the access control policies and responds with a yes/no answer. This is the synchronous response, or outsourced policy paradigm, as depicted in Figure 2.6.

ECA policies use the provisioned policy paradigm. The policy server has access to a set of policy rules, stored in a policy repository, and provisions network elements or software components (PEPs) with configuration information based on what the rules dictate. As an example, there may be a rule that dictates that at a certain time of day all network routers must be provisioned with certain DiffServ queuing mechanisms. The policy server must trigger this rule at the specified time of day and push the required configuration down to the network elements. This is the asynchronous pushing paradigm, or the provisioned policy paradigm described in the previous section, and is illustrated in Figure 2.7.

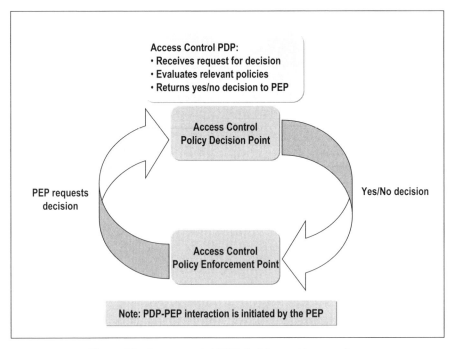

Figure 2.6. Access control PDP decision-making paradigm.

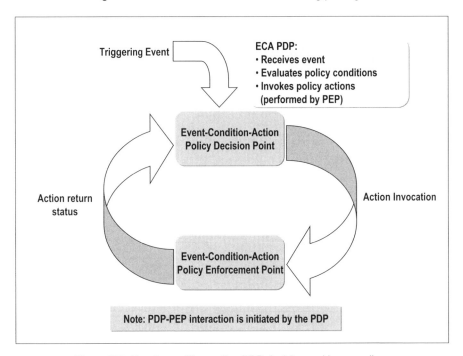

Figure 2.7. Event–condition–action PDP decision-making paradigm.

2.7 SUMMARY

This introductory chapter provided motivation for the use of policies in network management. A brief overview of the IETF Policy Framework was provided, along with associated terminology. The use of policies for automating network management and for increasing the flexibility available to network operators was discussed. Different types of policies were described, along with the associated decision-making paradigms.

3

POLICY LANGUAGES AND FRAMEWORKS

The previous chapter provided an introduction to policy-based management, the different types of policies used for network management, and policy decision-making paradigms. In this chapter, the focus shifts to policy languages and frameworks that have been developed recently. The chapter begins by looking at the policy-related efforts of various standards bodies, and then it describes some proprietary languages and frameworks that have been used in academic and experimental settings. Section 3.1 describes IETF efforts that are relevant to policy management, while Section 3.2 provides a brief overview of the DMTF (Distributed Management Task Force) policy standards efforts. In Sections 3.3.1 and 3.3.2, two policy languages, Ponder and PECAN, are described in order to give the reader a flavor of the state of the art in policy language implementations. Section 3.3.3 concludes this chapter with a brief summary.

3.1 RELATED IETF EFFORTS

A high-level discussion of the IETF work in the area of COPS and Policy Framework definition was provided in the previous chapter, in order to explain the genesis of different policy decision-making paradigms. In this section, details of the relevant standards developed by the IETF in the area of policies are provided. Descriptions of RSVP, COPS, DiffServ, COPS-PR, the IETF Policy Framework, and directory standards are provided. Although RSVP, COPS, and COPS-PR are not recommended for use in MANETs, it is instructional to take a look at these standards in order to understand why.

Policy-Driven Mobile Ad hoc Network Management, by Ritu Chadha and Latha Kant
Copyright © 2008 John Wiley & Sons, Inc.

3.1.1 COPS and RSVP

COPS [Durham et al., 2000] was the first policy-related protocol to be standardized by the IETF. This section provides some background about the motivation for this work and the resulting standards that were developed.

3.1.1.1 Overview. The service model of the Internet is based on *packet switching*; that is, data are inserted into *packets*, which are carried from the source to their destination based on information in the packet *header*. The IP packet header contains information such as the source and destination IP addresses of the packet, the source and destination ports, and other relevant data that are used to route the packet to its destination. In general, all packets are treated equally with respect to scheduling them for delivery; this paradigm came to be known as *best effort* delivery. However, in the mid-1990s, an effort at defining extensions to the basic best effort paradigm of the Internet began with the definition of the Integrated Services (IntServ) framework in the IETF [Braden et al., 1994]. This effort was motivated by the fact that new real-time services being supported over IP required certain service guarantees that were not supported by the existing best effort delivery paradigm. IntServ and its control protocol, RSVP (Resource reSerVation Protocol) are described in the remainder of this section.

3.1.1.2 A Quick Tour of RSVP and the Integrated Services Framework. The IntServ framework is based on the concept that per-flow reservation state must be maintained and resources must be reserved at every router in the network in order to provide different types of service for different applications.

Applications can be classified into two categories: *real-time* applications, which have stringent bounds on packet delivery delay, and *elastic* applications, which can wait indefinitely for packet delivery. In order to support the needs of these applications, the IntServ framework defined two new types of service, guaranteed and predictive service. Guaranteed service places a reliable upper bound on delay, whereas predictive service provides a fairly reliable, but not guaranteed, delay bound on packet delivery.

In order to maintain per-flow reservation state in every router in the network, every flow must request admission into the desired class of service. This is accomplished via the Resource reSerVation Protocol (RSVP) [Braden et al., 1997]. RSVP is a control protocol that is used to signal applications' QoS requirements, described via a *flowspec* (flow specification). An accompanying *filterspec* is used to describe the packets of the flow. Every router along the path of the flow processes an RSVP request and determines whether or not it can admit the flow. The decision about whether or not to admit a flow has two components to it. First, a decision needs to be made about whether a flow should be admitted or not based on criteria other than network resource availability. Such decisions are usually called *policy decisions*, and they are made based on criteria such as the identity of the flow requestor, the type of application, and so on. If the policy decision indicates that the flow should be admitted, then the second decision that needs to be made is whether the router has enough resources to admit the flow so that it can be given the requested level of service. Once a flow has been admitted, each router configures its packet classification mechanism based on the filterspec. Whenever a packet arrives at the router, the packet classification mechanism at the router tells it whether the packet should receive QoS treatment or not. Such treatment consists of placing the packet in appropriate queues and scheduling it for transmission in a manner that reflects the quality of service guarantees that were requested in the flowspec.

The IntServ framework includes four components: a packet scheduler, an admission control component, a classifier, and a reservation setup protocol. The packet scheduler manages packet forwarding using queuing and policing mechanisms, based on the flowspec. The classifier maps flows into one of several classes of service based on the filterspec; each class gets similar treatment by network routers. The admission control component is the decision-making component that decides which flows get admitted into which classes of service, based on the available network resources and policies, as described above. The last component is a reservation setup protocol that is used to signal flow requirements and sets up flow-specific state in every node along the path of a flow. An important point to note about RSVP is that it assigns responsibility for requesting QoS to the receiver of the flow rather than the sender. This is intended to help with accommodating multicast flows.

Now for a closer look at RSVP. RSVP includes seven messages: PATH, PathTear, RESV, ResvTear, ResvConf, PathErr, ResvErr. At a very high level, these messages perform the following functions:

- *PATH*: The PATH message initiates QoS signaling on behalf of the source of a flow. It is sent to the destination of the flow, and it carries information describing the flow. It also stores information about the path that it traverses from source to destination (as routed by the routing protocol), so that the RESV message can use this path in the reverse direction from the destination to the source.

- *RESV*: The RESV message is sent in the reverse direction of the PATH message, in response to receipt of the PATH message. The receiver of a PATH message sends the RESV message hop by hop, using information about which hop to send to from the PATH message. The RESV message contains information about the flow and the requested QoS for the flow (since the receiver is responsible for requesting QoS). At every hop, the RESV message is processed and a determination is made about whether to admit the flow or not.

- *ResvConf*: Receivers may request confirmation in their RESV messages; the resulting message is a ResvConf message.

- *PathErr*: A PathErr message is sent from receiver to sender to indicate an error in a received PATH message.
- *ResvTear*: A ResvErr message is sent to the sender to indicate a processing error anywhere along the path of a received RESV message.
- *PathTear, ResvTear*: The PathTear and ResvTear messages are used to tear down an existing flow reservation and are sent by the sender and receiver of the flow, respectively. Note that explicit teardowns of existing reservations are not strictly required, since flow state will eventually time out in the routers.

As mentioned above, since the receiver makes the request for QoS (via the RESV message), at every hop on the path of the RESV message, the RSVP process on that node makes two decisions:

1. *A Policy Decision*: The policy decision takes into account policy data in the RESV message and returns a policy decision about whether or not the flow should be admitted (more on this below).
2. *An Admission Control Decision*: The admission control decision determines whether the node has sufficient resources to provide the flow with the requested QoS.

If either of these decisions are negative, the reservation fails and an error message (ResvErr) is returned to the receiver. If both decisions are positive, then the node configures a packet classifier to select the packets for this flow (as described in the RESV message) and provide them with the requested QoS by interacting with the appropriate link layer mechanism.

The policy data contained in the RESV message is examined next. Every RSVP message has a common header, followed by a body with variable-length, typed objects. Each object header contains information about its length, along with a class number that identifies the object class. One of the defined object classes is the POLICY_DATA object class, which may appear in PATH, RESV, PathErr, and ResvErr messages. This object class was not fully defined in the original RSVP specification (although a placeholder was specified); it was later defined in RFC 2750 [Herzog, 2000]. The POLICY_DATA object contains the following options:

- FILTER_SPEC or SCOPE: Describes the set of senders.
- Originating RSVP_HOP: Identifies the node that constructed the policy object.
- Destination RSVP_HOP: Identifies the destination policy node. If a node's address doesn't match this, it should ignore the policy data.
- INTEGRITY: Used for securing policy communications.
- Policy refresh TIME_VALUES: Specifies how frequently policy data must be refreshed. Since policy data can be rather large in size, this parameter

can be used to reduce the refresh frequency (i.e., make refreshes of policy data less frequent than the refreshes needed for other RSVP data).
- Policy elements: One or more policy elements can be included in this message. These policy elements are opaque to RSVP.

Policy elements constitute the interesting part of the POLICY_DATA object. The intent is that these policy elements will be used by a policy decision point local to the node to make a policy decision about whether or not to admit the flow. To date, several policy elements have been defined in IETF RFCs:

- *Identity Representation for RSVP* [Yadav et al., 2001]: This policy element carries identity information that can be used to securely identify the owner and the application that is requesting QoS, so that a policy decision point can use this identity information in its decision-making process about whether or not to admit the flow. This is a very useful element, because any policy decision point should take into account the identity of the requestor before deciding to grant a request. The defined authentication policy element (AUTH_DATA) is inserted into the RSVP PATH message at the source node and carries information about the credentials of the user or application requesting QoS. This information is processed at each node by a policy decision point; the policy decision point may modify this information before forwarding the request to the next hop in the path. The AUTH_DATA policy element contains a list of authentication attributes. The currently defined attributes include:
  - Policy Locator: This is a unique string for locating the admission policy; an example is an X.500 distinguished name, which provides a handle that can be used to look up the appropriate authentication information in an X.500 directory [CCITT, 1993] (see Section 3.1.4).
  - Credential: Contains user credential information such as a Kerberos session ticket or a digital certificate (for public key-based authentication); or application credential information such as application identifier information.
  - Digital Signature: Contains the digital signature of the authentication data policy element.
  - Policy Error Object: If the policy decision point at an RSVP node encounters a request that fails to be admitted due to the contents of its AUTH_DATA policy element, it adds a policy error code containing information about the failure to the policy element and inserts the policy element into a PathErr or ResvErr message.
- *Signaled Preemption Priority Policy Element* [Herzog, 2001]: This policy element was introduced to carry information about the priority of a flow, so that in situations where a flow could not be admitted into the network, other flows of lower priority could be preempted to accommodate the new

incoming flow. The PREEMPTION_PRI policy element carries two 16-bit integer quantities that represent the preemption priority and the defending priority, respectively, of the incoming flow. The preemption priority is compared with the defending priority of existing flows; if the defending priority of an existing flow is lower than the preemption priority of the requesting flow, then the former can be preempted to accommodate the latter. Once a flow has been admitted, its defending priority is used to determine if it can be preempted to accommodate new flow requests.

- *Application and Sub-Application Identity Policy Element for Use with RSVP* [Bernet and Pabbati, 2000]: This policy element carries information that describes the application that is requesting QoS for its data flow. The AUTH_APP policy element carries a policy locator similar to that used by the AUTH_DATA policy element. The policy element also carries an attribute representing the application name.

- *Session Authorization Policy Element* [Hamer et al., 2003]: This policy element is used to support policy-based per-session authorization and admission control. The AUTH_SESSION policy element is used to convey information about the resources authorized for use by a session. The sending host inserts an AUTH_SESSION element into the resource reservation message to allow verification of the network resource request. Contrast this with the AUTH_DATA policy element described above, which provides information about user or application identity [Yadav et al., 2001]; the AUTH-SESSION policy element provides the ability to perform per-session admission control.

3.1.1.3 *Usage of COPS for Policy Decision-Making.* In the IETF Resource Allocation Protocol (RAP) Working Group, the need for using criteria for RSVP-based admission control that went beyond simple link capacities led to the standardization of the Common Open Policy Service (COPS) protocol for communicating policies from policy servers to RSVP routers. The underlying architectural model [Yavatkar et al., 2000] dictated that policy servers administer the network, communicating decisions to policy clients (e.g., network elements), where the policy decisions are enforced. Basically, the decisions concern who is authorized to access what resource in the network. In particular, if IP QoS is deployed, the users can access different transport services and this access must be administratively regulated. However, despite this initial application focus, the COPS policy model is general enough to be used for outsourcing policy decisions for technologies other than RSVP. The functional model originally considered for supporting outsourcing of policy decisions involved (1) routers or other network elements with a Policy Enforcement Point (PEP) module and (2) a Policy Decision Point (PDP) located in a separate policy server. The PEP and PDP communicate via the COPS protocol. This model works well, for example, for an IntServ/RSVP framework [Braden et al., 1997] for supporting Internet quality of service (QoS).

The RAP working group developed the COPS protocol with the objective of complementing the *resource-related* admission control defined in the IntServ model with *policy-related* admission control. The requirements for the initial definition of the policy-based admission control architecture and of the COPS protocol were mainly derived considering the IntServ RSVP signaling protocol. In this scenario (see Boyle et al. [2000]), the network nodes, running the RSVP protocol, represent the Policy Enforcement Points (PEPs), while a logically centralized element acts as a policy server and is called Policy Decision Point (PDP). The PEP makes *Requests* to the PDP for policy-related admission control, and the PDP provides the needed policy *Decisions*.

The COPS standard [Durham et al., 2000] describes a client-server model for supporting policy control over QoS signaling protocols. In this model, the clients are PEPs, or Policy Enforcement Points, and the server is a PDP, or a Policy Decision Point. As described earlier in this section, an example of a policy client is a router running RSVP, where policy-controlled admission control is provided by outsourcing policy decisions to a policy server. The COPS protocol allows the client to send requests, updates, and deletes to the policy server; the latter returns policy decisions to the client. In addition, the policy server can push unsolicited configuration information to the clients. COPS runs over TCP; clients are responsible for initiating a TCP session with the server, and this session is used for message exchanges between the client and server. COPS was designed to communicate self-identifying objects which contain the data necessary for decision-making. The data transmitted via these objects depend on the client type.

The different types of COPS operations are:

- *Request (REQ)*. This message is sent from the PEP to the PDP and is used to establish a handle for a client with the PDP. This handle is then used in subsequent messages to refer to the session and update its state. The REQ message includes client-specific information. When COPS is used in conjunction with RSVP, the client-specific information includes all the objects received in the RSVP message that results in a COPS decision request.
- *Decision (DEC)*. This message is sent from PDP to PEP in response to a REQ message. It includes the client handle (that was specified in the REQ message) and decision or error objects (if there were problems in the decision process).
- *Report State (RPT)*. This message is sent from the PEP to the PDP and is used by the PEP to report success or failure in implementing a PDP decision.
- *Delete Request State (DRQ)*. This message is sent from the PEP to the PDP and is used by the PEP to tell the PDP to delete the state associated with the specified client handle.

- *Synchronize State Request (SSQ)*. This message is sent from the PDP to the PEP and is used by the PDP to request the latest state information associated with a specified client handle from a PEP. If the client handle is not specified, the PEP sends all its state information to the PDP.
- *Client-Open (OPN)*. The Client-Open message is sent from the PEP to the PDP. It can be used by the PEP to tell the PDP the client types that the PEP can support, as well as the last PDP to which the PEP connected for the given client type.
- *Client-Accept (CAT)*. This message is sent from the PDP to the PEP. It is a response to the OPN message.
- *Client-Close (CC)*. This message may be sent from the PDP to the PEP, or vice versa. As an alternative to the CAT message, a PDP may respond with a CC message to an OPN message, if it chooses not to communicate with the client. It can also be sent from PEP to PDP to indicate that a particular type of client is no longer supported.
- *Keep-Alive (KA)*. This message may be sent from the PDP to the PEP, or vice versa. This message is sent by the PEP within the specified timer value. The PDP echoes this message back to the PEP upon receipt.
- *Synchronize State Complete (SSC)*. This message is sent from the PEP to the PDP. This message is sent as a response to the SSQ message to indicate that synchronization is complete.

An RFC [Boyle et al., 2000] defines the usage of COPS with RSVP for policy decision-making. A PEP on an RSVP router initiates communication with its PDP via an OPN message. When the PEP receives a PATH or RESV message, it sends an REQ message to the PDP and includes all of the objects included in that RSVP message in the client-specific information object. This includes information in the policy elements that were described in Section 3.1.1.2. The PDP responds with a DEC message that instructs the PEP to either accept or reject the RSVP message and allocate or deny the requested resources. When making a decision, the PDP considers the RESV as well as the associated PATH messages. The PEP then responds with a RPT message indicating successful installation of the PDP decision.

3.1.1.4 Summary.

As can be seen from the summary in the last few pages, RSVP and COPS provide a comprehensive framework for policy-based admission control for the purpose of guaranteeing QoS in an IP network. However, as mentioned earlier, RSVP has known scalability problems and therefore has not been widely deployed. It is more than likely that this is the reason why COPS has not been successful either. Even though COPS was designed to be a generic policy communication protocol, the IETF work on COPS only explored one use for COPS, and that use was closely coupled with RSVP. Although a lot of research and experimental work has been done on the use of COPS as the communication protocol between the PDP and PEP, vendors have not adopted this protocol, and to date there are hardly any commercial implementations of COPS in vendor equipment. Thus it is doubtful that COPS will ever become a significant player in the policy-based management field.

In the context of mobile ad hoc networks, RSVP is not a viable approach to providing QoS, for several reasons. Apart from the scalability argument, the most important of these reasons is the dynamic nature of ad hoc networks. RSVP relies on reserving resources for every flow on every router along the path of the flow. There is an implicit assumption that the path between the sender and the receiver will remain more or less static; recall that the RSVP PATH message encodes every hop along the path of the flow, and the RSVP RESV message is sent along the path specified in the corresponding PATH message. In MANETs, where the path from a sender to a receiver is expected to change frequently, the RSVP paradigm is not viable because RSVP messaging would have to be repeated every time the path from sender to receiver changes.

3.1.2   COPS-PR and DiffServ

The previous section described the IETF efforts in the area of policy-based admission control using RSVP as the QoS signaling protocol and COPS as the corresponding policy protocol. Following the RSVP work in the IETF, provisioned QoS and DiffServ [Blake et al., 1998] were introduced as a more scalable alternative to IntServ and RSVP; and COPS was extended to provision policies, instead of being used only to outsource decisions for network nodes. As a result, COPS-PR [Chan et al., 2001] was introduced and standardized. Below is a high-level overview of DiffServ, followed by a description of COPS-PR and its usage for policy provisioning.

3.1.2.1   A High-Level Overview of DiffServ. The Differentiated Services framework, or DiffServ, was introduced by the DiffServ working group in 1998 as a reaction to the lack of scalability of RSVP. As a contrast to the IntServ framework, where every router that processes the packets of a flow must maintain state related to that flow, the DiffServ framework aggregates packets into a small number of *behavior aggregates*, or traffic classes, which are accorded similar treatment by every router in the network. When packets enter the network, they are *classified* into one of these traffic classes based on any number of criteria, such as fields in the IP packet header (source IP address, destination IP address, source port, destination port, protocol), user information, and so on. The classification is encoded in the IP header of the packet using the first six bits of the TOS (Type of Service) octet [Nichols et al., 1998]. These six bits are called the DiffServ Code Point (DSCP) of the packet, and they are used by every router along the path of the packet to give the packet the appropriate treatment. The process of stamping an IP packet with a DSCP is called *marking*, or *classification*. The other function performed at the network edge is to *meter* (or *police*) the incoming traffic based on preconfigured traffic profiles. Metering at the edge allows the volume of traffic entering the network to be regulated to control the amount of traffic admitted into each traffic class. Every router along the route of an IP packet examines its DSCP and provides it the treatment corresponding to the traffic class represented by that DSCP. Thus every router must be provisioned ahead of time with configuration that tells it how to treat packets with different DSCP markings (called *conditioning*). Typically, this treatment consists of placing packets with different DSCPs into different queues with corresponding queuing behaviors that result in processing certain packets ahead of others, so that in times of congestion, packets with certain DSCP markings get preferential treatment over others. Much work has been done in the area of queuing treatments; the reader is referred to Jha and Hassan [2002] for an in-depth treatment of queuing in IP routers.

A high-level view of the process of classification, metering, and conditioning is shown in Figure 3.1. The classifier selects packets from a traffic flow. Packets may be sent to a meter, which (a) determines whether the actual rate of the flow conforms to the configured rates and (b) selects an appropriate action, such as sending the packet to another action or dropping it. The marker marks packets by setting the DSCP bits in the TOS octet based on its configuration. A shaper is used to buffer packets when the packets are arriving faster than allowed by their traffic profile. Buffers drop packets that exceed the buffer size. Finally, policers are used to drop traffic that exceeds a specified traffic profile.

The DiffServ standards describe a number of traffic classes that provide different per-hop behaviors in an IP network. The first of these is Expedited Forwarding (EF) [Davie et al., 2002], which is designed for traffic with stringent requirements on delay, and is suited for real-time applications such as voice and video, which are delay-intolerant. For such applications, packets that arrive later than a certain bound are of no use; also, such applications can tolerate some amount of loss. The second set of traffic classes is the Assured Forwarding (AF) group [Heinanen et al., 1999], which is designed for traffic

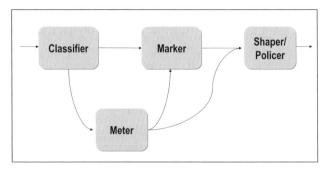

Figure 3.1. Traffic conditioning, marking, shaping.

that cannot tolerate loss but can tolerate some amount of delay. Examples of such applications are file transfer, e-mail, and so on which do not have stringent delay guarantees, but require that all sent packets are received at the destination. Packets that do not require QoS are not marked and are said to belong to the Best Effort class.

In order to provide Differentiated Services, routers must be configured with information that allows them to classify and condition traffic in accordance with traffic requirements. Edge routers must be configured with information about how to mark packets based on information in their IP headers. Note that this may require a large amount of configuration for every edge router, since every edge router must be configured with information about potentially every ingress flow that enters the network at that point. The reason this solution can scale, however, is that the number of flows entering the network at any given ingress router can be limited so that the router has sufficient resources to handle the flows entering the network at that point. Edge routers must also be configured with information about how to meter the traffic belonging to different traffic classes. This is particularly important for EF traffic, which is usually strictly rate-limited to ensure that it does not starve other traffic classes of bandwidth. Bandwidth is usually apportioned among AF traffic classes based on a percentage of the total available bandwidth, without strict rate limits. This apportionment typically reserves some percentage of bandwidth for Best Effort traffic as well.

Core routers in a DiffServ network must be configured with information that tells them how to treat packets that are marked with different DSCPs. Since the total number of DSCPs that may exist is 64 or fewer (recall that 6 bits are used to encode the DSCP, which therefore limits the total number of DSCPs to a maximum of 64), each core router will not have to maintain more state than that required for 64 different behavior aggregates. In practice, it has been shown that a much smaller number of DSCPs would typically be used, thus further lowering the amount of state required in each core router. The configuration information required for each DSCP is queuing and scheduling information. Packets belonging to different DSCPs are placed in different queues when they arrive at a core router, and they are scheduled for transmission based on different scheduling algorithms. For example, an implementation of the EF class may provide a priority queuing scheme that will service all EF packets ahead of any other packets, subject to a rate limit so that other traffic classes do not get starved.

Having described at a high level the configuration required for edge and core routers for supporting DiffServ, the next section takes a look at COPS-PR and its use for provisioning routers.

3.1.2.2 *An Introduction to COPS-PR.* COPS-PR [Chan et al., 2001] was introduced for the purpose of carrying provisioning information from a PDP to a PEP. The difference from the original COPS was that the PDP could initiate provisioning to the PEP(s) rather than be driven by events such as an

incoming RSVP request for resources that resulted in a request/connection from a PEP. But the most significant addition with COPS-PR was the addition of Policy Information Bases (PIBs). In order to allow for bulk provisioning of policy data, PIBs were introduced, providing an enhanced data model similar to that of SNMP Management Information Bases (MIBs) but tailored toward policy provisioning. COPS-PR was designed to be a generic policy provisioning protocol, not limited to QoS. The usage of the terminology "policy provisioning" is a confusing one because the protocol was touted as one that was used for installing *policies* in PEPs; a more apt description would have been installation of *configuration information* in PEPs. The policies that COPS-PR installs in PEPs are really just configuration data, such as that described in the previous section for DiffServ. These policies belong to the category of configuration policies, as described in the previous chapter. The use of the term "policy" for QoS provisioning data is very common and is something that the reader should be aware of.

For the DiffServ model, COPS-PR is used by a logically centralized management center acting as the PDP; the PDP "installs" the proper configuration (*decisions*) in the DiffServ network elements (routers) which represent the PEPs. The configuration is encoded based on objects defined in PIBs. A PIB can be viewed as a tree namespace where the branches of the tree represent structures of data or Provisioning Classes, while the leaves represent various instantiations of Provisioning Instances (PRIs). Each PRI is identified by a Provisioning Instance Identifier (PRID). A PRID is carried in the COPS client-specific object (see Section 3.1.1.3).

As far as the protocol itself is concerned, COPS-PR defines some extensions to the COPS protocol that are used by COPS-PR clients to populate client-specific data. Clients can include these new objects within the COPS Named Client-specific information object or the Named Decision Data object. Policy data are encoded based on the identifiers defined in the corresponding PIB. Thus variables defined in a PIB can be set on the PEP by the PDP using COPS-PR protocol exchanges. A small number of PIBs have been defined in IETF RFCs [Chan et al., 2003; Sahita, 2003]. The first of these is the DiffServ PIB, which defines objects used for provisioning DiffServ capabilities in routers. A central innovation developed in the DiffServ PIB was the use of *roles*. The role mechanism enabled policy data sets destined to groups of interfaces performing the same role to be downloaded only once from the PDP to the PEP, and then used by all interfaces in the PEP with the same role.

3.1.2.3 Summary. The DiffServ standards provide useful concepts for implementing traffic differentiation in IP networks. These concepts are especially valuable for MANETs. This is because bandwidth is typically scarce in MANETs, and therefore it is important to be able to differentiate different types of traffic based on traffic priorities. DiffServ provides the ability to mark IP packets with DSCPs that are used to provide differing levels of service to different types of traffic.

The static provisioning model for DiffServ networks, however, has some annoying limitations. For example, the preconfiguration of network elements may lead to underutilization of resources; it is difficult for the provider to adapt to changes in traffic demand; and the service offering of the provider is basically limited to the transfer of large and stable traffic aggregates. The static assignment of DSCPs to different traffic flows poses problems, since it may happen that too much traffic is marked with a given DSCP, causing congestion in that traffic class. The use of a Bandwidth Broker [Nichols et al., 1999] provides the capability to handle resource requests dynamically. Resources are requested by applications, and a Bandwidth Broker is used to dynamically admit or reject flow requests. Appropriate policies are then dynamically provisioned in ingress routers to allow admitted flows into the network. A higher utilization of network resources and the possibility of offering more advanced services are some of the benefits of Bandwidth Broker-based resource allocation. However, the design of a Bandwidth Broker (or indeed, any type of admission control algorithm) is especially challenging in MANETs due to the dynamic nature of the network. This dynamicity makes it difficult to accurately estimate whether new flows can be admitted to a network or not. Most existing Bandwidth Broker implementations assume a relatively static network topology for making admission control decisions. The assumption of a static topology breaks down in MANETs, where network topology is usually very dynamic. Thus a new admission control paradigm is needed for MANETs that makes effective use of the power of DiffServ. Chapter 8 provides a description of an Admission Control function that acts as a Bandwidth Broker for implementing DiffServ-based QoS in a mobile ad hoc network.

3.1.3 IETF Policy Framework

In 1997, the idea of modeling and using policies for network management resulted in the creation of the Directory-Enabled Networks (DEN) initiative [Strassner, 1999], which was later incorporated into the Common Information Modeling (CIM) effort in the DMTF (Distributed Management Task Force) [DMTF CIM Standards]. The IETF Policy Framework WG was created to leverage and extend standardization of policy information modeling from relevant working groups in the DMTF. Whereas the scope of the RAP policy networking model was limited to the communication protocol between PDPs and PEPs and data modeling of PIBs, which are associated with devices rather than any higher-level network policies, the charter of the Policy Framework WG included (a) standardization of a Policy Information Model for representing high-level network policies and (b) demonstration of its application to QoS policy representation. The information models developed in this WG included the Policy Core Information Model (PCIM) [Moore et al., 2001]; PCIM extensions (PCIMe) [Moore et al., 2003]; the QoS Policy Information Model (QPIM) [Snir et al., 2003]; and the QoS Device Information Model [Moore et al., 2004].

These information models are based upon DMTF CIM/DEN information models.

An interesting topic of debate during the development of the IETF policy information models was the issue of whether or not to define a standard policy language. Early in the history of the IETF Policy Framework working group, the charter of the group included the definition of a Policy Definition Language, and Internet Drafts were written describing such a language [Strassner and Schleimer, 1998]. The consensus within the IETF was that the IETF is not in the business of defining programming languages, and therefore was not going to define a policy language either. Also, the development of a policy language was viewed as being too ambitious a project for this working group. The development of a policy information model was deemed within the scope of the IETF, however, since information models for management information, such as SNMP Management Information Bases (MIBs), have always been accepted as being within the scope of the IETF.

3.1.3.1 Architecture. The IETF policy framework architecture was shown earlier in Chapter 2 and consists of four elements: the policy management tool, the policy repository, the policy decision point, and the policy enforcement point. A network operator uses the *policy management tool* to define the policies that are to be enforced within the network. A device that can apply and execute the different policies is known as the *policy enforcement point* (PEP). A *policy repository* is used to store the policies generated by the management tool. In order to ensure interoperability across products from different vendors, information stored in the repository corresponds to the policy information model specified by the Policy Framework Working Group. The *policy decision point* (PDP) is responsible for interpreting the policies stored in the repository and communicating them to the PEP. The PEP or PDP may be in a single device or different physical devices. The protocol to be used for communicating between the PDP and the PEPs is not dictated by the standards and could be COPS, SNMP, or a proprietary command-line interface (CLI). The policy repository could be a network directory server accessed using the LDAP protocol [Wahl et al., 1997], although this was never mandated by the standards. The IETF's preference for the use of an LDAP server for storing policies was underlined by the fact that LDAP schemas were defined for all of the developed IETF policy information models.

3.1.3.2 IETF Policy Information Model. The IETF Policy Framework working group defined a policy information model as an extension to the DMTF Common Information Model (CIM), known as the Policy Core Information Model, or PCIM [Moore et al., 2001]. An information model is an abstraction and representation of the entities in a managed environment, along with their properties, operations, and relationships. This is independent of any specific repository, application, protocol, or platform. The IETF also defined a mapping of the PCIM model to a directory schema so that an LDAP directory could be used as a policy repository [Strassner et al., 2004]. PCIM provides a high-level framework for the definition of policies in the following format:

If <condition> then <action>.

The condition set can be expressed in either disjunctive or conjunctive normal form. Policy rules may be aggregated into nested policy groups to define the policies pertaining to a department, user, and so on. Conditions and actions can optionally be defined once and reused by multiple rules; or they may be specific to a rule. Sophisticated time period conditions can be defined in terms of times, masks for days in week, days at the beginning or end of the month, months in a year, and so on. The actions can be defined as being sequential or in any order. Specific priority values are assigned to policy rules to resolve conflicts.

PCIM defines two hierarchies of object classes: (a) structural classes representing policy information and control of policies and (b) association classes that indicate how instances of the structural classes are related to each other. The policy classes and associations defined in this model are sufficiently generic to allow them to represent policies related to different applications. Policy rules may be aggregated into policy groups, and these groups may be nested to represent a hierarchy of policies. Policy groups are either (a) aggregations of policy rules or (b) aggregations of policy groups. The set of conditions associated with a policy rule specifies when the policy rule is applicable. If the set of conditions associated with a policy rule evaluates to true, then the set of actions associated with that rule is executed.

PCIM does not contain any structures that enable a user to define any concrete conditions or actions, with the exception of time-period-based conditions. The latter allow a user to specify temporal conditions that capture the semantics of a condition that is true if certain time constraints are satisfied— for example,

Day of the week = Saturday or Sunday

Time of day = 8 am to 5 pm

and so on.

The PCIM model went through several iterations and was finally accepted as a proposed standard in early 2001. However, it was felt that certain extensions to this information model were necessary, and this resulted in the PCIMe information model (PCIM extensions). The work on PCIM extensions focuses on incorporating mechanisms for describing certain kinds of conditions and actions that are of general utility and are not dependent on the domain of application. These extensions allow a user to specify the following:

- Conditions based on IP packet header values—for example, conditions of the form

$$\text{IP source address} = 128.92.3.1$$

$$\text{Port number} = 21$$

and so on.

- Conditions based on values of attributes of object instances in the repository: The policy repository will contain not only policies (the policy information model), but also object instances representing information about the system being modeled and operated upon (the device information model). For example, in an MPLS traffic engineering scenario, the repository would contain object instances representing traffic trunks, LSPs, and associations between a traffic trunk and its currently assigned LSP and its backup LSPs, if any. Policy conditions could then be formed that express a condition in terms of the attributes of these object instances—for example,

$$\text{LSP bandwidth} > 10\,\text{Mbps.}$$

- Generalized actions that allow a user to set the value of an attribute in an object instance—for example,

$$\text{Set LSP bandwidth to 20\,Mbps.}$$

Other domain-specific actions are not part of PCIM and are left to domain-specific information models to define. For example, the QoS Policy Information Model (QPIM) [Snir et al., 2003] contains a set of abstractions specific to IntServ and DiffServ management. The model includes abstract objects for classifiers, meters, shapers, droppers, and queues, which can be mapped onto the control elements provided by DiffServ routers.

3.1.3.3 *Policy Roles.* Policies are expected to be useful because they enable network operators to (a) specify general policies for an entire system and (b) have these policies be automatically applied to all the necessary network elements to realize the intent of the policies. This obviates the need to individually configure each involved resource. However, this necessitates some mechanism to be able to associate policies with the object instances to which they are applicable. As an example, suppose that a network operator wants to double the bandwidth of every gold traffic trunk on weekdays during business hours. The operator could specify a policy that states.

If (it is a weekday) and (time is 8 am to 5 pm) then (double bandwidth)

and apply this policy to all gold traffic trunks.

PCIM therefore provides a mechanism to associate policies with the resources to which they apply. This is done by associating resources with *roles* and by labeling policies with role combinations. Thus a policy could have role combination "silver" + "gold," which means that it applies to resources with role = "silver" and to resources with role = "gold" (called the *targets* of the policy). A traffic trunk could be associated with the role "gold" and thereby have the above policy applied to it. Thus rather than configuring and then later having to update the configuration of hundreds or thousands of resources in a network, a policy administrator assigns one or more roles to each resource, and then specifies the policies for each of these roles. A role combination is a set of attributes that are used to select one or more policies for a set of entities and/or components from among a much larger set of available policies. Roles and role combinations are especially useful in selecting which policies are applicable to a particular set of entities or components when the policy repository can store thousands or hundreds of thousands of policies. This use emphasizes the ability of the role (or role combination) to select the small subset of policies that are applicable from a huge set of policies that are available. This ability is critical for scalability of a policy-based management system.

3.1.3.4 *Weaknesses of the IETF Policy Information Model.* Although the IETF Policy Framework working group resulted in increasing the awareness of the networking community of the usefulness of policy-based management for increasing the automation of network management, the standards themselves suffered from several significant weaknesses that prevented widespread adoption. These shortcomings are summarized below:

• *Lack of an Execution Model*: One critical technical weakness of the IETF representation of policies is that the focus was solely on the definition of an information model for storing policies, while the execution model was left open to interpretation. To understand this, consider a sample IETF policy: *if* ⟨*conditions*⟩ *then* ⟨*actions*⟩. The intended semantics here are that the conditions are evaluated; and if they are true, the actions are executed. However, it is unclear *when* the conditions should be evaluated. Should they be evaluated at the time that the policy is created? Should they be evaluated every time the truth value of a condition changes? If so, how does the PDP know that the truth value of a condition has changed? If the policy conditions refer to certain variables, then it is safe to assume that the truth value of the condition cannot change unless the value of at least one of these variables changes. This means that the PDP must actively monitor the values of all the variables referred to in all the conditions in defined policies. This can be implemented using an active database system [Paton and Diaz, 1999; Paschke, 2005], which enables the triggering of actions whenever the content of the database changes. However, the realization of such a system requires specification of the appropriate rules regarding (a) the variables whose values need to be monitored and (b) the corresponding actions to be taken, thus adding to the complexity of implementation.

In contrast to the IETF policy model, ECA policies include one or more *events* that trigger policy evaluation. Here the execution model is clear: The PDP evaluates the conditions of a policy upon the occurrence of the events in the events portion of the policy. Thus there is a well-defined execution model for ECA policies.

• *Lack of Specificity*: Although the IETF policy model is extremely general and allows users to define their own conditions and actions by subclassing the generic classes for condition and action, this generality became a reason why the policy information model became largely useless. The fact that it became necessary to define proprietary extensions to the IETF policy information model for almost any meaningful policy management application completely defeated the purpose of defining a standard in the first place! There was no hope for interoperability between vendors because of the lack of specificity of the standards. Although PCIMe [Moore et al., 2003] did make a step in the right direction by defining a generic way to make use of variables in conditions and actions, it did not completely solve the problem.

• *Lack of Support from Other IETF Working Groups*: One major weakness of the IETF Policy Framework working group was that other working groups within the IETF provided little or no support. As an example, an attempt was made to define a policy information model for MPLS traffic engineering in 1999 [Isoyama et al., 2000]. This work could not progress because of lack of support from the IETF MPLS working group. The general feeling there was that policy was not part of the charter of the MPLS working group. However, without participation from the subject matter experts for MPLS, the Policy working group could not hope to achieve the required impact and thus this effort was dropped. The work was continued at the DMTF, but due to the much smaller membership and influence of this organization, the work did not have much impact.

3.1.4 Directory Standards

As mentioned in the previous section, although the IETF Policy Framework working group did not mandate any particular implementation for the policy repository, all of the developed IETF policy information models were specified in the form of Lightweight Directory Access Protocol (LDAP) schemas. In the late 1990s, there was a large amount of interest, comment, and speculation within the industry about directory services. In particular, LDAP was positioned as the standard capable of providing universal directory access, as well as a mechanism for integrating separate directories into large-scale, consolidated directory services. In order to get an understanding of LDAP, the place to start is the original directory standard, X.500 (the ISO/ITU international directory standards) [CCITT, 1993].

3.1.4.1 X.500. X.500 is the standard produced by the ISO/ITU defining the protocols and information model for a directory service that is independent of computing application and network platform. First released in 1988 and updated in 1993 and 1997, the X.500 standard defines a specification for a rich, distributed directory based on hierarchically named information objects (directory entries) that users can browse and search. X.500 uses a model of a set of Directory System Agents (DSAs), each holding a portion of the Directory Information Base (DIB). The DSAs cooperate to provide a directory service to user applications in a way that allows these applications to be unaware of the location of the information they are accessing. In other words, the user applications can connect to any Directory Server and issue queries to access information anywhere in the directory.

The X.500 standards address both (a) the way that directory information is structured and controlled within the directory service and (b) the protocols needed to provide access to this information. The original 1988 X.500 standards focused heavily on the protocols to be implemented; there were two protocols that were crucial in providing a truly distributed service:

- *Directory Access Protocol (DAP)*: Specifies how user applications access the directory information.
- *Directory Service Protocol (DSP)*: Specifies the protocol used to propagate user directory requests between Directory Servers when the request cannot be satisfied by the local Directory Server.

The 1993 standards refined the DAP and DSP protocols and addressed the key areas of the control and management of the data held in the directory. These standardized the administration of the directory, while simultaneously allowing maximum flexibility to the administrator of each portion of the directory. The 1993 standard also introduced a third protocol, the Directory Information Shadowing Protocol (DISP), for replication. Replication allows information mastered in one Directory Server to be shadowed to other Directory Servers. Several copies of directory data could then be used to satisfy user queries, resulting in greatly improved response times and greater resilience to failure.

X.500 specifies a distributed, tree-structured directory of information, along with allowed operations on the data. The following three subsections address the way information is distributed in an X.500 directory and the way X.500 entries are named; the way information is stored within individual X.500 entries; and X.500 directory operations.

3.1.4.1.1 The Architecture of an X.500 Directory. An X.500 directory is called a *Directory Information Base (DIB)*. It is structured as a tree, called the *Directory Information Tree (DIT)*. The entries in the database are stored as nodes in the tree. The X.500 Information Model defines the types of data and basic units of information that can be stored in the directory. An object class is used to group related information, and it has the following:

- A name that uniquely identifies the class (e.g., person, organization, router, etc.)
- An object identifier (OID) that also uniquely identifies the class
- A set of mandatory attributes and a set of allowed attributes

The basic unit of information is an *entry*, which is a collection of information about an object. An entry may belong to one or more object classes (e.g., a person, an organization, a network device, etc.) and contains *attributes*. Each attribute has a type, syntax, and one or more values. The collection of all information about object classes and their required and allowed attributes are called directory schemas.

As its name implies, a DIT is a hierarchical organization of data which lends itself naturally to representing containment relationships. Entries are stored in the DIT in a hierarchical fashion. A typical DIT is shown in Figure 3.2. This

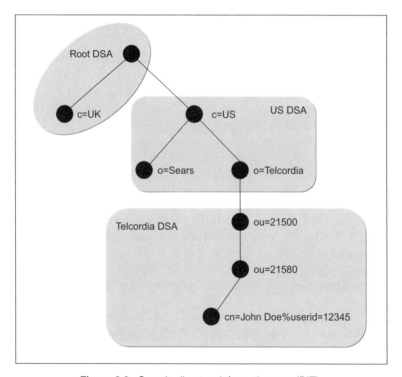

Figure 3.2. Sample directory information tree (DIT).

figure illustrates an employee directory, in which employees are organized under "company" and where companies are organized under "country." The countries represented in the database are the United Kingdom and the United States. Entries are composed of attributes and values. The countries in this database have the attribute `country` (abbreviated `c`). Thus the UK entry has country value "`c=UK`" and the US entry has country value "`c=US`." The organizations (Sears and Telcordia) appear as children of the entry "`c=US`," each having a single attribute `organizationName` (abbreviated `o`). Under the node "`o=Telcordia`," there is an organizational unit with attribute `organizationalUnit` (abbreviated `ou`) with value "`ou=21500`," and under this unit is a subunit with value "`ou=21580`." The leaf of this tree is an employee, with attributes `commonName` (abbreviated `cn`) and `ID`. The values are "`cn=John Doe`" and "`ID=12345`."

In a typical directory there would be other attributes as well, but the attributes shown in the figure are special attributes called *naming* attributes. The name of an attribute along with its value is called an attribute value assertion. The group of one or more attributes shown in the figure for each entry is called the *Relative Distinguished Name (RDN)* of the entry. This name must be unique among siblings. More than one attribute value assertion (AVA) may be used in an RDN. The notation used here to separate attributes in an RDN is the "`%`" character, as used in the employee entry at the leaf of the tree. Entries are named by concatenating the RDNs of all the entry's ancestors starting at the root of the tree to form the entry's *Distinguished Name (DN)*. The distinguished name uniquely names an entry. The "`@`" character is used to separate RDNs in a DN. For example, the DN for John Doe is

```
c=US@o=Telcordia@ou=21500@ou=21580@cn=John
Doe%userid=12345.
```

Directory System Agents (DSAs) are responsible for holding portions of the global tree. Besides containing the data for their portion of the tree, DSAs also have knowledge of other parts of the tree so that if an incoming request is not for data held by the local DSA, it will know where to send the request. The management of information within a particular domain is the responsibility of the DSA administrator. Furthermore, the underlying data may be held in any manner the administrator desires (e.g., a relational DB), so long as the X.500 interface is supported.

A Directory User Agent (DUA) is any process that accesses the Directory. If a DUA asks the US DSA in the figure to read the `c=UK` entry (which is not in the domain of the US DSA), the US DSA can either forward the request to the Root DSA (this is called *chaining* a request) or pass a *referral* for the Root DSA back to the DUA. But if the request is for the `o=Telcordia` entry, the US DSA can access that information locally and pass it back immediately to the DUA without need for chaining or referrals. Requests for entries under the Telcordia entry would have to be chained or referred to the Telcordia

DSA. The remaining examples here focus on Telcordia information, and therefore do not show the c=UK or c=US@o=Sears entries, although it is understood that those entries (and their subtrees) may also exist in the DIT.

3.1.4.1.2 Entries and Object Classes. The set of entries in the DIT comprise the DIB. X.500 entries belong to object classes and contain associated attributes. These object classes are organized in an inheritance hierarchy, such as the one shown in Figure 3.3. Each object class specifies a number of mandatory and optional attributes. In addition, the object class inherits all the mandatory and optional attributes from its ancestor classes. For example, an entry belonging to object class organizationalPerson also belongs to the object class person. Assuming all the attributes shown are mandatory, an organizationalPerson has mandatory attributes commonName, surname, employeeNumber, and telephoneNumber. Different object classes may also have the same attributes, such as the attribute telephoneNumber for both residentialPerson and organizationalPerson. Attributes may have multiple values.

An entry belongs to a collection of object classes, which do not have to be ancestors or descendants of each other. There is no correlation between the inheritance relationship in the object class hierarchy and the parent–child relationship in the DIT structure.

An entry representing a system administrator may be part of the object classes organizationalPerson (and thus person) as well as administrator, thus having mandatory attributes of all three of those object classes. Typically, entries representing the same real-world entities will have the same *set* of object classes. For example, all database administrators in the company might have object classes organizationalPerson, person, and administrator.

The following is an example of an entry:

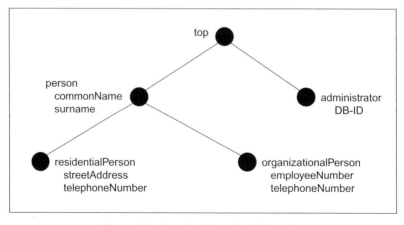

Figure 3.3. Sample object class hierarchy.

```
objectClass = person & organizationalPerson
objectClass = newPilotPerson & top
acl = others # read # entry
acl = self # read # entry
acl = manager # read # entry
surname = Doe
commonName = John Doe
userid = 12345
title = 2
telephoneNumber = 1-732-699-2000
roomNumber = RRC 1N-390
rfc822Mailbox = john.doe@telcordia.com
ou = 21580
```

There is a special entry type, called an alias, which must be a leaf node and which points to another entry. For example, information about Telcordia might be under `c=US@o=Telcordia` or under `c=US@o=Telcordia Technologies`. In order to support both names, `c=US@o=Telcordia Technologies` might be an alias entry from `c=US@o=Telcordia`. If the requestor asks to have aliases de-referenced, this means that any query for `c=US@o=Telcordia Technologies` will automatically be converted to a query for `c=US@o=Telcordia`. From the user's perspective, the entire set of Telcordia information will be under both entries. Furthermore, because it actually refers to the same subtree of the DIT, the information under the two entries will be identical.

3.1.4.1.3 Retrieving Data. There are a number of operations that can be used in the Directory. A description of the query operations that a user can invoke follows, with their salient arguments, along with a brief description of modification operations. Most operations have common arguments that specify service controls such as time limit (in seconds), size limit (in number of entries returned), whether or not to chain operations, and so on. Directory operations are connection-oriented, so there are two operations, `bind` and `unbind`, which are used to establish and tear down a connection to a DSA. All the other operations have the same common arguments, and therefore these are not listed below.

- **Read:** The parameters of the `read` operation include
  - *DN*: distinguished name—for example,

    ```
    c=US@o=Telcordia@ou=21500@ou=21580@cn=John
    Doe%userid=12345.
    ```

- *Entry Information Selection (EIS)*: The list of attributes of an entry that are to be returned. The user lists the attributes and specifies one of two options: Either return both attributes and values, or return just the attribute names of the attributes having defined values.
- *Modification Rights*: Boolean indicating whether the requestor's modification rights should be returned along with the request.

"Read" returns the selected information from the entry, if it exists and the user is allowed access to it.

- **List:** The `list` operation is used to list the immediate children of a particular entry. For example, listing under `c=US` will return `o=Sears` and `o=Telcordia` if those are the only two entries under `c=US`. Only the RDNs of the children are returned, since the full DN can be generated by concatenating the DN specified in the list operation with the RDNs returned (`c=US@o=Telcordia` is the DN of the Telcordia entry).
- **Search:** The `search` operation is used to search under a particular node for entries whose attributes match some filter. The parameters are:
  - *baseObject* (a DN): Only this entry and entries under it are considered. For example, one might specify the subtree under `c=US@o=Telcordia`.
  - *subset*: This parameter specifies what part of the subtree to search. Its possible values are:
    - *baseObjectOnly*: Search only the specified DN.
    - *oneLevel*: Search the baseObject and its immediate children.
    - *wholeSubtree*: Search the entire subtree under the base object.
  - *searchAliases*: Set to TRUE or FALSE, specifying whether aliases should be de-referenced in the search. Since the alias may reference other parts of the tree, this parameter could have a significant impact on the search scope.
  - *Entry Information Selection (EIS)*: The `search` operation also uses EIS as does `read`, to select which information from the matching entries should be returned.
  - *filter*: A Boolean expression used to select a subset of entries. Filters are evaluated using the values from the attributes of an entry to determine whether to return the entry. For example, the following filter finds entries with a last name that either starts with "Ch" or is exactly "Smith":

  ```
  ((surname=Ch*) | (surname=Smith))
  ```

  The DN and the portion of the entry specified by the EIS are returned for each entry that matches the specified filter.
- **Compare:** The `compare` operation is used to determine whether an attribute value assertion (AVA) holds for an entry. There are two parameters:

- *DN*: Of the entry, for example,

```
c=US@o=Telcordia@ou=21500@ou=21580@cn=John
Doe%userid=12345.
```

- *Purported Attribute Value*: AVA in question—for example,

```
Title=Security Group Manager.
```

The operation returns either TRUE or FALSE, depending on whether the specified AVA is true for the entry specified by the given DN or not.
- **Other Operations:** The `abandon` operation is used to halt an operation, and the other operations listed below are used for modification of the directory. The operations, with brief descriptions, are as follows:
  - `Abandon`: Used to abandon an operation in progress.
  - `AddEntry`: Used to add an entry to the DIT.
  - `RemoveEntry`: Used to remove an entry from the DIT.
  - `ModifyEntry`: Used to modify an entry's attributes.
  - `Modifiers`: Used to modify the name of an entry.
  - `ModifyDN`: Modify the DN or RDN of an entry.

3.1.4.2 LDAP.

The strengths of X.500 included a flexible information model, versatility, and openness, whereas its weakness was that the complexity of the standard led to buggy implementations. Furthermore, it was based on the OSI network protocols that were never widely deployed. LDAP was originally conceived as a way to simplify access to a directory service that was modeled according to X.500.

LDAP is an IETF standard protocol that was originally defined in order to provide a lightweight directory access protocol that was built on top of TCP/IP. Since the TCP/IP protocol suite was much more widely deployed in the mid-1990s than the OSI 7-layer stack, it was believed that a new, lightweight access protocol that could be used to access X.500 directory servers would help speed up the deployment of directory standards. The idea was to ease the burden of directory access on clients; servers were still implemented as full-fledged X.500 servers. In 1993, the first LDAP specification was published [Yeong et al., 1993] by the IETF. LDAP version 2 was defined in 1995 [Yeong et al., 1995] and was widely deployed. An accompanying standard C API was also standardized [Howes and Smith, 1995].

In 1995, implementors noticed that over 99% of directory accesses were via LDAP; this provided the incentive to develop and deploy stand-alone LDAP directory servers by late 1995. Thus LDAP became the foundation for a complete directory service, rather than simply a protocol for accessing X.500

directories. In 1997, LDAP version 3 was defined [Wahl et al., 1997], providing referrals, security, feature and schema discovery, and internationalization.

3.1.4.3 Summary. The X.500 directory standards, aided by the subsequent simplification of directory implementation via LDAP, provide useful capabilities for storing network management information. In particular, Chapter 9 describes the use of a Public Key Infrastructure (PKI) that makes use of directory services for distributing public certificates.

3.2 DMTF POLICY STANDARDS

The Distributed Management Task Force (DMTF) is a members-only standards body that develops management standards for enterprise and Internet management. Members pay a fee for membership and can participate in one or more DMTF working groups, where their degree of participation in standards development is restricted based on their membership grade. The standards documents produced by the DMTF are made available to non-members on their web site; however, non-members cannot participate in the standards development process.

The DMTF worked closely with the IETF to develop the IETF policy framework standards; in fact, the IETF Policy Framework working group was created to leverage and extend standardization of policy information modeling from relevant working groups in the DMTF. Given that the DMTF is a members-only organization, its membership is naturally much smaller and more restricted than that of the IETF. This is the reason why the DMTF decided to work closely with the IETF: to make its policy standardization efforts known to and accepted by a much larger audience than just DMTF members. The same core group of people authored the DMTF and IETF policy information model standards, and they are fundamentally the same. The DMTF, however, is still continuing its work on policy information modeling. This work is a subset of the DMTF CIM (Common Information Model), which seeks to provide information models for management of devices, applications, databases, security, IPSec [Kent and Atkinson, 1998], policy, users, and so on.

3.3 POLICY FRAMEWORKS

This section describes two policy frameworks. The first is the Ponder language and the associated toolkit, and the second is PECAN and its agent-based policy framework. Before getting into the details of these two frameworks, some of the important features that are required of such a framework are listed below. Note that this is not an exhaustive list of required features; the intent here is to highlight some of most critical high-level features that are required for MANET management.

- *Expressiveness of Policy Language*: The language should support the ability to specify ECA policies and AC policies.
- *Distributed Implementation*: Due to the distributed nature of MANET nodes and the need for localized management, there is a need to be able to enforce policies in a completely distributed manner. In other words, every MANET node should have the capability to evaluate and enforce policy decisions. Using the terminology defined earlier, every node should provide PDP and PEP functionality.
- *Survivability*: Again, due to the dynamic and unreliable nature of MANETs, it is critical that the policy framework be survivable; in other words, it should be able to handle random disconnections, mobility, high link loss rates, and so on. This has implications for the mechanisms used for policy dissemination and for information sharing among MANET nodes, as these mechanisms must be able to function correctly in the face of these challenges.

Sections 3.3.1 and 3.3.2 describe Ponder and PECAN, respectively; this is followed by an evaluation and comparison of their features with respect to the criteria listed above.

3.3.1 Ponder

Ponder [Damianou et al., 2001] is a declarative, object-oriented language for specifying security and management policies for distributed systems that was developed by a research group at Imperial College London in the United Kingdom. It is one of the better-known policy languages, primarily because it has been in existence for over a decade and because an implementation of the language is freely available for experimentation. Ponder supports a wide range of policies, including the constraint style of policies as well as the rule style of policies discussed in Chapter 2. Some of the salient features of the language are described in this section.

3.3.1.1 Architecture. The implementation of Ponder provided by Imperial College London assumes a centralized policy repository, in the form of an LDAP directory. Policies are created and modified via a centralized management console, which is responsible for storing policies in the LDAP directory. Policies are stored in the form of references to a Java RMI implementation of the policy. Using Ponder nomenclature, multiple "Ponder Management Components" (PMCs) can be distributed throughout the network. Although these entities are called "PMCs" in the Ponder terminology, they are really PDPs according to the definition of PDP in this book; that is, they are capable of receiving policies and implementing them. The architecture is depicted in Figure 3.4.

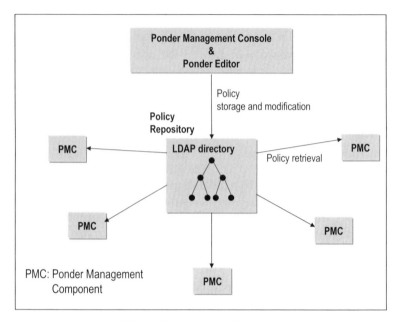

Figure 3.4. Ponder architecture.

3.3.1.2 Ponder Management Console and PMCs. As described in the previous section, the Ponder architecture uses a centralized management console for policy creation and manipulation. Policies are created via a policy editor hosted at this management console, and they are stored in an LDAP directory which is co-located with the console. The management console is responsible for loading the policies and activating them. Multiple PMCs may be distributed throughout the network. These PMCs implement policies that they retrieve from the LDAP directory. Actions that are performed as part of ECA policy implementation are written in Java, and they can be preloaded on the PMCs so that they can be executed when invoked by the PMCs.

3.3.1.3 Policy Dissemination. As can be seen in Figure 3.4, the Ponder management console stores policies in the LDAP directory, which serves as a policy repository. Policies are disseminated to the PMCs as follows: The Ponder management console pushes an LDAP reference to all the Ponder PMCs that need to implement the policy. The set of PMCs that need to implement the policy is determined based on the domain(s) to which the policy applies. These PMCs can then go and retrieve the referenced bytecode from the LDAP directory. As mentioned earlier, what is stored in the LDAP directory is really a collection of Java RMI references, which refer to TCP connections. The problem with this implementation is that if the TCP connection breaks—which is very likely to happen in MANET environments—the reference becomes invalid. This implementation is therefore extremely brittle and is not viable for a MANET environment. However, a more robust implementation of Ponder policy dissemination could be developed that is suitable for MANETs. An example of such an implementation is provided in Section 3.3.2.3, which discussed policy dissemination implemented by PECAN.

3.3.1.4 Ponder Policies. The Ponder language definition provides a powerful set of constructs for specifying ECA policies, AC policies, and other types of policies that will be described below. The implementation of the first version of Ponder, however, implements less than half of the language specification. AC policies are not implemented at all, and the event sublanguage (described below in Section 3.3.1.4.5) is not implemented either. This section describes the Ponder language specification; the reader should keep in mind that not all of the described features have been implemented.

Ponder policies specify the following:

- Subjects
- Targets
- Actions
- Constraints for authorizations and obligations

Policies can be of several different types and are described in more detail below. The remainder of this section discusses the Ponder language and its policy representation, including related concepts such as domains, subjects, targets, and different types of policies.

3.3.1.4.1 Ponder Domains. Ponder defines a domain as a collection of objects that have been explicitly grouped together for management purposes—for example, to apply a common policy. This makes it easier to apply policies to a group of objects, because it may be impractical to specify policies for individual objects in large systems with many objects. Another advantage of this approach is that one can change the domain membership of an object without changing policies; the policies for the new domain will thereby be made applicable to the object.

3.3.1.4.2 Subject and Targets. Using Ponder terminology, a *subject* refers to the managing entity or the entity that needs access to a resource; and a *target* refers to the managed resources. For example, if a certain user needs access to a database, then the subject is the user, and the target is the database. Policies are then used to specify which subjects can or cannot access a certain target. Policies can also specify which subjects cannot access certain targets. Examples of these policies will be provided later.

3.3.1.4.3 Authorization Policies. An authorization policy defines what a subject is permitted or not permitted to do to a target. It defines the permitted or forbidden operations on a specific target and thereby protects target objects from unauthorized management actions. Thus authorization policies in Ponder are used for specifying security policies and can provide target-based interpretation and enforcement. These policies are akin to the *outsourced* type of policy defined in Chapter 2 earlier.

Authorization policies can be either positive or negative. Positive authorization policies specify permitted actions, whereas negative authorization policies specify forbidden actions on the specified targets.

- **Example:** The following is an example of a positive authorization policy specified in Ponder:

```
type auth+ qosChat (subject s, string start, string
end)
{
    target chatServer;
    action setup;
    when time.between (start, end); }
```

This policy describes which subjects are authorized ("auth+") to perform a specified action ("setup") on a given target ("chatServer"), and any conditions that restrict this access ("when   time.between (start, end)").

The above policy can be viewed as a template; and instances of this policy can then be specified by instantiating this template—for example,

```
inst kidsChat = qosChat (/family/kids, "1400,"
"1900");
adultChat = qosChat (/family/adults, "2000",
"2400");
```

The first instance above creates a instance of the policy called "kidsChat" with subject /family/kids, a start time of 1400, or 2 pm, and an end time of 1900, or 7 pm. The second instance creates a instance of the policy called "adultChat" with subject /family/adults, a start time of 2000, and an end time of 2400.

Additional filtering provides the ability to perform various kinds of transformations on parameters of positive authorization policies, where it is not practical to provide different operations to reflect permitted parameters. The example below shows a positive authorization policy with subjects "engineers" and "managers," and where the target is the human resources database.

```
inst auth+ engineerAccess {
subject      engineers + managers ;
target       ⟨DB⟩ hrDB ;
action       getID (employeeID) ;
       if (subject = engineers)
       result = reject (result, salary); }
```

The idea here is that both engineers and managers are allowed access to the human resources database; however, certain restrictions can be placed on the subject "engineers," as shown by the conditional statement above. This conditional statement says that engineers are not allowed to access information about people's salaries.

· **Example:** The following is an example of a negative authorization policy:

```
type auth- noChat (subject s, string start, string
end)
  {
   target chatServer;
   action setup;
   when time.between (start, end); }
```

This policy states that subjects are not authorized ("auth-") to perform a specified action ("setup") on a given target ("chatServer") at certain times ("when time.between (start,end)"). Contrast this policy with the first one shown above, which was a similar policy framed as a positive authorization policy.

As before, an instance of this policy can be specified by instantiating this template—for example,

```
inst nokidsChat = noChat (/family/kids,
"1900","2400");
```

this instance would effectively deny the kids access to chat service between 7 pm and midnight.

3.3.1.4.4 Delegation Policies. Delegation policies are used to specify what kinds of actions can be delegated to other subjects by subjects who are permitted to perform certain actions on targets. In other words, if a subject S can perform an action A on target T, it may be able to delegate this action to another subject D, to allow it to perform the same action A on target T. Thus delegation policies are closely tied to authorization policies that authorize the original subject S to perform action A on target T. The original subject S is known as the *grantor* of access rights, and the subject to which the grantor delegates the access rights is known as the *grantee*.

Delegation policies can be either positive or negative. A positive delegation policy is used to delegate access rights to another subject, whereas a negative delegation policy is used to withhold such rights from another subject.

· **Example:**

```
type auth+ accessServices (subject s, target t) {
   action t.accessFTP, t.accessHTTP;
}
deleg+ sDelegT (accessServices a) (subject grantor,
grantee granteeD) {
   action accessFTP;
}
```

In the above example, the subject delegates a subset of its access rights (namely, accessFTP) to the grantee. The access rights obtained by the grantee are derived from the associated authorization policy, which authorizes accessFTP and accessHTTP.

3.3.1.4.5 Obligation Policies. Obligation policies are used in Ponder to define the actions that a subject is required to perform on a target. These policies are event-driven, and they correspond to the *provisioned* type of policies that were described earlier in Chapter 2. When the specified event or events occur, the specified actions must be performed by the subject on the target. An example is shown below.

· **Example:** The following is an example of an obligation policy:

```
type oblig restartServer (subject s, target t) {
   on [t.alarm];
   do t.restart;
}
inst oblig R1 = restartServer(/mgmtAgents/agent1,
/servers/dnsServer);
```

In this example, an obligation policy states that when a target server emits an alarm (which is the triggering event for this obligation policy), the subject must perform a server restart. An instance of the policy R1 is shown with specific values for the subject, a management agent that can perform the action, and the target, a DNS server.

Ponder specifies a sub-language for describing events that are used within Ponder policies. Event expressions can be defined, and they can be combined in various ways to form more complex expressions. Event composition operators for Ponder are defined below:

- e_1 && e_2: Occurs when both $e1$ and $e2$ occur, in any order.
- e + time_period: Occurs a specified period of time after event e occurs.
- $\{e_1; e_2\}$! e_3: Occurs when e_1 occurs followed by e_2, without no occurrence of e_3 in between e_1 and e_2.
- $e_1 \mid e_2$: Occurs when either e_1 or e_2 occur, in any order.
- $e_1 \rightarrow e_2$: Occurs when e_1 occurs before e_2.
- $n*e$: Occurs when e occurs n times.

As noted earlier, the above event language has not been implemented in Ponder version 1.

3.3.1.4.6 Refrain Policies. The flip side of an obligation policy is a refrain policy, which dictates what a subject must not do to a target. Refrain policies are very similar to negative authorization policies; the immediate question that arises is, What's the difference between the two? The difference is a subtle one: A refrain policy is enforced by the potential performer of the action (i.e., the subject), whereas a negative authorization policy is enforced by the target. So when do you know which one to use? The rule of thumb is to use refrain policies if the control, or the policy enforcement, is better suited for the subject side rather than the target side. As an example, a subject may decide to hold off on certain actions on a target under certain circumstances; this could be expressed by a refrain policy, since the target may not have access to the decision-making algorithms or data to feed these algorithms. On the other hand, if the idea is to prevent access to a particular resource and the access control is implemented by the resource itself (e.g., a file system), then the access control is better represented by a negative authorization policy.

3.3.1.5 Policy Conflicts in Ponder. There are several obvious ways in which Ponder policies can give rise to conflicts. A simple example is in the case of positive and negative authorization policies. A positive authorization policy may allow a subject to perform a given action on a target, whereas a negative authorization policy might forbid the subject from that very action on the same target. One way that such conflicts could be dealt with is by assigning priorities to the policies, so that the policy with the higher priority takes precedence over a policy with lower priority if they happen to be in conflict. Ponder defines a meta-policy language that can be used to specify conflict resolution strategies. A detailed treatise of how conflicts for policies expressed in Ponder can be resolved using the Event Calculus approach is provided in the next chapter.

3.3.2   PECAN (Policies Using Event–Condition–Action Notation)

This section describes PECAN [Chadha et al., 2003], a policy-based mobile ad hoc network management system developed by Telcordia, which addresses the special needs posed by mobile ad hoc networks. The system provides the capability to express networking requirements at a high level and have them automatically realized in the network by agents, without requiring further manual updates. Network management functionality is realized by policy agents that are organized in a hierarchy to provide both scalability and autonomy. Survivability is achieved by enabling any component to take over the management role of another component in the case of failure. It should be noted that PECAN provides much more than just a policy language, as Ponder does; PECAN also includes a framework for using policies for managing MANETs.

At a high level, PECAN provides the following capabilities:

- A policy engine (implemented by *policy agents*) that evaluates and enforces policy rules
- A distributed, cooperative set of policy agents that self-organize themselves into a hierarchy for the purpose of network management
- A reliable, survivable policy dissemination mechanism
- A flexible, pluggable framework for inserting new management functionality into policy agents

These capabilities are described in more detail in the following sections.

3.3.2.1   Architecture.   The high-level architecture of PECAN is shown in Figure 3.5, and it is described in detail in Chadha et al. [2004]. As shown here, a collection of Policy Agents manage all the nodes in the mobile ad hoc network. These policy agents are organized into a number of clusters. A cluster is a collection of entities subject to the same set of policies and managed by a policy agent, known as the Domain Policy Agent. Clusters are organized in a hierarchical structure. At the highest level, the Global Policy Agent (GPA) manages multiple Domain Policy Agents (DPAs). A DPA can manage multiple DPAs or Local Policy Agents (LPAs). An LPA manages a node. LPAs perform local policy-controlled configuration, monitoring, filtering, aggregation, and reporting, thus reducing management bandwidth overhead. Policies are disseminated from the GPA to DPAs to LPAs, or from DPAs to LPAs. Policy Agents react to network status changes on various levels (globally, locally, domain-wide) by automatically reconfiguring the network as needed to deal with fault and performance problems. In this architecture, any node can dynamically take over the functionality of another node to ensure survivability. A flexible agent infrastructure allows dynamic insertion of new management functionality.

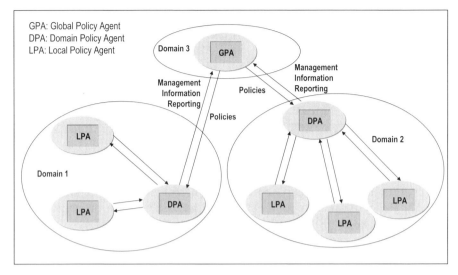

Figure 3.5. PECAN architecture.

The clustering mechanism used in PECAN is configurable and can be altered depending on the management requirements. Chapter 5 provides an overview of clustering mechanisms for management purposes in an ad hoc network; these mechanisms are implemented within PECAN. The particular clustering mechanism to be used in a given situation can be selected from these mechanisms.

3.3.2.2 PECAN Policy Agents. A Policy Agent in PECAN is the entity responsible for managing and enforcing policies on a network node. The Policy Agent on a node is referred to as a Local Policy Agent (LPA) if it does not manage any other nodes (i.e., if it is not the leader for a cluster). The Policy Agent at the top of the clustering hierarchy is referred to as the Global Policy Agent (GPA). Intermediate Policy Agents are referred to as Domain Policy Agents (DPA). A GPA, DPA, or LPA have the same basic structure and consist of the same code base. They have similar functionality but different scope.

The main function of a Policy Agent in PECAN is to enforce policies. Policies consist of events, conditions, and actions, and they conform to the description given in Section 2.4.1. Policy events are named entities that may contain zero or more parameters. Policy conditions can specify boolean operations on event parameters and database parameters, which typically contain management-related data. When an event occurs, the Policy Agent looks up policies containing that event, evaluates their conditions, and, if they evaluate to *true*, enforces the actions of the policies.

An interesting feature of PECAN is the way policy actions are specified. Defining how policy actions will be specified is always tricky business for a policy language. Define the actions at too high a level (as done by the IETF), and you have a useless construct; define the actions too narrowly and precisely, and you have destroyed flexibility and constrained the capabilities of your policy language. The solution used in PECAN is to specify an action as a general-purpose piece of code and to supply certain actions that are useful for network management as part of the PECAN system. The idea is that users of PECAN can write their own actions to customize PECAN for their purposes; and PECAN provides a well-defined mechanism for creating actions that can plug into the PECAN framework. All that a user needs to do is to use the PECAN-defined Java action class and implement its base methods. Policies can then be written that use this action and trigger it based on different events and conditions. The standard interface implemented by all actions allows PECAN Policy Agents to communicate with the actions in order to query them and invoke them as needed when policies are being enforced. Thus a Policy Agent can launch actions, set/modify their parameters, and terminate them. Management policies can therefore be defined in terms of the actions that could be performed by these agents. The standard interface also enables other systems (such as a Policy Editor GUI) to query the actions to determine their functionality and associated parameters.

3.3.2.3 *Policy Dissemination.* PECAN provides a mechanism for distributing policies in a reliable and survivable manner. Policies created at the GPA are distributed to the DPAs, which, in turn, distribute them to the LPAs. The dissemination of policies (and, in general, the dissemination of any data) in a reliable and survivable fashion in a mobile ad hoc network poses special problems. Although reliability of delivery can be provided by the use of a reliable transport protocol such as TCP or its MANET-appropriate variants,[1] survivability is a challenge that cannot be addressed by TCP. Survivability means the following: A survivable dissemination mechanism ensures that every node in the network to which data has to be disseminated will eventually receive the data, provided that it eventually becomes reachable. Survivability is usually not of concern in highly reliable, static wireline networks; however, in MANETs where nodes typically have intermittent connectivity and may be disconnected from the network for long periods of time, survivability of dissemination of critical data such as policies for network management becomes a key requirement. Note that a reliable transport protocol such as TCP does not provide this survivability guarantee; indeed, all that such protocols guarantee is that either the data will reach its destination or the sender of the data will be informed about failure to deliver the data. A survivable mechanism, however, must guarantee eventual delivery, even in the face of intermittent disconnection.

---

[1] Note that TCP has been shown to provide poor performance in high-loss environments that are typical of MANETs (see Holland and Vaidya [1999]); however, other reliable transport protocols have been proposed as replacements [Liu and Singh, 2001; Argyriou and Madisetti, 2003].

In order to provide survivable policy dissemination, PECAN implements a mechanism that disseminates data using the management hierarchy. Policies are assumed to be created at the GPA and must be disseminated to all the nodes in the network. The rationale behind this is that the entire network must be managed in a coherent fashion and therefore the same policies must be used everywhere. The exception to this is configuration policies, which are usually specific to a given node and provide configuration directives for that node.

The following mechanism is used for policy dissemination:

- When any policy update (e.g., policy creation, modification, deletion, activation, deactivation) is performed at the GPA, the GPA assigns a sequence number n to this policy update and *periodically sends* this sequence number n to all its children. This can be done using unicast UDP messaging, or multicast if available. Policy update sequence numbers are assigned sequentially, starting from 1, and are strictly increasing. As soon as a new policy update operation is performed, it is assigned the next number ($n + 1$ if the last sequence number was n), and the GPA begins to send the sequence number $n + 1$ to all its children periodically. In other words, the GPA is periodically sending to all its children a single positive integer, which represents the sequence number of the most recent policy update performed at the GPA.

- When a policy sequence number arrives at a remote Policy Agent, this Policy Agent determines whether it is missing any policy update operation by comparing the transmitted sequence number with its last received sequence number. If the new sequence number is larger than the last received sequence number, the Policy Agent requests the missing policies from its parent, using a reliable transport protocol. The parent responds with all the missing policy updates to the requesting node.

- When a policy sequence number arrives at a remote Policy Agent, if this Policy Agent has any children (i.e., if it is a DPA), it forwards this policy sequence number to its children. As soon as a new policy update sequence number is received, the Policy Agent begins to send the newly received sequence number to all its children periodically in place of the previous one. In other words, every DPA periodically sends to all its children a single positive integer, which represents the sequence number of the most recent policy update received from its parent.

Now, let us look at what happens if a node N is temporarily disconnected from the network. Let us assume that the node had received n policy updates while it was connected to the network. While it was disconnected, assume that m policy updates occurred. Now the node rejoins the network and automatically joins an existing cluster. One of the following two situations can arise:

- The parent of node N has received all policy update operations or is the GPA: In this case, the parent of node N will be periodically sending the sequence number $n + m$ to all its children, and node N will eventually receive it. Node N will then request the missing m policy updates from its parent and will receive all the policy updates that it missed while it was disconnected from the network.
- The parent of node N has *not* received all policy update operations: In this case, whenever the parent of node N reestablishes connectivity with its parent in the management hierarchy, it will receive all the missing policy updates (using the same argument as above), and therefore this case reduces to the previous one. Since the hierarchy is finite, this argument shows that the dissemination algorithm is survivable.

3.3.2.4 PECAN Policies. The PECAN policy language by and large conforms to the description given in Chapter 2 earlier. A PECAN policy rule is composed of *events, conditions*, and *actions*. A policy rule has zero or more events, an optional condition, and one or more actions. When the events of a rule occur, the conditions are evaluated; if they evaluate to *true*, the actions are executed. The general format of a policy rule is as follows:

$$\textbf{When } (events) \textbf{ if } expression \rightarrow action$$

Events, conditions, and actions are described in more detail below.

3.3.2.4.1 Events. The *event*s section of a policy contains either

- an event sequence (if order of events is relevant) or
- an event set (if order of events is not relevant)

The following syntax is used:

$$\textbf{when } (event\_seq|event\_set)$$

A special type of events are *time-based* events, which allow the specification of a time at which a policy should be triggered, or a time interval that results in periodic triggering of the policy at the specified time interval. For the former, these policies are akin to one-time events and are triggered at a specified time that may be absolute or relative to the policy's activation. For the latter, periodic events are generated at the specified intervals.

3.3.2.4.2 Conditions. The condition section of a policy contains an expression with the following syntax:

$$\textbf{if } expression$$

The expression language provides infix notation for arithmetic and relational operators on the supported scalar types, which include:

- Support for integers, floating point numbers, boolean values, dates, records with named fields, and lists
- Single element lists (considered scalars)
- Named variables

A named variable is a named reference to a value that can be accessed from within a policy. Policy variables can be used to transfer information between triggered policies or between multiple occurrences of a triggering of the same policy. When used in conditions, variables are accessed for condition expression evaluation. Three basic variable access methods are supported, one each for scalars, records, and lists. Access to records in a database is provided by specifying a list of records to be selected from a relational database by specifying column names and named fields; the result is a list of all the records whose named fields match the values in the query.

3.3.2.4.3 Actions. As described earlier, PECAN actions consist of *code* that is part of the network management system. The values of policy variables can be passed to actions as action parameters. Different types of actions can be defined for different needs. In addition to allowing users to define actions, PECAN also provides certain predefined action skeletons that are delivered with the system. The idea is that these actions can be extended and/or modified as needed by the operator. These actions are described below:

- *Configuration Actions*: These actions perform configuration on a local managed element, or node. Configuration Actions may invoke local system commands to perform the configuration. They may also perform the configuration by writing to the database or by sending commands via a standard interface such as SNMP.
- *Monitoring Actions*: These actions collect information at configurable intervals from the local node and store it in the PECAN database. A monitoring action may collect information about the local platform using various mechanisms, including SNMP, operating system calls, or log files or via an Element Management System.
- *Reporting Actions*: Reporting actions report two types of information: monitoring and configuration information. The information is reported at configurable intervals to the parent Policy Agent of the node in the management hierarchy. Thus LPAs report information to their DPA; DPAs report information to their DPA or GPA, depending on the identity of their parent. Information monitored by Monitoring Actions or produced by Aggregation Actions (see below) may be reported on demand by Reporting Actions. Configuration information, however,

is reported regularly to verify that nodes are configured correctly. If the configuration has changed, the new configuration is reported. However, if configuration remains the same since the last reporting, a "hash value" of the configuration is computed and reported in order to conserve bandwidth.

- *Aggregation Actions*: These actions perform aggregation of locally collected data before the information is reported to the parent Policy Agent. For example, an Aggregation Action may compute the average value of an attribute over a given period of time using the last five monitored values of the attribute. The enables pre-processing of information at the node where this information is being monitored, and it reduces the amount of information that needs to be reported. It also provides flexibility in the type of information that could be reported.

- *Filtering Actions*: Filtering actions report information to the parent Policy Agent only when certain conditions are met. For example, a policy can be created that specifies that battery power should only be reported when it drops below a certain threshold. These actions perform intelligent filtering of monitored data, thus reducing processing effort of management stations and the communication overhead due to transmission of monitoring information.

In addition to the above types of actions, general-purpose actions can be written that perform or invoke other specialized management functions—for example, root cause analysis, and so on.

3.3.2.4.4 Manipulating PECAN Policies. PECAN implements an API that allows external applications (such as GUIs) to manipulate policies and their attributes. Policies contain the following attributes:

- *Policy Identifier*: Provides a unique identifier for each policy.
- *Description*: A human-readable description of the policy.
- *State*: A policy is either active or inactive. An active policy is one that Policy Agents enforce; an inactive policy is one that is ignored by Policy Agents until its state is changed to active.
- *Date and Time of Last Modification*: Indicates when a policy was last modified.
- *Date and Time of Last Activation*: Indicates when a policy was last activated.
- *Date and Time of Last Deactivation*: Indicates when a policy was last deactivated.
- *Policy Scope*: A policy has a dissemination scope, which dictates the targets to which it is applicable.

The following operations are provided via the PECAN policy management interface:

- *Create*: Create a new policy. Policies are created in the inactive state.
- *Delete*: Delete an existing policy. Policies that are deleted are permanently removed from the policy repository.
- *Modify*: Modify an existing policy. An existing policy is modified using this operation.
- *Activate*: An inactive policy is activated using this operation.
- *Deactivate*: An active policy is deactivated using this operation.
- *Retrieve*: An existing policy can be retrieved from the policy repository using its unique identifier using this operation.

3.3.2.5 *Policy Conflicts in PECAN.* PECAN includes a simple conflict detection system that detects a well-defined set of conflicts, as described in Chadha et al. [2005]. Conflict detection is currently limited to compile-time detection of *actual* conflicts. Typically, the term *policy conflict* refers to undesirable consequences of combined execution of actions prescribed by several policies. Such consequences ultimately include (a) concurrent setting of the same variable by two different actions to two different values and (b) the resulting system behavior. Other forms of conflict include *incompleteness* of the policy set, where, for instance, the condition part of a policy rule includes a variable whose value is supposed to be produced as a result of another rule's action, but such a policy does not exist. Other constraints, such as those based on the freshness of data explicitly set in policy conditions or implicitly in the policy type, can also be violated, leading to *inconsistent* policies.

PECAN uses a limited implementation of the event calculus (EC) (discussed in much more detail in the next chapter) as a formal language to represent PECAN policies and conflicts, as well as to detect actual conflicts using abductive reasoning. As shown below, it is possible to identify a set of generic constraints on policies containing monitoring, aggregation, and reporting actions (referred to below as monitoring policies, aggregation policies, and reporting policies). Domain-specific constraints, which are described below, are expressed in EC. Once a combined representation is obtained, actual conflict detection is performed with an abductive reasoning procedure.

3.3.2.5.1 Conflict Types for Monitoring Policies. A monitoring policy rule is specified as follows:

- *Event*: Time event, arriving with periodicity T_m
- *Condition*: None
- *Action*: Obtain current value of variable v, store in database.

The following conflict types can be identified for monitoring policies:

1. *Identical Rules*: Rules that differ only by rule name and/or the name of the database object in which v is stored.
2. *Redundant Rules*: A rule to monitor v with periodicity T_m and a rule to monitor v with periodicity nT_m, where n is any natural number.
3. *Data Overwriting Rules*: Two rules to monitor v with periodicities T_m and T_k, where $m > k$ and m mod $k \neq 0$, assuming v is stored as a scalar. If v is stored as a vector indexed by periodicity T, this will not be a conflict.

3.3.2.5.2 Conflict Types for Aggregation Policies. A *spatial* aggregation policy rule is specified as follows:

- *Event*: Time event, arriving with periodicity T_a.
- *Condition*: A Boolean expression on multiple variables $\{v_i\}$, whose current values are available in the database.
- *Action*: If condition evaluates to true, compute the value of an expression on one or more variables $\{v_j\}$, whose values are assumed to be available in the database, and store that value in the database.

In spatial aggregation, the variables v_k are treated as scalars. In addition to the conflict types described for monitoring policies above, the following types can be identified for spatial aggregation policies:

1. *Nonexistent Rules*: There exists v_k that is assumed to be refreshed by a periodic policy (i.e., a monitoring or aggregation policy), but there is no corresponding policy. Identification of this conflict type depends on the existence of an attribute for each database object that identifies it as the output of a policy.
2. *Incompatible Periodicity*: For any v_k that is assumed to be refreshed by a periodic policy (i.e., a monitoring or aggregation policy), the following two constraints must hold: $T_k \leq T_a$ and T_a mod $T_k = 0$. Any violation of these constraints constitutes a conflict.
3. *Incompatible Phase*: The maximum difference Δt_s between any two *start* times for any two policies that refresh the value of any v_k used in either the condition or the action part of an aggregation rule exceeds a predefined value $\Delta t_{s\text{-}max}$. The latter is a configurable system-wide parameter.

A *temporal* aggregation policy rule is specified as follows:

- *Event*: Time event, arriving with periodicity T_a.
- *Condition*: A Boolean expression on multiple frames of the variable v_i, whose values (indexed by time) are assumed to be available in the database.

- *Action*: If condition evaluates to True, compute the value of an expression on multiple frames of v_j, whose values are assumed to be available in the database, and store that value in the database.

In temporal aggregation, the variable v_i is treated as a vector. The following conflict types can be identified for temporal aggregation policies, in addition to those described for monitoring policies:

1. *Nonexistent Rules*: v_i is assumed to be refreshed by a periodic policy (i.e., a monitoring or aggregation policy), but there is no corresponding policy. Identification of this conflict type depends on the existence of an attribute for each database object that identifies it as the output of a policy.
2. *Incompatible Periodicity*: v_i is assumed to be refreshed by a periodic policy (i.e., a monitoring or aggregation policy), and the following constraint must hold: $T_i \leq T_a$. Any violation of this constraint is a conflict.

3.3.2.5.3 Conflict Types for Reporting Policies. A reporting rule can be specified as follows:

- *Event*: Time event, arriving with periodicity T_r.
- *Condition*: A Boolean expression on multiple variables $\{v_j\}$, whose current values are available in the database.
- *Action*: If condition evaluates to True, report the value of a variable v_i, which is available in the database.

In reporting, the variables v_k are treated as scalars. The following conflict types can be identified for reporting policies:

1. *Identical Rules*: Rules that differ only by rule name and/or the name of the database object in which v is stored.
2. *Redundant Rules*: A rule to report v with periodicity T_m and a rule to report v with periodicity nT_m, where n is any natural number.
3. *Incompatible Periodicity*: For any v_k that is assumed to be refreshed by a periodic policy (i.e., a monitoring or aggregation policy), the following two constraints must hold: $T_k \leq T_r$ and $T_r \bmod T_k = 0$. Any violation of these constraints constitutes a conflict.
4. *Incompatible Phase*: The maximum difference Δt_s between any two *start* times for any two policies that refresh the value of any v_k used in either the condition or the action part of a reporting rule exceeds a predefined value $\Delta t_{s\text{-}max}$. The latter is a configurable system-wide parameter.

The set of constraints identified above is used as an initial set for PECAN and may be modified and extended based on new actions. The above constraints are clearly generic to the monitoring domain and independent of the actual policies.

3.3.3 Summary: Policy Frameworks

This section described the Ponder and the PECAN frameworks for policy management. To summarize the capabilities of these two systems, it is useful to refer back at the requirements that were listed at the beginning of this section. Table 3.1 summarizes the capabilities of Ponder and PECAN with respect to these features. Since the specification of Ponder diverges significantly from its implementation, the Ponder specification as well as its implementation are evaluated in Table 3.1.

In terms of expressiveness of the policy language itself, the Ponder specification provides a much richer set of features and constructs for expressing different types of policies, including both ECA policies and AC policies. The Ponder implementation, however, does not implement many of these features. The PECAN language is restricted to ECA policies, and has a much more limited set of constructs for expressing policies than specified by Ponder. However, as far as the framework is concerned, PECAN provides a much more comprehensive set of features for distribution and for survivability, since it was built with a MANET environment in mind. The Ponder framework relies on centralized components and cannot handle an unreliable network environment where nodes connect to and disconnect from the network in a random manner.

TABLE 3.1   Comparison of Ponder and PECAN Features

Feature	Ponder	PECAN
Expressiveness of policy language	Ponder specification: Excellent; supports ECA and AC policies, with a rich language for event definition. Actual Ponder (version 1) implementation: Fair; supports only ECA policies. Many of the features in the language specification have not been implemented.	Good; supports ECA policies but not AC policies.
Distributed implementation	Poor; relies on centralized policy definition point, which cannot be easily modified at run-time.	Excellent; completely distributed implementation with self-forming information-sharing structure in the form of a management hierarchy.
Survivability	Poor; completely dependent on reliable communications between distributed nodes for operations.	Excellent; provides mechanisms for automatically dealing with mobility and intermittent connectivity.

3.4 SUMMARY

This chapter described a number of policy standards and languages. The IETF policy-related standards were chiefly targeted at QoS applications; IETF standards in this area include IntServ, COPS, DiffServ, COPS-PR, and LDAP. The RSVP resource reservation paradigm is not suitable for use in MANETs because the dynamic nature of MANETs makes it difficult to reserve resources on a per-flow basis. The DiffServ paradigm of providing differentiated services to a small number of traffic classes, however, is very useful in MANETs, due to the scarcity of bandwidth in such networks. Mechanisms for QoS assurance based on DiffServ will be described in Chapter 8.

Two general-purpose policy languages, Ponder and PECAN, were also described, along with a description of the current implementations of these languages. Further information about these languages is available in the references cited in this chapter. More details about policies for mobile ad hoc network management will be provided in Chapter 5. From the point of view of suitability for managing MANETs, both Ponder and PECAN have sufficient expressive power to be able to specify the required network management policies. However, PECAN provides a framework for deployment of policy management functionality that is much more suitable for a MANET environment.

4

POLICY CONFLICT DETECTION AND RESOLUTION

As seen in the preceding chapters, policy-based network management promises to deliver a high degree of automation for ad hoc network management. A policy-based network management system provides the capability to express networking requirements in the form of policies and have them automatically realized in the network, without requiring further manual updates. However, as with every technology, these benefits come at the expense of certain obvious risks. The biggest risk associated with policy-based management is that the policies themselves can interact in undesirable ways, by causing conflicting actions to be taken by the management system. Thus it is essential that policies be analyzed for conflicts, and that mechanisms be put in place for determining how to resolve these conflicts. A number of policy conflict resolution techniques have been described in the literature; however, they often concentrate on the abstract problem of formal policy analysis and have very little to do with practical policy conflict resolution in live management systems. This chapter provides an overview of the state of the art in policy conflict detection and resolution, followed by a critical look at what is really needed to resolve practical policy conflicts in network management systems. In particular, although policy conflict and resolution tools can often be used as debugging aids and can help in automatically verifying certain system properties, it will be shown that application-specific policy conflict detection and resolution requires a great deal of manual analysis and can often be addressed by careful policy writing (or rewriting), rather than via cumbersome and unrealistically complex policy conflict resolution solutions.

4.1   INTRODUCTION

The subject of policy-based network management has received a great deal of attention in the recent past. Ad hoc networks are highly dynamic and pose stringent requirements for security, reliability, and, above all, operations automation. Policy-based network management shows a great deal of promise due to its potential for providing a many-fold increase in automation of network operations. However, as with any technology, policy-based management is a double-edged sword. The ability to define policies provides a great deal of power to network operators but, at the same time, puts a dangerous tool in their hands. It is essential that policies be analyzed before and during deployment to ensure that they do not give rise to undesirable or inconsistent behavior.

Policy conflict detection and resolution is not a new topic, and several approaches have been suggested in the literature for detecting and resolving various types of policy conflicts. There are two problems here: The first is that there is a danger that people will assume that policy conflict resolution strategies will magically correct all problems with poorly expressed policies, and the second is that much of the work in this field provides academic solutions with oversimplified examples and often tends to substitute complex conflict resolution techniques for common sense and practical policy-writing guidelines. Furthermore, some of the approaches are extremely cumbersome to use and require extensive modeling of the managed system and of the effect that the policies have on the managed system, which itself is extremely error-prone, leading to the obvious next question: How can anyone guarantee that these models themselves are conflict-free? Is there a need for conflict resolution systems for conflict resolution systems?

First, conflict detection should be distinguished from conflict resolution. Conflict detection is a necessary precursor to conflict resolution, since conflicts must be *detected*, or discovered, before they can be *resolved*, or removed. It is important to note that automated policy conflict resolution techniques require a great deal of manual analysis in order to correctly model the policies and the management system. This must happen before any automation of the conflict detection and resolution task can be achieved. Also, another point to remember is that such analysis has to be performed each time a new policy is added to the system, in order to specify the intent of the policy and to specify any undesirable interactions of this new policy with existing policies. As an example, there exist techniques for conflict detection that consist of specifying invariants, or safety properties, of the system; such techniques require manual specification of these invariants, and they make use of tools for automated reasoning to verify that the invariants are not violated. These techniques are only as good as the invariants themselves: If the specification of invariants is erroneous or incomplete, then so is the obtained result. Furthermore, invariant checking only provides conflict detection and not resolution.

In this chapter, it is shown that if policies are not going to be altered (i.e., created, modified, or deleted) after system deployment, then conflict resolution mechanisms do not need to be applied on an ongoing basis for run-time policy conflict resolution. In other words, if policies are not altered after system deployment, then *application-specific run-time conflict resolution is largely unnecessary*, and *effective conflict resolution can be achieved by careful inspection and rewriting of policies*. The argument that has been put forward against this approach is that when a user wants to add a new policy, the user would have to examine the entire set of existing policies to be able to write the new policy correctly. This argument is refuted as follows: If application-specific conflict resolution is being used, the user may be able to write a new policy without looking at other policies, but will still need to look at all the existing policies to uncover potential conflicts with these policies! This is because any mechanized conflict resolution approach still requires the user to manually develop conflict resolution rules. Thus manual analysis is still required, regardless of the approach. The question is, Which approach is better? In order to answer this question, one must examine the pros and cons of each approach.

If policies are going to be altered (i.e., created, modified, or deleted) after system deployment, then some form of mechanized conflict resolution is required at run time. The question here becomes, What are the realistic scenarios where policies are modified at run time in a network management context, and what are the types of policy conflicts that can arise? This is discussed later in the chapter, following a discussion of policy types.

This chapter is structured as follows. The next section provides a brief definition of the structure of a policy, to set the stage for discussions about policy conflicts. Section 4.3 provides an overview of related work. Section 4.4 describes types of policy conflicts, and Sections 4.5 and 4.6 describe the proposed approach to conflict resolution. In Section 4.7, two case studies that compare conflict resolution strategies from the literature with the proposed approach are presented. Section 4.8 provides a summary of the chapter.

4.2 ANATOMY OF A POLICY

In order to have a meaningful discussion about policies, it is necessary to first establish a common understanding of what a policy is. For the sake of simplicity, the discussion in this section will largely be based on Ponder policies, as described in the previous chapter [Damianou et al., 2001]. To recapitulate, using Ponder terminology, a *subject* refers to the managing entity or the entity that needs access to a resource, and a *target* refers to the managed resources. Subjects perform *actions* on targets. Furthermore, Ponder policies are either obligation/refrain or authorization policies. An authorization policy defines what actions a subject is permitted (positive authorization) or not permitted (negative authorization) to perform on a target. Obligation policies are used in Ponder to define the actions that a subject is *required* to perform on a target. The flip side of an obligation policy is a refrain policy, which dictates what a subject must not do to a target. From a network management point of view, obligation policies are the most commonly used type of policies and are used to define actions that network management systems must automatically take under different circumstances. Refrain policies are less useful because it is less common to specify what a manager cannot do, although it is conceivable that there may be some actions that only certain components within a network management system are allowed to perform, for security reasons. An obligation policy has the following syntactical elements: a *triggering event*, which defines what events can trigger the policy; a subject and target as defined earlier; an optional set of *conditions* that are checked before triggering the actions; and, finally, one or more actions to be performed by the subject on the target.

4.3 RELATED WORK

In Moffett and Sloman [1994], the authors provide a taxonomy of policy conflicts for obligation (referred to as "imperitival") and authorization policies, along with a very cursory discussion of detection and resolution methods. They define overlap in the subjects of obligation policies as a "multiple managers" conflict when the goals of the policies are semantically incompatible. These definitions are further refined in Lupu and Sloman [1997], where a Prolog-based tool for detecting modality conflicts (see Section 4.4.1.1 for a definition of modality conflicts) at compile time (i.e., at policy definition time) is described.

An approach to conflict detection that has received some attention recently is based on use of the Event Calculus [Bandara et al., 2003; Kowalski and Sergot, 1986] for policy conflict resolution, which provides a formal language for reasoning about policies and applies abductive reasoning (described in more detail later in this chapter) to analyze policy specifications and identify conflicts. This approach will be examined in some detail in Section 4.6.2.2.3. The use of this approach for policy conflict resolution requires the following components: (i) a set of time points, fluents, event types, base predicates (*initiates, terminates, holdsAt, happens*), and domain-independent axioms with respect to *holdsAt*; (ii) a model of policy rules themselves; (iii) a model of policy enforcement, that is, how the system behaves when policies are enforced; (iv) a model of the managed system, which includes representation of the effects of actions, and so on; and (v) a model of policy conflicts. One can then run queries on the system to derive conflicts. Charalambides et al. [2006] illustrates the application of this approach to a management system for MPLS networks, along with the detection and resolution of application-specific policy conflicts that arise due to the problem of inconsistent configurations. The drawback of this approach is that it is, for the most part, extremely complex.

As an example, the user is required to formally specify the effect of every policy action on the managed system by specifying predicates that describe the pre-conditions and post-conditions for every action. While theoretically sound, the approach frequently places undue burden on the user (due to the high level of complexity of this task) and makes the process of conflict detection very prone to errors. This chapter shows that the conflicts presented in Charalambides et al. [2006] can be easily resolved by the methods outlined here, rather than by using the event calculus.

Chomicki et al. [2003] present an interesting approach to conflict resolution that is based upon modeling the effects of policy actions, manually identifying application-specific conflicts, and explicitly blocking either conflicting actions, or events that will give rise to conflicts. They introduce the notion of event monitors and action monitors that are used to filter the output of policies to eliminate conflicts. Conflict rules must be specified in the form of constraints; these constraints are then used to specify *blocking* rules that either (a) eliminate actions that would be triggered by policies or (b) eliminate events that would trigger policies. The notion is elegant and can be implemented using a logic programming language such as Prolog. However, again, the approach rapidly becomes difficult to manage because the specifications of the monitors themselves may have conflicts that require resolution; furthermore, the effort required to write conflict rules could be applied instead to modifying the policies so that they are conflict-free. In Section 4.6, a simpler approach based on policy rewriting is shown that can achieve similar effects.

4.4 TAXONOMY OF POLICY CONFLICTS

In order to be able to discuss policy conflict resolution mechanisms, it is useful to classify policy conflicts into categories so that appropriate mechanisms can be applied to each category. Policy conflict taxonomy has been discussed in Lupu and Sloman [1999]. The literature on policy conflicts has identified certain types of policy conflicts, which can be classified into the following two categories: application-independent conflicts and application-specific conflicts. These are described separately below.

4.4.1 Application-Independent Conflicts

Application-independent conflicts, as the name suggests, are independent of the policy application, and therefore they lend themselves most easily and naturally to mechanized policy conflict resolution. For this type of policy conflict, conflict resolution tools are very useful and can provide valuable automation aid. Application-independent conflicts are the easiest types of conflicts to detect and resolve, because there is no manual policy analysis required on a per-application basis. A generic policy detection tool can be used, provided of course that it has a built-in understanding of the policy syntax in use. Application-independent conflicts include the following.

4.4.1.1 Modality Conflicts. Modality conflicts arise when there is a refrain and an obligation policy with the same subjects, targets, and actions. In other words, one of these policies requires that certain actions be taken and the other forbids the same set of actions, leading to a contradiction. Such conflicts can be detected by simple inspection—that is, by an automated syntactical scan of the policies. In order to resolve such conflicts, all policies could be tagged with a priority, and the policy with the lower priority could be deleted; or all conflicting policies could be displayed to the user, who would be required to manually delete either the obligation or the refrain policy for each of the conflicts. Lupu and Sloman [1999] provides an overview of conflict resolution strategies for modality conflicts. It should be noted that modality conflicts are good candidates for automated conflict detection tools, since such tools can be easily built and applied to any policy management system, regardless of application.

4.4.1.2 Redundancy Conflicts. Redundancy conflicts arise when there are two or more identical policies in the system. Such conflicts can be easily detected by syntactic analysis of the defined policies, and any duplicated policies can be deleted. As above, such policies are good candidates for automated conflict resolution since they are not application-dependent.

4.4.2 Application-Specific Conflicts

Application-specific conflicts require some preliminary manual work that involves manual inspection of the defined policies, and, based on a semantic understanding of the policies and the managed system, a formulation of the types of conditions that could arise that would result in undesirable actions being performed on the management system. This manual step is required because it is not possible to formally describe an application-specific conflict without an understanding of how the policies and the managed system should behave. Although it is difficult to characterize in a generalized fashion the types of application-specific conflicts that can arise, some important sources of conflict in policies for network management systems are outlined below.

4.4.2.1 Redundancy Conflicts. Redundancy conflicts arise when there is duplication in two or more policies. The simplest case of redundancy was described in the previous section, where two or more policies are found to be identical. Another form of redundancy is when one policy subsumes another. Subsumption here means that one policy achieves everything that another does, and possibly something more. A simple example of the latter is shown below:

Policy 1:   <Event:E, Subject: S, Target: T, Actions: A1, A2>
Policy 2:   <Event: E, Subject: S, Target: T, Actions: A1>

If action A1 is idempotent (i.e., performing it more than once at a given instant in time has the same effect as performing it exactly once), then Policy 1 is said to subsume Policy 2. In other words, Policy 2 is redundant and can be eliminated. Such a determination is necessarily application-specific, since it requires a semantic understanding of the effects of the various actions in policies. An example of an action that is not idempotent is the action of incrementing a counter (incrementing a counter twice is not the same as incrementing it once). Another example of subsumption was described in Chadha et al. [2005] and is summarized as follows: Suppose that an action Set_Monitoring_Interval sets the time interval at which a management system polls a certain variable (e.g. setting the variable to 5 seconds would mean that the variable would be polled every 5 seconds). Then the action Set_Monitoring_Interval(5) would subsume the action Set_Monitoring_Interval(10).

4.4.2.2 *Mutually   Exclusive   Configurations   Conflicts   (Mutex Conflicts).* This type of conflict arises when a parameter on a target is set to two different values simultaneously by two different actions. Depending on the implementation of the policy engine, the result will be that the parameter will be set to one of the two values. However, this may be an indication of a conflict. As an example, consider the following two policies:

Policy 1:   <Event: Intrusion detected; Subject: Manager1; Target: Router1, Interface1; Action: Shut down target interface>
Policy 2:   <Event: Provisioning Request; Subject: Manager1; Target: Router1, Interface1; Action: Bring up target interface and assign customer to this interface>

Here the action in Policy 1 will attempt to shut down an interface on a router, whereas the action in Policy 2 will assign a customer to this interface and bring the interface up. These two states for the router interface (up and down) are mutually exclusive, and therefore simultaneous firing of both these policies gives rise to an unpredictable situation. Note that such a conflict will normally not be detectable at policy compile-time (i.e., at policy definition time), because the triggering of these two policies happens in response to two different events, and these two events may or may not occur at the same point in time. Thus run-time detection is required for all such conflicts, except the most trivial case where both policies are identical except for the value assigned to a target parameter. In that case, the conflict can be statically detected based on a compile-time static analysis of the policies.

4.4.2.3 Inconsistent Configuration Conflicts. This class of conflicts, described in Chadha [2006a], is related to the previous one, but is a little more subtle. Here, two (or more) different parameters may be the targets of configuration in two different policies, so there may be no conflict of the above type (mutually exclusive configuration). However, the parameters may be related in some way and the two policies' actions may be trying to configure them in ways that result in an inconsistent configuration. Consider the following example drawn from Charalambides et al. [2006]. Here there are two actions, *incrAlloc(Link, PHB, BW)* and *setBwMax(Link, PHB, BW)*, that can potentially lead to inconsistent configurations. The first action increases the bandwidth allocation for the specified PHB (Per Hop Behavior) on a link, while the second specifies an upper bound on the bandwidth allocation for the PHB on a link. If *incrAlloc* attempts to increase the bandwidth for a certain PHB on a given link to a value greater than that allowed by *setBwMax*, an inconsistency will arise. *IncrAlloc* gets invoked every time an alarm is raised in the system, based on certain thresholds. Again, as before, such a conflict cannot be detected at compile time because the conflict only arises if *incrAlloc* gets invoked enough times to exceed the specified threshold.

4.4.2.4 Other Conflicts. A conflict of duty arises when a subject performs two operations on the same target, such as submitting a voucher and approving it. A conflict of interest arises when the same subject performs operations on different targets, such as providing advice to two clients who are competitors.

The above types of conflicts are not directly relevant to network management. First, note that these conflicts are applicable only to obligation policies. The underlying theme is that there are certain actions that subjects really should not be allowed to perform (such as approving their own vouchers or providing advice to clients who are competitors). Such situations are not likely to arise in network management applications. For network management obligation policies, the subject is typically the network management application itself; the target is a network device; and the action is usually a reconfiguration of some network device parameters. Here there is less emphasis on multiple subjects performing different management actions on network devices; normally there is just one subject, namely the network management application. For this reason, these types of conflicts are not considered further in this chapter.

4.5 WHAT TYPES OF POLICY CONFLICT RESOLUTION ARE REQUIRED IN A NETWORK MANAGEMENT APPLICATION?

Recall that, as described before in Chapter 2, network management policies can be classified into the following three categories:

- ECA policies
- AC policies
- Configuration policies

Now consider what types of conflict detection and resolution are needed for each of the above types of policies for network management applications.

- ECA policies will probably not be altered at run time (except maybe for some parameters of conditions), so static, manual policy analysis or static tool-supported policy analysis may be sufficient, as will be demonstrated later. Policies can be modified so that there are no conflicts at run-time as specified in this chapter.
- Access Control policies will probably be altered at run time and therefore need run-time conflict detection. These are basically just modality conflicts and are straightforward to resolve, as will be described in the next section.
- Configuration policies will definitely be modified at run time. The only conflicts that can arise here are those among inconsistent configuration parameters. Thus some run-time conflict resolution will be required to ensure that inconsistent configuration conflicts do not arise.

4.6 CONFLICT DETECTION AND RESOLUTION STRATEGIES

Conflict detection and resolution strategies fall into two major categories: compile time, or static conflict detection and resolution at policy definition time, and run time, or dynamic conflict detection and resolution in a deployed system. It would be ideal if all conflicts could be resolved at compile time, when policies are defined, because there would be no need to perform time-consuming conflict detection at run time. Also, it is easier to have a human in the loop (if necessary) for conflict resolution at compile time, before policies are deployed, than at run time, when a system is fielded, and up and running. However, for obvious reasons, it is not possible to perform all conflict detection and resolution at compile time, as described in examples in the previous section. The sequences of events that lead to the triggering of policies may or may not occur in a fielded system. Furthermore, policy events and actions may contain variables that depend on system state, which is unknown at compile time. For these reasons, it is necessary to incorporate both compile-time as well as run-time conflict resolution strategies into any realistic policy-based network management system. Strategies for resolving the conflicts described in the previous section are examined next.

First, to summarize, the following types of conflicts were described in the previous section:

- Application-independent conflicts:
  - Modality conflicts
  - Redundancy conflicts
- Application-specific conflicts:
  - Redundancy conflicts
  - Mutually exclusive configuration conflicts
  - Inconsistent configuration conflicts

Modality conflicts are easy to detect, and resolution can be performed based on a number of different strategies, as described later in this chapter. Only the most trivial redundancy conflicts are application-independent, namely when there are two or more policies that are identical in every respect; the conflict resolution strategy for this is to syntactically inspect every policy and delete redundant policies. Thus the only remaining conflicts that need to be considered are application-specific conflicts, which can be resolved statically (compile-time conflict resolution) or dynamically (run-time conflict resolution). These are discussed separately below.

4.6.1 Application-Independent Conflict Resolution

As mentioned above, tools for application-independent conflict resolution are easier to build than tools for application-specific conflict resolution. This is because the tool can be built without requiring a detailed analysis of the application. Furthermore, such tools can be reused for resolving conflicts for any application, by definition. This section discusses application-independent conflict resolution.

4.6.1.1 Modality Conflicts. Modality conflicts are relatively simple to detect. Recall that such policy conflicts arise for authorization policies, where there are policies that simultaneously authorize and deny access to a resource. When the Policy Decision Point receives a request for access to a resource, it consults the relevant policies, and if the policies do not provide a definite answer (i.e., the response is both "authorize" and "deny"), there is a policy conflict. In order to resolve the conflict, the user typically defines a conflict resolution strategy that determines which policy "wins" (i.e., whether the authorization or the denial prevails). Commonly used conflict resolution strategies include the following:

- *Deny Overrides*: Any policy that denies access to a resource overrides any policy that authorizes access to the resource.
- *Permit Overrides*: Any policy that permits access to a resource overrides any policy that denies access to the resource.
- *First Applicable*: The result of the first relevant rule encountered is the final authorization decision.

In XACML, which is a language for describing access control policies (see Chapter 9 for a description of XACML), conflicting results are resolved by what is called a combining algorithm. A combining algorithm determines which rules or sub-policies are to be evaluated, and how various combinations of results are to be resolved. A combining algorithm may also make use of optional parameters supplied to it in the policy. XACML defines several standard combining algorithms, but the mechanism is extensible to allow users to define additional algorithms. For example, one standard XACML combining algorithm says "return Deny if any sub-policy (or rule) returns a result of Deny." Another algorithm says "return the result of evaluating the first sub-policy that applies to the request." Yet another algorithm says "permit access if at least one sub-policy or rule returns a result of Permit." A fourth says "permit access only if none of the sub-policies or rules returns a result of Deny."

XACML sub-policies or rules can return a result of Indeterminate if some error occurs during evaluation such that the policy engine is unable to determine a valid result for evaluating the sub-policy or rule. Each combining algorithm specifies how to combine every possible combination of Permit, Deny, Not Applicable, and Indeterminate.

4.6.1.2 Redundancy Conflicts. Redundancy conflicts where one policy is identical to another are trivial to resolve, as any duplicate of a policy can simply be deleted. Tools for detecting such conflicts can perform a simple syntactic check to detect duplicate policies.

4.6.2 Application-Specific Conflict Resolution

This section describes approaches for application-specific conflict resolution. Such approaches can be further categorized as *compile-time* conflict resolution and *run-time* conflict resolution strategies. Each of these are described in turn below, with respect to the conflict types identified earlier.

4.6.2.1 Compile-Time Conflict Resolution. Conflict resolution may be performed in a static manner for a subset of the above-listed types of conflicts. For each type of conflict, there are some simple cases where a simple static syntactical inspection will reveal policy conflicts, as discussed below.

4.6.2.1.1 Redundancy Conflicts. Certain types of redundancy may be detected based on a static analysis of policies. Consider the case where subsumed policies need to be detected. This can be done by looking for policies that have the same events, conditions, and targets. For this set of policies, the actions for each pair of policies *P1* and *P2* need to be analyzed to determine whether the set of actions in *P1* is a subset of the set of actions in *P2* and whether the set of actions in *P1* is idempotent. If this is true, then the policy *P1* is subsumed by the policy *P2* [Chadha, 2006a] and can be deleted.

The above method can be further extended to detect redundancy of policy actions. As described in Section 4.4.2, an action in one policy may accomplish everything that another action in an otherwise identical policy accomplishes. This can be detected by again looking for policies that have the same events, conditions, and targets. For this set of policies, the actions for each pair of policies *P1* and *P2* need to be analyzed to determine whether the set of actions in *P1* is subsumed by the set of actions in *P2* [Chadha, 2006a]. This means that the actions in policy *P2* accomplish everything that the actions in *P1* accomplish. This check is necessarily application-specific. Again, the resolution strategy is to delete the subsumed policy.

4.6.2.1.2 Mutually Exclusive Configuration Conflicts. Only very trivial conflicts can be identified at compile time for mutually exclusive configuration conflicts, namely those where two policies are identical except for the value assigned to a target parameter. Since this type of conflict is application-specific, there is a need to model the effect of actions, in order to be able to reason about them. As an example, consider the following two policies, which each assign a value to a variable:

P1: <Event: QoS Provisioning Request; Subject: Manager1; Target: Router1, Interface1, EF; Action: Assign token bucket rate of 1Mbps to target>

P2: <Event: QoS Provisioning Request; Subject: Manager1; Target: Router1, Interface1, EF; Action: Assign token bucket rate of 2Mbps to target>

P1 and *P2* are in conflict because both of them will be triggered at the same time and will attempt to assign different values to the same parameter. The two policies can be translated into the following Prolog code:

```
assignedValue(p1, assignTokenBucket,1).
assignedValue(p2, assignTokenBucket,2).
target(p1, [router1, interface1, ef]).
target(p2, [router1, interface1, ef]).
events(p1, [qosProvReq]).
events(p2, [qosProvReq]).
```

The following code can be used to detect the conflict:

```
mutexConflictStatic(P1, P2):-
    identicalTargets(P1, P2),
    identicalEvents(P1, P2),
    assignedValue(P1, assignTokenBucket, Value1),
    assignedValue(P2, assignTokenBucket, Value2),
    Value1=\=Value2).
```

```
identicalTargets(P1, P2):-
    target(P1, T1),
    target(P2, T2),
    T1==T2.

identicalEvents(P1, P2):-
    events(P1, E1),
    events(P2, E2),
    E1==E2.
```

The problem that arises here, however, is that although the conflict can be detected very easily via mechanical means, resolution is a different matter. There is no easy way to specify which of the conflicting actions should "win"; a manual inspection of the policies and their intended effects will have to be conducted in order to determine which of the two different values should be assigned to the target parameter. Thus the above code can be used as a tool to assist the policy writer in writing a consistent set of policies, but cannot be used as an automated conflict resolution tool.

4.6.2.1.3 Inconsistent Configuration Conflicts. The final category of conflicts to examine is inconsistent configuration conflicts. Let's say you have multiple policies, whose actions modify a set of parameters $p1$. Suppose that these parameters are not independent, and that there exists some relationship that must always hold between the parameters in $p1$, expressed as $rel(p1)$. Then if a policy has an action to modify some of the parameters in $p1$, the invariant $rel(p1)$ may be violated, giving rise to a conflict. Again, trivial conflicts belonging to this category can be detected at compile time, namely in the case where

   (i) two or more policies are identical in every way except in their actions;

  (ii) their actions manipulate variables in $p1$; and

 (iii) the parameter values being assigned to the variables in $p1$ are known at compile time.

Then a check can be performed to see whether the constraint $rel(p1)$ is true or not. Again, this provides a conflict *detection* strategy, but resolution requires manual inspection of the conflicting policies.

4.6.2.2 Run-Time Conflict Resolution. Conflict resolution that must be performed at run time is examined next. Run-time conflict resolution is generally more complex than compile-time conflict resolution, as explained earlier; this section provides an overview of methods that can be used for this purpose.

4.6.2.2.1 Mutually Exclusive Configuration Conflicts. Assume that there are multiple policies that configure a certain network parameter. As an example, assume that one policy is a reaction to an intrusion and, upon receiving an intrusion event, it shuts down the interface being attacked on a router. Also, suppose that there is another policy that configures the same interface on the router to be up and to originate an IPSec tunnel for a VPN customer. These actions are in direct conflict, since the first action shuts down an interface, and the second one brings up the interface and configures it. Suppose that the two policies are:

Policy 1:   Event: Intrusion Detected on interface I1
            Action: Shut down interface I1
Policy 2:   Event: Provisioning VPN Event on interface I1
            Action: Bring up interface I1 and provision VPN.

Existing policy conflict detection systems [Bandara et al., 2003; Chomicki et al., 2003] would detect this as a conflict, and they would resolve it via different means. Bandara et al.'s Event Calculus approach would allow the specification of the conflict itself and of policies that indicate how the conflict should be resolved. The approach in Chomicki et al. [2003] would block one or the other of the actions if they occurred at the same time.

Stepping back to look at the problem, the question really is, What are the intended semantics? The intent of shutting down an interface in response to a detected intrusion is to prevent the attack. A manual analysis should determine whether prevention of the attack is a higher business priority than provisioning a new VPN in the system. If this is the case, then clearly, provisioning a new customer by bringing up the interface should only be allowed if the interface has not been shut down to prevent an intrusion. Thus there needs to be some indication in the system about the "intrusion state" of the interface, so to speak; and this state must be taken into account in any policy that attempts to bring up the interface. Thus the policies should be modified as follows:

Policy 1:   Event: Intrusion Detected on interface I1
            Actions: Shut down interface I1, and set state of interface to
                "under attack."
Policy 2:   Event: Provisioning VPN Event on interface I1
            Condition: Interface I1 is not under attack.
            Action: Bring up interface I1 and provision VPN.

Generalizing the above, either the intended semantics is that one action prevents the other one from modifying the parameter (in this case the operational status of the interface), or it is that the most recently performed action "wins." As another example, consider two policies, each with an action that increments a counter by different amounts. If the intended semantics is that the counter should be incremented by multiple policies, then there is no additional conflict resolution required, because the policies will result in the intended behavior.

4.6.2.2.2 Inconsistent Configurations Conflicts. Two approaches for handling inconsistent configurations conflicts are described below. The two methods achieve identical results, but differ in the implementation. The first modifies policies so that the inconsistent configuration conflict cannot arise, whereas the second leaves policies intact but augments the policy decision process to include a constraint check prior to the execution of any policy actions.

4.6.2.2.2.1   POLICY MODIFICATION.   One approach for dealing with the problem of inconsistent configuration conflicts is to modify the policies before they are deployed to ensure that this problem does not occur. As before, say you have multiple policies, whose actions modify a set of parameters *p1* with a relationship *rel(p1)* defined between them. Then if a policy has an action to modify some of the parameters in *p1*, the invariant *rel(p1)* may be violated, giving rise to a conflict. Again, such a situation has been described in Charalambides et al. [2006], where Event Calculus and conflict resolution policies are used to deal with the situation. However, a simpler solution is to insert the constraint *rel(p1)* within any policy with actions that alter parameters in *p1*. This can be done in one of two ways:

1. *Manually*: When authoring a new policy that has an action that alters parameters in *p1*, insert the constraint *rel(p1)* as a condition within the policy.
2. *Automatically*: When a new policy is created, the system should automatically check its actions to determine whether any parameters in *p1* are modified by the actions. If so, the policy should be automatically modified to include the constraint *rel(p1)* as a condition within the policy.

In the first approach, policies must be manually edited to ensure consistency; whereas in the second approach, a compile-time check can be used to automatically insert constraint-checking into the appropriate policies so that, at run time, the appropriate constraints are verified for consistency. If the constraints are found to be false, then the actions of the policy will not be triggered. Note that here the assumption is that this is the desired resolution; that is, if there is an action that attempts to modify parameters in a certain way that results in violation of a specified constraint, the action will not be performed.

4.6.2.2.2.2 POLICY ACTION CANCELATION. The second approach for dealing with the problem of inconsistent configuration conflicts is to introduce a run-time *action cancelation module*. The concept is analogous to that described in Chomicki et al. [2003]. At a high level, the idea is that that when a PDP is about to trigger an action, it first invokes the action cancelation module. This module checks whether the action would violate any existing constraints on the allowed values of system variables; if so, the action would be canceled (i.e., the action is never triggered), possibly with a warning to the user. Thus, as above, if you have multiple policies whose actions modify a set of parameters *p1* with a relationship *rel(p1)* defined between them, then *rel(p1)* is included in the set of policy constraints. The action cancelation module has access to a list of all such constraints in the system. Whenever the PDP invokes the action cancelation module, the latter checks the relevant constraints. Note that since there may be a large number of constraints defined in the system, only those that are relevant to a given action need to be checked before invoking that action. This can greatly improve the efficiency of the system. The list of constraints that are relevant to a given action are determined by looking at what system variables the action modifies. Suppose that an action A modifies a set of variables *V*. Then any constraint C that includes a variable belonging to *V* is said to be relevant to the action A.

The advantage of using this approach rather than embedding the check for constraint violation within the policies themselves is that it decouples policy definition from constraint definition. Policies can be written without knowledge of the constraints on the allowed values of system variables, and the action cancelation module is responsible for canceling actions that would violate the specified constraints. The results achieved are the same as the method described in the previous section.

4.6.2.2.3 General-Purpose Conflict Resolution. This section looks at a general-purpose conflict resolution mechanism based on the Event Calculus, which can be applied to any type of conflict. Event Calculus is based on abductive reasoning. In order to provide a basis for understanding the reasoning process behind the Event Calculus, an intuitive understanding of abductive logic is required, which is provided in the next section. For the purpose of this section, it is assumed that the reader has a working knowledge of logic programming, although an attempt is made to explain notation when it is introduced. The following section provides an overview of the approach and is largely based on Bandara et al. [2003].

4.6.2.2.3.1 OVERVIEW OF THE APPROACH. To set the stage, step back for a moment and think about what would be required to reason formally about the effects of policies on the system, so that conflicts can be detected and resolved. First consider obligation policies. For such policies, the following items need to be formally represented:

1. *The Effects that an Action Has on the Managed System*: In order to figure out whether two policies conflict, there is a need to be able to reason about the interactions between actions that are triggered by different policies at the same time.
2. *The Components of a Policy*: Policies can contain triggering events, conditions, and actions. Each of these needs to be formally represented.
3. *The Functioning of the PDP*: In addition to the above, there is also a need to model how the PDP works; in other words, when an event occurs, any policy containing that event will be triggered, and if its condition is true, its actions will be executed.
4. *The Policies*: Finally, a model of the policies themselves is needed, using a representation that allows automated reasoning about the policies.

4.6.2.2.3.2 OVERVIEW OF ABDUCTIVE LOGIC. The philosopher Peirce came up with the notion of abduction in 1958, which is described in Peirce [1958]. He describes the three forms of reasoning:

- *Deduction*: An analytical process that is based on the application of general rules to particular cases, with the inference of a result.
- *Induction*: A synthetic reasoning process that infers generalized rules from specific cases and the result.
- *Abduction*: A synthetic reasoning process that infers specific cases from a rule and a result.

Peirce wrote that abduction is "a weak kind of inference, because we cannot say that we believe in the truth of the explanation, but only that it may be true." Abduction is used in order to reason from effect to cause. Consider the following example:

```
congestion   ← link_failure
congestion   ← too_much_traffic
packet_drops ← congestion
```

For the reader not familiar with the above logic programming notation, the above statements can be read as follows:

```
There is congestion if there is a link_failure.
There is congestion if there is too_much_traffic.
There are packet_drops if there is congestion.
```

Thus if a network management system observes that packets are being dropped, there are two possible explanations based on the above statements: The first is that there is a link failure, and the second is that there is too much traffic in the network.

Kakas et al. [1993] defined abduction as follows. Given a set of statements T and a statement G (observation), abduction is defined as the problem of finding a set of sentences S (an *abductive explanation* for G) such that

$$T \cup S| = G(\text{read T union S satisfies G}), \quad \text{and} \quad T \cup S \text{ is consistent.}$$

Other authors (e.g., Cox and Pietrzykowski [1992]) have identified other desirable properties of abductive explanations. Explanations should be *basic* (not explainable in terms of other explanations) and *minimal* (not subsumed by other explanations).

4.6.2.2.3.3 THE EVENT CALCULUS. Event Calculus is a formal language for reasoning about event-based systems. A number of formulations of the Event Calculus have been presented. In this section, formulation described in Bandara et al. [2003] is presented, which consists of:

1. *A Set of Time Points (Mappable to the Set of Nonnegative Integers)*: A description of time is needed to be able to reason about the properties of the system at various points in time.
2. *A Set of Properties Called Fluents*: Fluents are used to describe properties of a system. A property either holds or does not hold at a certain point in time. Knowledge about which properties hold in a system at any given point in time allows one to reason about the system.
3. *A Set of Event Types*: Events are used to trigger policies.
4. *A Set of Predicates*: Predicates are statements that are either true or false. The following predicates are defined for the Event Calculus in order to be able to reason about the policy system. These predicates are used to express concepts that are needed for policy description. The concept of events occurring and causing system fluents (properties) to become true or false is captured by the following predicate definitions:
   initiates(A,B,T): Event A initiates fluent B for all time > T.
   terminates(A,B,T): Event A terminates fluent B for all time > T.
   happens(A,T): Event A happens at time point T.
   holdsAt(B,T): Fluent B holds at time point T.
   initiallyTrue(B): Fluent B is initially true.
   initiallyFalse(B): Fluent B is initially false.
   clipped(T_1,B,T_2): Fluent B is terminated sometime between time point T_1 and T_2.
   declipped(T_1,B,T_2): Fluent B is initiated sometime between time point T_1 and T_2.
5. *A Set of Domain-Independent Axioms with Respect to HoldsAt*: The following axioms are included in the system to enable an automated reasoning system to derive conclusions from system properties and events

occurring in the system. The following notation is used below: The arrow "←" should be read as "if."

- If B holds at a time T, and it is not terminated between T and T_1 where T_1 is some time after T, then B holds at time T_1:

$$holdsAt(B, T_1) \leftarrow$$
$$holdsAt(B, T) \wedge$$
$$\neg\, clipped(T, B, T_1) \wedge$$
$$T < T_1.$$

- If A happens at time T and initiates B at that time, and B is not terminated between T and T_1 where T_1 is some time after T, then B holds at time T_1:

$$holdsAt(B, T_1) \leftarrow$$
$$initiates(A, B, T) \wedge$$
$$happens(A, T) \wedge$$
$$\neg\, clipped(T, B, T_1) \wedge$$
$$T < T_1.$$

- If B does not hold at time T, and is not initiated between times T and T_1, where T_1 is some time after T, then B does not hold at time T_1:

$$\neg holdsAt(B, T_1) \leftarrow$$
$$\neg\, holdsAt(B, T) \wedge$$
$$\neg\, declipped(T, B, T_1) \wedge$$
$$T < T_1.$$

- If A happens at time T and terminates B at that time, and B is not initiated between T and T_1 where T_1 is some time after T, then B does not hold at time T_1:

$$\neg holdsAt(B, T_1) \leftarrow$$
$$terminates(A, B, T) \wedge$$
$$happens(A, T) \wedge$$
$$\neg\, declipped(T, B, T_1) \wedge$$
$$T < T_1.$$

4.6.2.2.3.4 POLICY COMPONENT MODEL. The next step is to create a formal model of the components of policies. These include events, actions, access requests, and permit and deny decisions. Events are used to represent the fact that an action is performed, that access is requested, and that a request is rejected.

Using the Ponder representation for policies, the required function symbols are:

- *state(Obj, V, Value)*: Represents the value *Value* of a variable *V* of an object *Obj* in the system. It can be used in an *initiallyTrue* predicate (see above) to specify the initial state of the system and also as part of rules that define the effect of actions.

- *operation(Obj, Action(V))*: Used to denote the operations specified in a policy function or event (see below). The function denotes the fact that object *Obj* can perform the action *Action(V)* and is used to specify all of the operations that objects can perform.

- *systemEvent(Event)*: Represents any event that is generated by the system at runtime and is used to trigger enforcement of obligation or refrain policies. The *Event* argument specified in this term can be any application-specific predicate or function symbol.

- *doAction(Obj_{Subj}, operation(Obj_{Targ}, Action(V)))*: *doAction* is an event that indicates the occurrence of an operation. Here the operation that has been performed is *Action(V)*, performed by Obj_{Subj} on target Obj_{Targ}.

- *requestAction(Obj_{Subj}, operation(Obj_{Targ}, Action(V)))*: *requestAction* is an event that indicates the occurrence of a request to perform an operation. Here Obj_{Subj} is requesting permission to perform operation *Action(V)* on target Obj_{Targ}. Note that when using the Ponder framework, this event will trigger a permission (or denial) decision to be taken by the target object's access controller.

- *rejectAction(Obj_{Subj}, operation(Obj_{Targ}, Action(V)))*: *rejectAction* is an event that indicates the occurrence of a denial of a request to perform an operation. Here Obj_{Subj} is denied permission to perform operation *Action(V)* on target Obj_{Targ}.

- *permit(Obj_{Subj}, operation(Obj_{Targ}, Action (V)))*: *permit* is a fluent, or property of the system, that represents the fact that permission was granted to a subject, Obj_{Subj}, to perform the operation *Action(V)* on the target, Obj_{Targ}.

- *deny(Obj_{Subj}, operation(Obj_{Targ}, Action (V)))*: *deny* is a fluent, or property of the system, that represents the fact that permission was denied to a subject, Obj_{Subj}, to perform the operation *Action(V)* on the target, Obj_{Targ}.

- *oblig(Obj_{Subj}, operation(Obj_{Targ}, Action (V)))*: *oblig* is a fluent, or property of the system, that represents the fact that the subject, Obj_{Subj}, is obliged to perform the operation *Action(V)* on the target, Obj_{Targ}.

- *refrain(Obj_{Subj}, operation(Obj_{Targ}, Action (V)))*: *refrain* is a fluent, or property of the system, that represents the fact that the subject, Obj_{Subj}, is to refrain from performing the operation *Action(V)* on the target, Obj_{Targ}.

The predicate symbols used with this approach are described below:

- *object(Obj)*: Used to specify that *Obj* is an object in the system.
- *attr(Obj, V)*: Specifies that *V* is an attribute of the object, *Obj*.
- *method(Obj, Action(V))*: Represents the fact that the action *Action(V)* is supported by object *Obj* in the system. It is used to define a separate ground term for every operation specified in the system.
- *isValidSpec(Obj$_{Subj}$, operation(Obj$_{Targ}$, Action(V))) ←*
  - *object(Obj$_{Subj}$),*
  - *object(Obj$_{Targ}$),*
  - *method(Obj$_{Targ}$, Action(V)).*

The above predicate defines what it means for an operation to be valid. Many of the function definitions above contain the tuple *(Obj$_{Subj}$, operation(Obj$_{Targ}$, Action(V))).* The *isValidSpec* predicate is defined to be true if the members of this tuple are consistent with the specification of the managed system. As such, it is used in the body of any rule where functions with the tuple *(Obj$_{Subj}$, operation(Obj$_{Targ}$, Action(V)))* are specified in the head, or in the portion of the rule before the "if" occurs.

4.6.2.2.3.5 SYSTEM BEHAVIOR MODEL. The next thing that needs to be modeled is the effect of actions on the managed system. This is a nontrivial task and requires an intensive, application-specific modeling effort. The effects of actions on the managed system are specified formally by capturing the state of the system after each action is performed, in the form of *post-conditions*. Any predicates that must be satisfied before an action can be performed must be captured in the form of *pre-conditions*. The following rules capture the relationship between performing an action (via *doAction*) and the corresponding pre-conditions and post-conditions. The first rule states that the event *doAction(Obj$_{Subj}$, operation(Obj$_{Targ}$, Action(Parms)))* results in the fluent *Post-True* being true for all time after time T, provided that the *doAction* parameters are valid (as defined at the end of the previous section) and that specified pre-conditions are true. In other words, the effect of performing the action *Action(Parms)* by subject *Obj$_{Subj}$* on target *Obj$_{Targ}$* is to make true the system properties represented by the fluent *PostTrue*. The second rule similarly states that the event *doAction(Obj$_{Subj}$, operation(Obj$_{Targ}$, Action(Parms)))* results in the fluent *PostFalse* ceasing to be true for all time after time *T*, provided that the *doAction* parameters are valid and that specified pre-conditions are true. In other words, the effect of performing the action *Action(Parms)* by subject *Obj$_{Subj}$* on target *Obj$_{Targ}$* here is to make false the system properties represented by the fluent *PostFalse*.

$$initiates(doAction(Obj_{Subj}, operation(Obj_{Targ}, Action(Parms))), PostTrue, T)$$
$$\leftarrow validSpec(Obj_{Subj}, operation(Obj_{Targ}, Action(Parms))) \wedge PreCondition.$$

$terminates(doAction(Obj_{Subj}, operation(Obj_{Targ}, Action(Parms))),$

$\quad PostFalse, T) \leftarrow validSpec(Obj_{Subj}, operation(Obj_{Targ}, Action(Parms)))$

$\quad \wedge PreCondition.$

Note that since *PreCondition* and *PostCondition* are application-specific, they cannot be specified in more detail here. In practice, each of these predicates would be replaced with application-specific conditions composed using the *state* function defined earlier. The *PreCondition* fluent would be defined using the *holdsAt* function defined earlier. Taken together, the above rules can be used to specify the effect that actions have on the system being modeled.

4.6.2.2.3.6   POLICY ENFORCEMENT MODEL. The following set of rules formally describes the effect of policy enforcement on the system. It describes events that are generated within the policy system (*requestAction, rejectAction, doAction*) as a result of fluents that hold in the system at a given time (*oblig, refrain, permit, deny*). The next section will describe how the *oblig, refrain, permit*, and *deny* fluents are initiated/terminated (i.e. become true/false) in a system.

Obligation/Refrain Enforcement Rules: The rules below describe the enforcement model for obligation and refrain policies. The first rule states that a request to perform a given action by a subject on a target is generated at time T_n if the fluent *oblig* for the same operation holds at some time T_m before T_n. The second rule states that a rejection of a request to perform a given action by a subject on a target is generated at time T_n if the fluent *refrain* for the same operation holds at some time T_m before T_n.

$happens(requestAction(Subj, operation(Targ, Action(ParmList))), T_n) \leftarrow$

$\quad holdsAt(oblig(Subj, operation(Targ, Action(ParmList))), T_m) \wedge (T_m < T_n).$

$happens(rejectAction(Subj, operation(Targ, Action(ParmList))), T_n) \leftarrow$

$\quad holdsAt(refrain(Subj, operation(Targ, Action(ParmList))), T_m) \wedge (T_m < T_n).$

Access Control Rules: The rules below describe the enforcement model for access control policies. The first rule states that an event indicating that an action is performed (*doAction*) is generated at time T_n if the fluent *permit* for the same action holds at some time T_m before T_n. The second rule states that a rejection of a request to perform a given action by a subject on a target is generated at time T_n if the fluent *deny* for the same operation holds at some time T_m before T_n.

$happens(doAction(Subj, operation(Targ, Action(ParmList))), T_n) \leftarrow$

$\quad holdsAt(permit(Subj, operation(Targ, Action(ParmList))), T_m) \wedge (T_m < T_n).$

$$happens(rejectAction(Subj, operation(Targ, Action(ParmList))), T_n) \leftarrow$$
$$holdsAt(deny(Subj, operation(Targ, Action(ParmList))), T_m) \wedge (T_m < T_n).$$

4.6.2.2.3.7 MODEL OF POLICIES. Every policy in the system must be modeled in terms of the Event Calculus. The rules below specify the semantics of policies in terms of events and the fluents that are initiated or terminated by these events. The function symbols used below were defined earlier in Section 4.6.2.2.3.4. Note that the previous section described what happened when certain fluents (*oblig, refrain, permit, deny*) were true in the system; this section describes how those fluents are initiated and terminated. In a nutshell, these fluents are initiated/terminated upon occurrence of the events *requestAction, rejectAction, doAction*, and *systemEvent*. These relationships are explained below and are then formalized in the form of rules.

- When an action is requested (via *requestAction*), one of the fluents *permit* or *deny* are initiated. This is because when a subject requests permission to perform an action, the request will either be permitted or denied.
- When an action is rejected (via *rejectAction*), one of the fluents *deny* or *refrain* are terminated. This is because rejecting an action is done due to one of the fluents *deny* or *refrain* being initiated; and once the action is rejected, the corresponding fluent should be terminated.
- When an action is performed (via *doAction*), one of the fluents *permit* or *oblig* are terminated. This is because performing an action is done due to one of the fluents *permit* or *oblig* being initiated; and once the action has been performed, the corresponding fluent should be terminated.
- When a system event occurs (via *systemEvent*), the one of the fluents *oblig* or *refrain* are initiated. This is because when a system event occurs, the event causes either an obligation policy or a refrain policy to be triggered.

Recall that in Ponder, policies can be authorization policies (positive or negative), obligation policies, refrain policies, or delegation policies. Delegation policies are a form of authorization policies and are not addressed separately here. For each of these policies, the rules below express the policy semantics.

Positive Authorization: The following two rules capture the semantics of positive authorization policies. The first rule states that given a request to perform an action by a subject Obj_{Subj} on a target Obj_{Targ} (as indicated by the occurrence of the *requestAction* event), the request will be permitted (i.e., the fluent *permit* will be initiated) if the operation specification is valid (using the "*isValidSpec*" predicate) and any specified constraints are satisfied. These constraints refer to rules that constrain what actions are permitted/denied by given subjects on given targets and should be specified via policies. The second rule simply states that the fluent *permit* should no longer hold (i.e., it will be terminated) once the requested action has been performed (i.e., the *doAction* event has occurred, indicating that the action was performed).

$$initiates(requestAction(Obj_{Subj}, operation(Obj_{Targ}, Action(ParmList))),$$
$$permit(Obj_{Subj}, operation(Obj_{Targ}, Action(ParmList))), T_m) \leftarrow$$
$$isValidSpec(Obj_{Subj}, operation(Obj_{Targ}, Action(ParmList))) \wedge Constraint.$$

$$terminates(doAction(Obj_{Subj}, operation(Obj_{Targ}, Action(ParmList))), permit$$
$$(Obj_{Subj}, operation(Obj_{Targ}, Action(ParmList))), T_m) \leftarrow$$
$$isValidSpec(Obj_{Subj}, operation(Obj_{Targ}, Action(ParmList))).$$

Negative Authorization: The following two rules capture the semantics of negative authorization policies. The first rule states that given a request to perform an action by a subject Obj_{Subj} on a target Obj_{Targ} (as indicated by the occurrence of the *requestAction* event), the request will be denied (i.e., the fluent *deny* will be initiated) if the operation specification is valid (using the "*isValidSpec*" predicate) and any specified constraints are satisfied. As above, these constraints refer to rules that constrain what actions are permitted/denied by given subjects on given targets and should be specified via policies. The second rule simply states that the fluent *deny* should no longer hold (i.e., it will be terminated) once the requested action has been rejected (i.e., the *rejectAction* event has occurred, indicating that the request to perform the action was rejected).

$$initiates(requestAction(Obj_{Subj}, operation(Obj_{Targ}, Action(ParmList))), deny$$
$$(Obj_{Subj}, operation(Obj_{Targ}, Action(ParmList))), T_m) \leftarrow$$
$$validSpec(Obj_{Subj}, operation(Obj_{Targ}, Action(ParmList))) \wedge Constraint.$$

$$terminates(rejectAction(Obj_{Subj}, operation(Obj_{Targ}, Action(ParmList))), deny$$
$$(Obj_{Subj}, operation(Obj_{Targ}, Action(ParmList))), T_m) \leftarrow$$
$$validSpec(Obj_{Subj}, operation(Obj_{Targ}, Action(ParmList))).$$

Obligation Policies: The following two rules capture the semantics of obligation policies. The first rule states that upon the occurrence of some system event E, the corresponding obligation fluent will be triggered (*oblig*), provided the operation specification is valid (using the "*isValidSpec*" predicate) and any specified constraints are satisfied. These constraints refer to policies that dictate what events trigger what actions under what conditions (in other words, ECA policies). The second rule simply states simply that the fluent *oblig* should no longer hold (i.e., it will be terminated) once the requested action has been performed (i.e., the *doAction* event has occurred, indicating that the required action was performed).

$$initiates(systemEvent(E), oblig(Obj_{Subj}, operation(Obj_{Targ},$$
$$Action(ParmList))), T_m) \leftarrow validSpec(Obj_{Subj}, operation(Obj_{Targ},$$
$$Action(ParmList))) \wedge Constraint.$$

$$terminates(doAction(Obj_{Subj}, operation(Obj_{Targ}, Action(ParmList))), oblig$$
$$(Obj_{Subj}, operation(Obj_{Targ}, Action(ParmList))), T_m) \leftarrow$$
$$validSpec(Obj_{Subj}, operation(Obj_{Targ}, Action(ParmList))).$$

Refrain Policies: The following two rules capture the semantics of refrain policies. Recall that refrain policies dictate what a subject must not do to a target. The first rule states that upon the occurrence of any system event, the corresponding refrain fluent will be triggered (*refrain*), provided that the operation specification is valid (using the *"isValidSpec"* predicate) and any specified constraints are satisfied. These constraints refer to refrain policies. The second rule simply states simply that the fluent *refrain* should no longer hold (i.e., it will be terminated) once the requested action has been rejected (i.e., the *rejectAction* event has occurred, indicating that the requested action was rejected).

$$initiates(systemEvent(\_), refrain(Obj_{Subj}, operation(Obj_{Targ}, Action$$
$$(ParmList))), T_m) \leftarrow validSpec(Obj_{Subj}, operation(Obj_{Targ},$$
$$Action(ParmList))) \wedge Constraint.$$

$$terminates(rejectAction(Obj_{Subj}, operation(Obj_{Targ}, Action(ParmList))),$$
$$refrain(Obj_{Subj}, operation(Obj_{Targ}, Action(ParmList))), T_m) \leftarrow$$
$$validSpec(Obj_{Subj}, operation(Obj_{Targ}, Action(ParmList))).$$

4.6.2.2.3.8 SPECIFICATION OF POLICY CONFLICTS. The final specification step in this process is to specify the conflicts that must be detected and resolved. This requires a manual analysis of the managed system and the policies, in order to determine what constitutes potential policy conflicts in the system. Once the conflicts that need to be detected have been specified, an abductive proof procedure can be used either to show that conflicts do not exist or to construct a sequence of events that lead up to the occurrence of a conflict. This sequence should help the user to determine the source of the conflict and to resolve the conflict in an appropriate manner. Examples of such policy conflict specifications are shown below:

- Recall that modality conflicts arise when there exist policies that simultaneously permit and deny the same operation. As specified in [Bandara et al. 2003], such a conflict can be specified as follows:

$$holdsAt(authConflict(Obj_{Subj}, operation(Obj_{Targ}, Action$$
$$(ParmList))), T) \leftarrow holdsAt(permit(Obj_{Subj}, operation(Obj_{Targ},$$
$$Action(ParmList))), T) \wedge holdsAt(deny(Obj_{Subj}, operation$$
$$(Obj_{Targ}, Action(ParmList))), T).$$

This specification states that an authorization conflict ("*authconflict*") occurs for the operation that a subject wants to perform on a target if this operation is simultaneously permitted and denied.

- Another type of modality conflict is specified below. Here the conflict is caused by the simultaneous triggering of rules that are obligation and refrain policies for the same operation.

$$holdsAt(obligConflict(Obj_{Subj}, operation(Obj_{Targ}, Action$$
$$(ParmList))), T) \leftarrow holdsAt(oblig(Obj_{Subj}, operation(Obj_{Targ},$$
$$Action(ParmList))), T) \wedge holdsAt(refrain(Obj_{Subj}, operation$$
$$(Obj_{Targ}, Action(ParmList))), T).$$

- A third type of modality conflict is specified below. Here the conflict is caused by (a) an obligation policy that requires a certain operation to be performed and (b) an access control policy that denies permission for performing that operation.

$$holdsAt(unauthObligConflict(Obj_{Subj},$$
$$operation(Obj_{Targ}, Action(ParmList))), T) \leftarrow$$
$$holdsAt(oblig(Obj_{Subj}, operation(Obj_{Targ}, Action(ParmList))), T) \wedge$$
$$holdsAt(deny(Obj_{Subj}, operation(Obj_{Targ}, Action(ParmList))), T).$$

Note that the above conflicts are all application-independent, since they apply to any policy system regardless of the application area.

4.6.2.2.3.9 DISCOVERY OF POLICY CONFLICTS. After having modeled the system as described in the last few sections, policy conflicts can be discovered by using an abductive logic procedure to reason backwards from the assumed occurrence of a conflict back to the causes of the conflict. Recall that abduction is used in order to reason from effect to cause. Therefore if the assumption is made that a certain conflict has occurred (the *effect*), then it is possible to reason about the sequence of events that led to the occurrence of this conflict and can pinpoint the *cause* of the conflict. In other words, the process used is to (a) provide as input to the system a predicate that states that a given conflict holds and (b) allow the system to reason about the sequence of events that could lead to the occurrence of the conflict. If the system fails to find such a sequence, then one can deduce that the system is free of the specified conflict. If the system does succeed in finding such a conflict, then the returned sequence of events will provide information about the source of the conflict. By analyzing this sequence of events, one can then pinpoint which policies led to the occurrence of the conflict and then determine how to handle the conflict. As will be shown in the first example in the Case Studies section (see Section 4.7), conflict resolution policies can be put in place to deal with different types of conflicts.

4.6.2.2.3.10 SUMMARY. The main drawbacks of the Event Calculus approach to policy conflict detection and resolution include the following:

1. *Complexity*: It is a highly complex process that involves creating the above-listed models and running queries on the identified potential conflicts. The process of specifying application-specific predicates to define the effects of policy actions on the system requires a high level of expertise and is extremely error-prone.
2. *Conflict Resolution*: The approach does not provide any solution for conflict resolution. It only provides an approach for conflict detection.

However, the approach could be applied for detecting conflicts such as those described in the previous section, which are not application-specific, since there is no need for application-specific modeling in this case.

4.7 CASE STUDIES

In this section, a couple of case studies are provided that examine examples from the literature, and contrast the use of formal policy conflict resolution computations with the approach suggested in this chapter.

Example 1. The following example is derived from Charalambides et al. [2006], but has been modified somewhat to simplify presentation. The paper describes a policy-based management module, DRsM (Dynamic Resource Management) for an MPLS network that supports DiffServ Quality of Service (QoS) [Blake et al., 1998] and can dynamically adapt to varying traffic demands by adjusting bandwidth allocations for different traffic classes. Some of the relevant policies implemented in this system can result in the inconsistent configurations conflict described earlier. As an example, the system includes the following policies for dealing with bandwidth allocation changes. When the upper threshold for bandwidth for a PHB (Per Hop Behavior) is exceeded because the traffic on a link *Link1* currently assigned bandwidth *BW* exceeds this threshold, the management system raises an alarm (event: *drsmAlarm-Raised*) that triggers a policy whose action *incrAllocRel* increases the bandwidth allocation for the specified traffic class by 20%:

> Policy p1:   Event: drsmAlarmRaised(upprTh, Link1, ef, BW)
> Subject: drsmPMA
> Target: mainMO
> Action: incrAllocRel(Link1, ef, BW,20)

The above policy is specified in the Event Calculus notation as follows:

> *initiates(systemEvent(drsmAlarmRaised(upprTh,Link1,ef,BW)), oblig(p1, drsmPMA,operation(mainMO,incrAllocRel(Link1, ef, BW,20)), T).*

Furthermore, there exists a constraint in the system that specifies an upper bound for the amount of network resources that can be allocated for each PHB. These constraints are put in place by policies that install the constraints in the system. As an example, the following policy has an action that sets the maximum allocation for the PHB EF (Expedited Forwarding) to 50 on all links:

> Policy p2:   Event: polReceived
> Subject: drsmPMA
> Target: mainMO
> Action: setBWMax(Link, ef, 50)

which is specified in Event Calculus as follows:

> *initiates(systemEvent(polReceived), oblig(p2,drsmPMA,operation(mainMO, setBWMax(Link,ef,50)), T).*

Here *polReceived* is an event that occurs when policies are received in the system. It can be seen that there is a potential conflict between the two actions *incrAllocRel* and *setBWMax*, since the latter sets the value of a system parameter that represents the maximum bandwidth allocation for EF to 50, and the former modifies the bandwidth allocation for EF. The conflict is specified formally as follows:

> Policy d1:   holdsAt(conflict(ndMaxConflict,conflictData([PolID1,PolID2, Link, PHB, BW2, BW3])),T) ←
> holdsAt(oblig(PolID1, Subj, op(Targ, incrAllocRel(Link, PHB, BW1,Incr))), T) ∧
> holdsAt(oblig(PolID2, Subj, op(Targ, setBWMax(Link, PHB, BW2))), T) ∧
> increasedBW(Link, PHB, BW1, Incr, BW3) ∧
> BW3 > BW2.

The above rule states that a conflict of type *"ndMaxConflict"* holds at a given time *T* if both *incrAllocRel* and *setBWMax* are triggered at the same time *T*, and *incrAllocRel* is attempting to increase the bandwidth allocated to link *Link* to *BW3*, where *BW3 > BW2*. Here *BW2* is the upper bound on the bandwidth for this link for the specified PHB, as specified by *setBWMax*. The function *increasedBW* above computes the new bandwidth setting that would be set by *incrAllocRel* by increasing the current bandwidth allocation for the link for this PHB by 20%.

Note that *conflict*, as defined above, is a fluent; an additional policy can be defined so that an event *conflDetected*, with the conflict parameters, is generated whenever this fluent holds. This policy can be triggered by a periodic check:

Policyc1:   initiates(systemEvent(periodicCheck),genEvent(conflDetected
            (ConflictType, conflictData(Parms))),T) ←
            holdsAt(conflict(ConflictType, conflictData(Parms)),T).

The above policy states that whenever the *periodicCheck* event occurs at some time *T*, if the fluent *conflict* holds at that time, then the operation *genEvent(conflDetected(ConflictType, conflictData(Parms)))* is performed, where *genEvent* is the action of generating an event.

The next step is to define a policy for conflict resolution; the relevant policy to resolve the above potential conflict is as follows:

Policyr1:   Event:conflDetected(ndMaxConflict,conflictData([PolA,PolB,
            Link, PHB, NDMaxBW]))
            Subject: drsmPMA
            Target: mainMO
            Action: setNDMax(Link, PHB, NDMaxBW)

The above policy is triggered by the event *conflDetected*, which indicates that a conflict of type *ndMaxConflict* has been detected in the system. Recall that this conflict indicates that an attempt is being made to assign a bandwidth allocation *NDMaxBW* to a given PHB, which is higher than the allowed upper threshold for that PHB. The resolution here is to invoke the operation *setNDMax*, which changes the upper threshold on bandwidth allocation permitted for the specified PHB to *NDMaxBW*. This policy is specified in Event Calculus as follows:

initiates(systemEvent(conflDetected(ndMaxConflict,conflictData([PolA,
    PolB,Link,PHB,NDMaxBW]))),oblig(r1,drsmPMA,operation(mainMO,
    setNDMax(Link, PHB, NDMaxBW)), T).

Given all of the above specifications, the Event Calculus approach can be used to detect the above conflict. As mentioned above, the implementation must generate a periodic event (*periodicCheck*) that triggers the conflict detection rule *d1*. When a conflict is detected, the system generates the *conflDetected* event that triggers policy *r1* above, which resolves the conflict.

Now let us take a look at how the above conflict could be dealt with using the approach described in the previous section. First, upon analysis of the managed system, it is clear that policy *p1* has a potential conflict with policy *p2* because their actions modify related parameters (giving rise to the "Inconsistent Configurations" conflict). In order to formalize the parameters and their relationships, let MAX_EF_BW be the system variable that represents the maximum bandwidth allocation for EF; and let EF_BW be the system variable that represents the actual bandwidth allocation for EF. Then the action *setBWMax* modifies MAX_EF_BW, and the action incrAllocRel modifies EF_BW; and the two are related via the following constraint:

$$\mathbf{EF\_BW \leq MAX\_EF\_BW.}$$

The proposed conflict resolution strategy could be used to either (a) manually modify the policies to address the above conflict or (b) automatically insert the constraint into the policies. With the latter approach, the resulting policies *p1′* and *p2′* would be as follows:

    Policy p1′:   Event: drsmAlarmRaised(upprTh, link1, ef)
                  Subject: drsmPMA
                  Target: mainMO
                  Action: incrAllocRel(link1, ef, 20)
                  Condition: EF_BW ≤ MAX_EF_BW.
    Policy p2′:   Event: polReceived
                  Subject: drsmPMA
                  Target: mainMO
                  Action: setBWMax(link1, ef, 50)
                  Condition: EF_BW ≤ MAX_EF_BW.

The effect of these policies would be the following. Policy *p2′* is triggered at installation time, where presumably EF_BW is initialized to some value lower than the maximum allowed bandwidth MAX_EF_BW. Following this, whenever *p1′* is triggered, it will only increase the EF allocation if the EF_BW remains less than or equal to the specified maximum (MAX_EF_BW). Note that any conflict is avoided automatically; however, the resolution strategy is different from that in Charalambides et al. [2006] because in the latter, the resolution strategy involves modifying MAX_EF_BW to match with EF_BW if the latter exceeds MAX_EF_BW. However, intuitively the resolution strategy proposed here is better aligned with the stated goals of the system, since this strategy causes the system to act in accordance with the maximum thresholds set by system policies. Furthermore, note that the resolution strategy in

Charalambides et al. [2006], which increments the upper bound on resource allocation whenever it is exceeded, could be achieved in a much simpler way by simply eliminating policy *p2* from the system! If the upper bound on an allocation is always going to be ignored whenever it is exceeded, then it doesn't serve a useful purpose and its removal has no impact on the system (other than reducing the number of potential policy conflicts!).

Example 2. The following example of policy conflict resolution is taken from Chomicki et al. [2003]. The following policies are specified (note that the policies have no subjects and targets; these are assumed to be implicit in the policy system):

Policy p1:	Event: defectiveProduct
	Action: stop
Policy p2:	Event: orderReceived
	Action: mailProduct

If both the events *defectiveProduct* and *orderReceived* occur together, then the two actions *stop* and *mailProduct* will be triggered at the same time. However, this is a conflict, since it is not possible to stop shipment and to mail a product at the same time. The conflict resolution strategy is defined via the constraint: *never stop AND mailProduct.* The authors in Chomicki et al. [2003] describe an action-cancelation monitor that cancels the action *mailProduct* when *stop* and *mailProduct* are being triggered together. The action-cancelation monitor is modeled as follows:

- A constraint of the form *never $a_1 \wedge a_2 \wedge \ldots \wedge a_n$ if C* is captured as a conflict rule:

$$block(a_1) \vee block(a_2) \vee \ldots \vee block(a_n) \ if \ exec(a_1) \wedge \ldots \wedge exec(a_n) \wedge C.$$

- For each action *a* in a policy, an accepting rule is defined:

$$accept(a) \ if \ exec(a) \wedge \neg \ block(a);$$

this accepting rule means that action *a* is executed ("accepted") if *exec(a)* is invoked and the action *a* is not blocked.
- Thus the constraint *never stop $\wedge$ mailProduct* is expressed as the following conflict rule:

$$block(stop) \vee block(mailProduct) \ if \ exec(stop) \wedge exec(mailProduct).$$

The implementation of this conflict rule will result in one of the two actions *stop* or *mailProduct* being blocked, thus resolving the conflict.

Next, consider an alternative to the above strategy. First, the model of the managed system is unclear. From the specified policies, it appears that the event *defectiveProduct* indicates that all products in the inventory are defective (and therefore none should be shipped to a customer). It is probably more realistic to assume that there is a certain inventory of products, some of which are defective and some of which are not. Thus the *defectiveProduct* event should really specify which products are defective. Second, the second policy should only result in mailing of the product if there exists a nondefective product in the inventory. The resulting policies are as follows:

Policy p1′:   Event: defectiveProduct(prod)
               Action: markDefective(prod)
Policy p2′:   Event: orderReceived
               Action: mailProduct(prod)
               Condition: existsProductInInventory(prod) ∧ ¬defective
               (prod)

In other words, an analysis of the policies shows that the policies needed more details and the semantics needed to be further expanded to include a parameter identifying the defective product(s). Once the policies are rewritten as above, the potential for conflicts is removed.

4.8 SUMMARY

In the previous section, it was shown via two case studies that the usage of complex policy conflict resolution mechanisms is not always warranted and that careful inspection and rewriting of policy rules can achieve similar effects, without imposing the burden of having to implement a formal policy conflict resolution system. Rather than writing policies and then writing rules about how to resolve policy conflicts, it is often simpler to rewrite the policies themselves. The argument that policy writers cannot be trusted to write correct policies does not hold water, because ultimately human effort is required to write correct rules for conflict resolution. Furthermore, every time a new policy is added, human effort is required to check all existing policies and determine whether new conflict resolution rules are required to accommodate the new policy! In other words, manual examination is required for both policy rewriting and conflict detection policy creation. The question that must be addressed is, Which one is easier and less error-prone? As was seen in this chapter, there are situations where it is simpler to rewrite the policies themselves. There may be cases where an automated policy conflict resolution system can provide value, especially when there are a large number of policies in the system, and compact separate models for system behavior and conflicts can be developed without much effort and without a need for constant updates. This may be the case when the addition of new policies does not introduce new conflict types and therefore does not require manual conflict analysis each time a new policy is added.

Additionally, it should be noted that the above conclusion holds only for run-time application-specific policy conflict resolution. Conflicts such as modality conflicts that are application-independent should be resolved automatically using conflict resolution tools, because no human effort is required to write complex policy conflict resolution rules for every application and for every new policy that is added to a set of policies. Furthermore, conflict detection tools that rely on syntactic checks for compile-time application-specific policy conflict detection may be useful for flagging potential conflicts to the user; however, even with such tools, human involvement is required to determine the appropriate conflict resolution strategy.

5

POLICY-BASED NETWORK MANAGEMENT

Having looked at policy terminology, policy languages and frameworks, and policy conflict resolution, this chapter focuses on the use of policies for network management. An overview of policy-based network management is provided by contrasting policy-based network management with current network management paradigms. The requirements for network management in MANETs are listed, and a reference MANET architecture is introduced. This is followed by a description of the architecture of a policy-based network management system for MANETs. The chapter closes with some practical usage scenarios.

5.1 OVERVIEW

This overview begins with a brief look at the state of the art of existing network management, and examines the benefits that can be provided by policy-based management. This is followed by a listing of management requirements for MANETs.

5.1.1 What's Missing in Current Network Management Systems?

The traditional network management functions are well known in telecommunications literature as FCAPS (Fault, Configuration, Accounting, Performance, Security). A 1998 Yankee group study [Yankee Group, 1998] found that configuration management makes up 45% of network operations cost for IP networks, whereas network monitoring accounts for only 6% of this cost.

Policy-Driven Mobile Ad hoc Network Management, by Ritu Chadha and Latha Kant
Copyright © 2008 John Wiley & Sons, Inc.

Another Yankee group report [Kerravala, 2004] quantified the impact of the lack of adequate configuration management tools by examining the different sources of errors in network configuration. In this report, human error was found to be the leading cause of network downtime, constituting 62% of all configuration errors. One reason for this is that while interfaces to network elements and/or the appropriate element management systems have been largely standardized for fault and performance management, this is not the case for configuration management of IP devices. Today, there is no standard interface that allows a service provider to configure IP routers from different vendors via the same configuration management interface. On the other hand, many standard SNMP MIBs (Management Information Bases) have been defined that allow the collection of fault and performance statistics from network elements provided by different vendors via the same standard interface. This disparity can be partly explained by the fact that versions of SNMP preceding SNMPv3 [Case et al., 1990] did not provide sufficiently robust security mechanisms to allow configuration via SNMP.

In order to address this need, a number of configuration management systems have emerged on the market in recent years that provide advanced, vendor-independent configuration management functionality. Such configuration management systems automatically translate high-level service specifications into appropriate configuration commands for network elements provided by different vendors. One aspect that has not been addressed by currently available commercial network management systems is the *feedback loop between configuration and fault/performance management*. Statistics collected for fault and performance management can be processed and analyzed by currently available operations support systems, leading to diagnoses of network problems. In order to fix such problems, there may be a need for repairing or replacing faulty equipment (in the case of network faults); or it may be necessary to reengineer the network to add more capacity to deal with severe network congestion problems; and so on. Such requirements need to be addressed manually, due to the need to physically install/repair equipment in the field. However, many performance and some fault problems can be handled by *network reconfiguration*. For example, if a network link is severely congested, it may be possible to alleviate the problem by sharing the traffic load with another underutilized link. This can be accomplished simply by appropriately reconfiguring the network. Today, this is done manually by experienced network operators who examine outputs from fault and performance management systems and decide on how to appropriately reconfigure the network. This is the fundamental problem with network management today: There is too much human intervention required to run a network. In order to reduce the cost of network operations, it is necessary to take the human out of the loop by creating a feedback loop between fault/ performance monitoring systems and configuration systems and by specifying *policies* that regulate how the system should be reconfigured in response to various network events.

Another aspect that is missing from current configuration management systems is the ability to state *long-term, network-wide* configuration objectives. Examples of such objectives include the following:

- All gold traffic trunks should have dynamically routed backup Label Switched Paths (LSPs).
- Halve bandwidth of gold traffic trunks during off-peak hours.
- If link utilization exceeds 80%, reduce by half the bandwidth allocated for best-effort traffic; and so on.

The above objectives can be achieved manually using current configuration management systems; for example, the first objective could be achieved by manually searching for all gold traffic trunks configured in a network and creating dynamically routed backup LSPs for each of them. The problem with doing this manually is that first, it is a tedious process to do this for each gold traffic trunk individually; and second, the process has to be repeated for every new gold traffic trunk that is added to the network.

5.1.2   How Can Policy-Based Management Help?

The need for a new management paradigm for MANETs was discussed in the first chapter of this book. As was discussed there, the advantage of policy-based management is that it allows the network administrator to define all the device provisioning and configuration at a single point (the management tool) rather than provisioning and configuring each device itself. In a system with a large number of machines, the control of configuration within a single logical entity reduces the manual effort required of an administrator. The administrator inputs the policies needed for network operation into the management tool which populates the repository. The information in the repository is specified in terms of the technology being deployed within the network. The PDP retrieves policies and converts them into the appropriate configuration of the PEP that can enforce the desired policies. The benefits of centralization on reducing manual tedium can easily be seen. In a network that requires configuring 1000 machines where configuring each machine requires 10 minutes, an administrator would need to about a week working continuously to configure the devices manually. In a policy-based solution, the administrator would need to spend only about 15 minutes populating the repository with the appropriate policies. The PDP would generate configurations for the devices automatically. This approach is likely to get the configuration done much faster, with a smaller likelihood of erroneous configuration.

Business-level abstractions make the job of a policy administrator simpler by defining the policies in terms of a language that is closer to the business needs of an organization rather than in terms of the specific technology needed to deploy it. The administrator need not be very familiar with the details of the technology that supports the desired business need. As an example, consider the case of a network operator who needs to define two levels of service, premium and standard. It is fairly simple for an administrator to identify customers and to define which service level they map to. One way to support the business need for two different levels of service is to use DiffServ within the network. If the administrator wants to define policies using the technology-level abstractions, he or she needs to be familiar with the details of the technology—for example, be aware that differentiation is obtained by assigning traffic to different PHBs (Per Hop Behaviors) and that the premium customer should be mapped to a specific PHB (e.g., the EF PHB, or Expedited Forwarding) and then define some parameters for this PHB. This process requires special knowledge that is harder to find than the familiarity with the business needs of the organization.

As mentioned above, traditional network management functionality is divided into the following components, commonly known as FCAPS: Fault, Configuration, Accounting, Performance, and Security Management. As will be described in the next four chapters, typical network management solutions provide vertically stovepiped products that address one or two of the above areas, with, at best, some loose coupling between the products via shared databases. The fundamental problem with this approach is the inability to:

- Tie together the operation of these systems; and
- Do so in a way that can be changed as needed, depending on management goals.

The first problem above needs to be addressed by closing the feedback loop between network status and network reconfiguration. A policy-based management system allows the network operator to enter management objectives as *policies* into the management system, and it ensures automatic enforcement of these policies so that no further manual action is required on the part of the network operator. The second problem indicates that the interactions between network status gathering and network reconfiguration need to be expressible in a flexible, user-friendly manner, so that these interactions can be customized as needed based on possible changing management goals. This problem is accentuated in military scenarios, where missions are typically short-lived (72 hours or less) and the requirements of each mission require that the network and its mode of management be tailored to the needs of the mission [Chadha 2006b]. For example, certain missions in enemy territory require stronger encryption for over-the-air communications than missions in friendly territory; the nature of communications in different missions may dictate different bandwidth allocations for QoS queuing mechanisms for each mission; certain missions may be particularly sensitive to jamming and interception; and so on.

In order to tie together the operations of FCAPS, there is a need to add another component to FCAPS: the *Policy Management* component, used to provide the glue between FCAPS functions. The Policy Management component is used to:

- Initiate the operations of FCAPS components, by kicking them off as needed.
- Control the operations of FCAPS components, by conditionally altering their behavior.
- Tie the operations of one FCAPS component to another, by capturing the desired interactions between these components.
- Allow the specification of all of the above in a user-friendly manner via policies.

Policy-based network management provides the capability to express networking requirements at a high level and have them automatically realized in the network by configuration agents, without requiring further manual updates. This approach provides the network administrator with the capability to specify high-level policies that:

- Specify *long-term, network-wide* configuration objectives—for example:
  - All control traffic must get the highest level of QoS priority.
  - All private communications must be encrypted.
- Provide an *automated feedback loop* so that information reported by monitoring agents can be used to automatically trigger correction of network problems based on policies—for example:
  - If server response time >5 seconds, determine whether to relocate the server.

Once policies such as those described above are defined, they are automatically enforced by the management system. These capabilities can provide military personnel as well as commercial network operators with very powerful tools to configure and control their network and to reconfigure their network in response to network conditions, with the highest possible level of automation. Some examples of added functionality that would be enabled by policy-based management include:

- Dynamically changing the role of a node to act as a server (e.g., DNS server), based on relevant capabilities such as computing power, battery power, signal strength, hardware, and so on. Capabilities that are important in a certain environment may not be relevant in others (e.g., battery power may be important at night but not during the daytime for solar-powered nodes); such constraints are captured as policies.
- Switching between proactive (e.g., Optimized Link State Routing [Clausen and Jacquet, 2003]) and reactive (e.g., Ad hoc On-demand Distance Vector [Perkins, 2003]) routing protocols to optimize performance depending on the known density of nodes in the network.
- Setting network-wide values for configuration parameters; and so on.

As described in Chapter 3, several languages and frameworks have been developed by various organizations for describing network policies. In the ECA (Event–Condition–Action) paradigm, policies are represented as rules with events, conditions, and actions. These rules can be used to express high-level business goals that are automatically translated into detailed device configurations that achieve the desired business goals. Examples of such high-level business rules are:

1. Specification of network-wide configuration actions to achieve a certain objective, which involves more than one device: For example, all routers in a service provider's network need to be configured with the same parameters for weighted fair queuing for a class of service to provide quality of service guarantees for that class. This rule can be expressed as follows:

   Event: Configuration event

   Condition: None

   Action: Configure specified WFQ parameters.

2. Automating actions based on time of day: For example, provide quality of service guarantees for customer A from 8:00 am to 5:00 pm by reserving sufficient resources at each router between these times. Two policies are used to express the desired behavior:

   Policy 1:

   Event: Timer event at 08:00 UTC

   Condition: None

   Action: Begin policing and marking customer A's traffic at the customer's ingress router.

   Policy 2:

   Event: Timer event at 17:00 UTC

   Condition: None

   Action: Stop policing and marking customer A's traffic at the customer's ingress router.

3. Automating actions to react to changes in network state: For example, drop packets to alleviate congestion if congestion is detected in the network.

   Event: Link utilization crosses threshold of 80%

   Condition: None

   Action: Drop best effort traffic exceeding 10 Mb/s.

Chapter 1 explained why policy-based management becomes even more critical for MANETs than for static wireline networks. Essentially, MANETs require much more frequent reconfiguration than static wireline networks.

This is a major factor in the drive towards more automated network management for MANETs.

5.1.3 MANET Management Requirements

In order to be able to effectively use the previously described policy rule-based approach to network management, a policy-based management system for MANETs must support a number of essential features, which are listed below. It should be noted that the list below is by no means an exhaustive list of requirements for a network management system for MANETs; rather, this list attempts to enumerate the salient features of a network management system for MANETs that *differ* from the management requirements for wireline networks. In other words, requirements that apply to wireline network management have not been repeated here, because the emphasis is on the *additional* functionality required for managing MANETs.

- *Minimal Use of Bandwidth*: Given that MANETs are characterized by scarce bandwidth, it is imperative that a management system for MANETs minimize its use of wireless bandwidth to be effective, so that bandwidth is available for mission-critical applications. In general, management traffic should utilize no more than 2% and 5% of network bandwidth.
- *Survivability*: Given that MANET nodes are often disconnected from the network due to a variety of reasons, such as mobility, adverse weather conditions, and limited power, it is important that the network management system be survivable. In other words, it should continue to function even when one or more nodes are missing from the network, and it should be able to automatically handle frequent entry and exit of nodes to and from the network. This requirement implies a distributed architecture that can automatically recover from the loss of any node.
- *Automated Reconfiguration*: Given the need for frequent reconfiguration in MANETs, the network management system must support automated reconfiguration as far as possible. Automation is required both to (a) improve network performance by enabling rapid responses to network problems and (b) reduce the amount of human intervention required to manage the network.
- *Quality of Service Management*: The scarcity of bandwidth in MANETs dictates that traffic be accorded differentiated treatment based on the type of traffic and its priority. Unlike wireline networks where bandwidth is typically plentiful, applications in a MANET must compete for network resources. This means that a network management system for MANETs must be capable of differentiating between high-priority traffic and low-priority traffic and ensuring that the most important traffic gets preferential treatment.

• *Scalability*: The architecture of the network management system must be scalable in order to be able to support a large number of nodes.

5.2 ARCHITECTURE

This section begins by describing a reference network architecture for a mobile ad hoc network, along with the accompanying security considerations. This is followed by a description of an architecture for a policy-based network management system for ad hoc networks that addresses the requirements outlined above.

5.2.1 MANET Reference Architecture

This section describes a reference architecture for a MANET. At the outset, it should be noted that the network architecture for MANETs can vary widely, based on the capabilities of the hardware used, the purpose of the network deployment, the scale of the network, and so on. Given these possible variations, it is clear that there is no single MANET architecture that fits all possible deployments. So one may ask the question, Why bother defining a reference network architecture? The intent here is to provide a network architecture that can be referenced in the remainder of this book when network management functions are discussed. The architecture described here is sufficiently general that most actual deployments should be derivable from this architecture, with minor variations.

Figure 5.1 shows the MANET reference architecture. Nodes of different capabilities are deployed in an ad hoc network. Nodes are differentiated into the following major categories:

1. *Manned Platform Nodes*: These nodes contain a wired network within the node and have a wireless interface to the other nodes in the MANET, as described in [Poylisher et al., 2006]. The wired network contained within the node is comprised of one or more wired LANs and may contain equipment such as hosts, routers, switches, firewalls, and so on. Examples of such nodes are military tanks or commercial vehicles that carry passengers who need access to computing equipment on board the vehicle while it is moving. Platform nodes are characterized by considerable computing capabilities and typically unlimited power. They may have one or more LANs; one scenario where multiple LANs are required is when a platform must host information at different security levels (unclassified, secret, etc.) and needs to physically separate the information residing at each of these levels on different LANs.

2. *Unmanned Platform Nodes*: These nodes may contain a wired network within the node and have a wireless interface to the other nodes in the MANET. These nodes are very similar to manned platform nodes, except that they do not have passengers on board and therefore cannot rely on

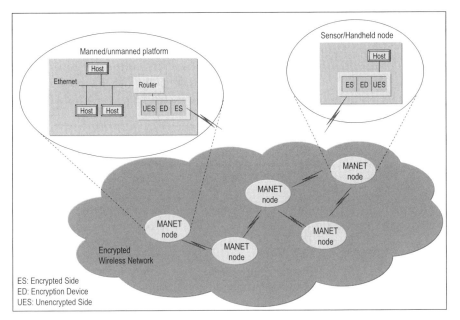

Figure 5.1. MANET reference architecture.

any manual operations. Examples of such nodes are unmanned vehicles such as Unmanned Aerial Vehicles (UAVs) that provide relay capabilities for ad hoc networks on the ground.

3. *Sensor Nodes*: These nodes contain sensing equipment and are unmanned. The sensing equipment is provided by special-purpose hardware. These sensor nodes communicate with other nodes in the MANET, both sensor and nonsensor nodes, via wireless links. Sensor nodes are characterized by constrained computing capabilities and limited power.

4. *Handheld Nodes*: These nodes are typically small handheld devices such as PDAs (Personal Digital Assistants) that are carried by a human being. These nodes are used to communicate with other MANET nodes and have some kind of simple user interface. Like sensor nodes, handheld nodes are also characterized by constrained computing capabilities and limited power.

The different types of nodes described above can be networked together, using gateways where necessary to connect different types of wireless technologies.

5.2.2 MANET Security Considerations

In order to secure wireless communications against eavesdropping, all over-the-air communication must be encrypted. The type of encryption, algorithms used, keying mechanisms, and so on, depend on the degree of security that is required; for example, military operations may have much more stringent requirements for security than, say, web browsing.

In order to understand the implications of encryption requirements on the nodes in a MANET, consider a manned platform node for the sake of illustration. Figure 5.2 shows a simple LAN on a manned platform, connected to other nodes via a radio. Host computers on the LAN are connected to an encryption device, which is a router that supports IPSec [Kent and Seo, 2005]. The router has a wired interface on the LAN side and has a wireless radio interface for over-the-air communications. All IP packets originating on the wired LAN that are destined for other MANET nodes arrive at the router, get encrypted and encapsulated within a new IP packet (with new IP source and destination addresses), and are sent out via the wireless interface over the air. The new IP source and destination IP addresses are addresses of wireless interfaces and are known as "black" IP addresses; the original source and destination IP addresses are known as "red" IP addresses. Over-the-air communications are encrypted and are therefore usually referred to as "black" communications. On the other hand, any information on the wired platform LAN is unencrypted and is commonly referred to as "red" communications. When the IP packets arrive at the destination MANET node, their contents are decrypted and passed to the red side, to the destination host. All MANET applications other than some essential management functions are hosted on the red side for maximum security.

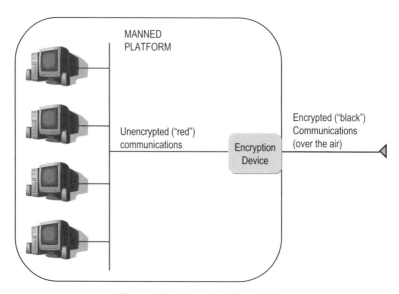

Figure 5.2. Manned platform LAN.

5.2.3 Policy-Based Network Management System for Ad hoc Networks

The policy-based management architecture defined by the IETF that was described in Chapter 2 is a very high-level architecture and does not address a lot of issues that must be answered before such a system can be deployed. First, the policy management system must be tied in with the traditional components of network management (FCAPS); second, the issue of deployment must be addressed. How should a network management system for ad hoc networks be distributed across the network?

These questions are addressed in the following sections.

5.2.3.1 Policy-Based Management System Architecture. As discussed earlier in this section, the motivation behind the use of policy-based management is that it can help to tie together the operations of FCAPS management. Figure 5.3 shows these components and their relationships, and it provides a high-level view of the components of a network management system for ad hoc networks. At the highest level is the Policy Management component, which is used to create and edit policies. The Policy Management component also interfaces with an optional Network Planning component, which could generate policies for the Policy Management component (shown as "Policy Definition" in the figure). Policies are stored in a Policy Repository. The Policy

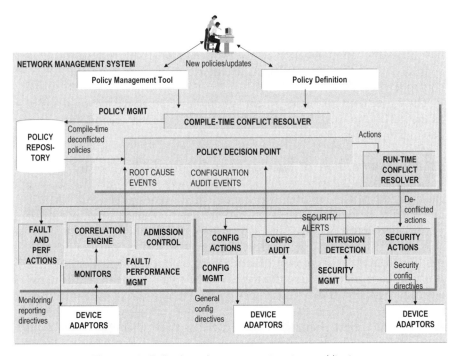

Figure 5.3. Policy-based management system architecture.

Management component includes compile-time as well as run-time conflict resolvers. The core functionality of the Policy Management component is performed by the Policy Decision Point, or PDP. The PDP controls the functioning of the remainder of the network management components. These include Fault and Performance management components, which monitor the network and receive fault notifications from network elements; the Configuration management component, which configures the network elements and verifies that they are correctly configured; the Security management component, which configures security devices and performs intrusion detection and generates security alerts; and finally, the QoS management component, which is a subset of the Performance management functionality and performs admission control on network traffic flows that require QoS.

Each of these FCAPS components implements actions that can be triggered by the Policy Management component. They also generate different types of events that can trigger policies. The Fault Management component generates root cause events; these events are sent to the Policy Management component, where they may trigger policies that reconfigure the network. This reconfiguration is accomplished by invoking actions implemented by the Configuration Management component. The Configuration Management component periodically conducts audits of the network elements, to verify that they are configured according to the latest stored information; if there is a discrepancy, configuration audit events are generated. The Security Management component performs intrusion detection and generates security alerts when intrusions are detected; these alerts are sent to the Fault Management component for further root cause analysis and correlation with other network alarms.

Finally, any communication with network elements is performed via network adaptors that hide vendor-specific interface details from the FCAPS components.

The entire network management system depicted above will be referred to as the NMS (Network Management System). The next section discusses the deployment of the NMS.

5.2.3.2 *Distribution of Management Functionality.* The first and foremost question is that of distribution. The architecture in Figure 5.4 shows one centralized NMS and multiple network elements managed by that NMS. This architecture is commonly used in small and medium-sized wireline networks, where one central management station uses SNMP polling to gather device status and also listens for traps from network nodes. This may not be the best solution from the point of view of survivability and robustness. For a MANET, in particular, a centralized solution with one NMS managing the entire network would bring the network to its knees due to the amount of over-the-air communications required between the NMS and network elements! Furthermore, this architecture is definitely not survivable, as it has a single point of failure.

As mentioned in Section 5.1.3, one of the foremost issues to be addressed by a network management system for ad hoc networks is the scarcity of band-

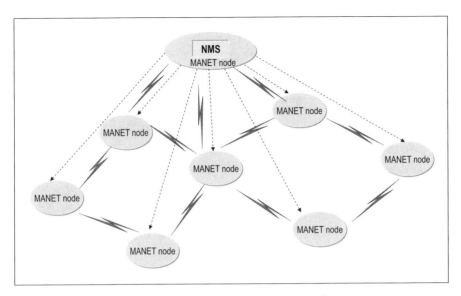

Figure 5.4. Centralized management solution.

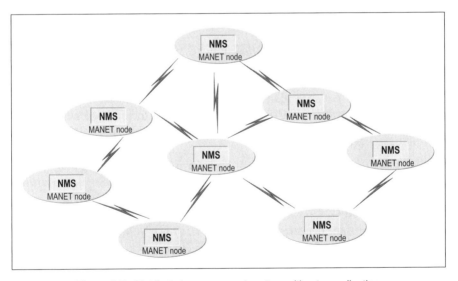

Figure 5.5. Distributed management system without coordination.

width available for application and management traffic. Thus management must be localized as much as possible. An alternate to the above picture is the one shown in Figure 5.5, where an NMS resides on every MANET node.

The above picture does not show any communication between NMSs residing on different nodes; however, in practice, it is often necessary to exchange

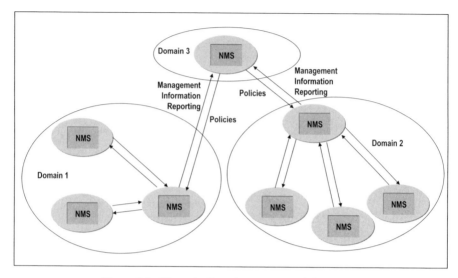

Figure 5.6. Dissemination of management information.

some subset of management information among network nodes. The two categories of information that must be disseminated (see Figure 5.6) are:

- *Policy/Configuration Information*: Policies and certain types of configuration information need to be defined by operators and disseminated to all nodes in the network. Policies are either (a) created manually or (b) generated from a network plan that is created manually. The manual step of creating a network plan or policies is typically performed at a central location. Thus policies must be disseminated from that central location to all nodes in the network.
- *Monitoring Information*: Monitoring information is collected at all nodes and stored locally. It may be necessary to summarize the collected information in some way and report it to other nodes. There could be several reasons for this:
  - A human operator at a centralized location may want to keep an eye on the high-level status of all the nodes in the network.
  - It may be necessary to collect information from several nodes and correlate the collected information to diagnose network problems.
  - It may be necessary to take certain global actions across the entire network if certain conditions are reported.

As was shown earlier in Figure 5.4, disseminating information directly to and from one centralized location is not a scalable solution, since as the number of nodes in the network grows, the centralized node will get overwhelmed by the amount of traffic it needs to handle, and the network will not be able to support the large volumes of data going to and coming out of this centralized node. Furthermore, such an architecture is not survivable in the face of node loss and intermittent connectivity, both of which are common occurrences in MANETs. An alternative that has been explored in depth in the context of routing [Toh, 2002] is the concept of *clustering*, which organizes nodes into small groups called *clusters* in a hierarchical fashion and is discussed in the next section.

5.2.3.3 Clustering for Network Management. Many of the concepts of clustering for the purpose of routing can be used for policy-based management, including building and restructuring a clustering hierarchy. The important difference to note here is that clustering is being used for the purpose of network management, rather than for routing. The concept has also been used for MANET management in Chen and Jain [1999], where a hierarchy of three levels is used for management. The principal limitation of the management architecture described in this paper is the restriction of the number of levels in the management hierarchy, which is limited to three. It is important to allow a management hierarchy of arbitrary depth, since limiting the depth of the hierarchy intrinsically limits the scalability of the system and only provides improvement by a constant factor over the centralized management solution. The system should be able to scale to the size of the network by expanding or shrinking the number of tiers in the hierarchy and appropriately assigning management functionalities to nodes at different tiers in the hierarchy. One way to form a hierarchical structure in such management systems is to associate it with the human reporting hierarchy so that policies can be formulated to serve the management needs originating from the organization of human administration. Another way to form a hierarchical structure is to align the structure with the physical network links between nodes so as to make information dissemination as efficient as possible. These two approaches are discussed in more depth below. First, the relevant terminology is provided below:

- A *cluster* is a collection of one or more nodes in the network.
- A cluster includes a single cluster *leader* and a set of children, known as *cluster associates* or *cluster members*. Leaders of several clusters can form another cluster (see Figure 5.7 for a sample set of clusters).

Given that a multi-tier management hierarchy is required for scaling a distributed management system, it is critical for this hierarchy to *systematically and autonomously self-form*, especially if the target networking environment is dynamic by nature. The reason for this is that nodes may disappear or become unreachable, thus requiring frequent reorganization of this management hierarchy. The basic idea for formation and maintenance of the management hierarchy is to form "clusters" and then link clusters into a tree.

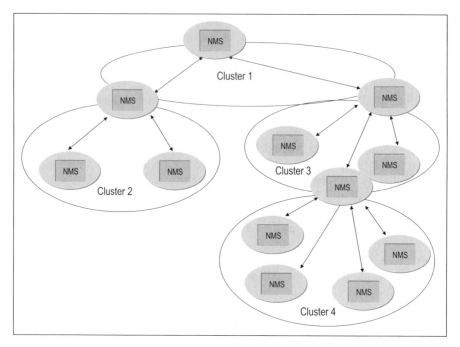

Figure 5.7. Hierarchy of network management clusters.

The next few sections describe formation and maintenance of this management hierarchy. The terminology used is borrowed from PECAN (see Chapter 3 for a description of the PECAN policy-based management system). In particular, the concepts of GPA (Global Policy Agent), DPA (Domain Policy Agent), and LPA (Local Policy Agent) are used.

5.2.3.3.1 Initial Formation of Management Hierarchy. The initial formation of the previously described management hierarchy can be achieved by advance planning; in this section, an implementation of this concept is described. A planned management hierarchy is loaded into a *policy naming server* on the GPA node when the system is launched (note that the GPA is the root of the management hierarchy). In this architecture, the policy naming server is replicated at every policy agent; its function is to distribute and store data about the management hierarchy. The policy naming server on the GPA distributes the naming data, which essentially includes <child, parent> pairs, to all children under the GPA. Nodes receiving naming data store the data in their local store, and distribute the data to all their children, if any.

During the initialization phase on each node, a policy agent uses the policy naming service to find out whether it has a parent in the predefined hierarchy. If it has no parent, it assigns itself the role of GPA; if it has a parent, it checks whether it has any children to determine whether its initial role should be a

DPA or an LPA. In case no naming information is available from the naming service, which could happen if the distribution of naming data by the GPA is unsuccessful due to intermittent connectivity, a policy agent will assign itself the role of GPA. This could result in more than one GPA for the network; this is handled via a merge after network connectivity is restored, as described below in Section 5.2.3.3.3.

In what follows, the parent of a cluster will be called the *leader* of the cluster, and the children of this leader will be called its cluster *associates*.

5.2.3.3.2 Cluster Maintenance. In a dynamic networking environment, after the initial formation of the management hierarchy, it is necessary to maintain this hierarchy for distributing policies and collecting management information. Cluster maintenance is achieved as follows. First, the leader of a cluster periodically broadcasts heartbeat messages intended for all its cluster associates, to keep them aware of the cluster that they belong to. These heartbeat messages contain cluster information such as cluster leader, cluster identifier, cluster settings, and so on. A cluster associate acknowledges the receipt of heartbeat messages from its leader in order to maintain its membership in the cluster. Note that it is possible for nodes that are not children of this cluster leader to hear the heartbeat broadcast by this leader; however, such children do not acknowledge receipt of heartbeat messages from any cluster leader other than their own. If a cluster associate misses a number of consecutive announcement messages from its leader, it assumes that the leader no longer exists and tries to join a different cluster. If no other cluster leader can be reached, it changes its role to become a GPA. Furthermore, if a cluster becomes too big (where "too big" is a configurable metric), it splits into two or more clusters.

For cluster signaling, two types of messages are used:

- When a node wants to join a cluster (when two clusters are merging), it sends a "join request" message to the cluster leader. If the request is granted, the leader sends a "join granted" message to the child and records the node as its child when an acknowledgement message is received.
- When a cluster leader splits its cluster into two or more clusters, it appoints one of its associates as the leader for a group of associates in its cluster. The new leader remains a child in the current cluster, while the associates to be split out join the new cluster by signaling their new leader.

5.2.3.3.3 Management Hierarchy Maintenance. To exercise global control of a policy system, it is crucial to keep one single management hierarchy in the system as far as possible, so that policies can be distributed in a top-down fashion and information can be collected from the bottom up. To ensure that the policy system has one single management hierarchy whenever possible, the following signaling mechanisms are used:

- When a node detects the existence of another GPA, it notifies its own GPA.
- When two GPAs detect each other, they negotiate with each other to merge into one hierarchy.

When a node receives a heartbeat message from a GPA other than its own, it forwards this message to its GPA. When a GPA learns the existence of another GPA, it initiates a negotiation session with the other GPA. The other GPA signals the session initiator its decision to remain a GPA or to step down and become a child of the session initiator. Once the session initiator acknowledges this decision, the two hierarchies merge and one of the GPAs steps down to become a child of the other. The two hierarchies also merge their contents to complete the merge process.

5.3 USAGE SCENARIOS

An oft-repeated complaint about policy-based management is that the basic concepts are relatively easy to grasp, but people have difficulty coming up with real usage scenarios for policy-based management and have trouble formulating policies for network management. For this reason, the ideas presented in this and the past few chapters are motivated by presenting some real-world scenarios that illustrate the use of policies in a MANET. Note that this section is by no means a comprehensive collection of policies required for managing a MANET; rather, it illustrates some of the concepts discussed so far via examples.

The policy-based management concepts introduced in this chapter are illustrated below via by a number of usage scenarios relating to managing a MANET using a policy-based network management system based on PECAN, which was described in Chapter 3. A sample MANET testbed is used as the basis for the usage scenarios and is shown in Figure 5.8. The testbed consists of three OLSR wireless ad hoc routing domains, numbered 1, 2, and 3. Servers providing various application-level services are placed on mobile nodes throughout the network; these nodes are labeled MOBILE-1 through MOBILE-7 in the figure. Two nodes (MOBILE-6 and MOBILE-7) are used as border routers to connect the ad hoc routing domains; MOBILE-6 connects domains 1 and 3, and MOBILE-7 connects domains 2 and 3. A GPA is running on a node in OLSR Domain 3, and LPAs are running on all other machines in the network.

Below is a description of some of relevant usage scenarios.

5.3.1 Network Monitoring

In order to manage the MANET nodes, vital network statistics must be collected by the policy agent on each node. Monitoring policies are created for

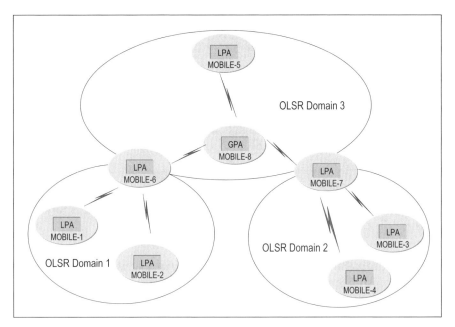

Figure 5.8. Mobile ad hoc network testbed.

this purpose at the GPA and distributed throughout the network. As an example, the following monitoring policy (expressed as an ECA policy below) is created at the GPA and distributed to all nodes:

Monitoring Policy

- *Event*: None
- *Condition*: None
- *Action*: Monitor CPU utilization every 30 seconds.

Upon receipt of this policy, monitoring agents on all nodes began to monitor the local CPU utilization and store it in a local data store every 30 seconds.

   The above monitoring policy is extremely simple; in practice, it would probably be more realistic to create a set of monitoring policies that adaptively monitor various network statistics based on current network status and reporting requirements. This is further elaborated upon in the next section.

5.3.2   Reporting Management Information

The previous section described collection of node statistics at each node. In order to provide a view of the network health, it is sometimes desirable to report management information back up the management hierarchy. This is achieved by creating a reporting policy at the GPA and distributing it throughout the network:

Reporting Policy

- *Event*: None
- *Condition*: None
- *Action*: Report CPU utilization every 30 seconds to parent node.

Upon receipt of this policy, all nodes begin to report the local CPU utilization every 30 seconds to their parents. Since this testbed has a two-level hierarchy, all information gets reported to the GPA. The policy agent on the GPA stores this reported information in a local data store.

As discussed above, the above policy is rather simple; in practice, reporting of management information should be adaptive and might vary based on several factors, which include external factors, network status, and threshold crossings. These are discussed in the following three sections.

5.3.2.1 *Adaptive Reporting Based on External Events.*

In certain situations, it is desirable to adjust reporting of management information based on external factors such as business needs, military situation, and so on. An example of a business need is the need to report SLA statistics to customers periodically; reporting of data from network nodes may need to be adjusted at certain times to collect the required data based on the reports that have been promised to customers. Policies can therefore be put in place that trigger reporting of the required data from the appropriate network nodes at the required times.

A military example is the following: The military adjusts various network and other parameters based on the prevalent *information condition* (INFOCON). The INFOCON level is determined for the U.S. Department of Defense by the Secretary of Defense; this level depends on the current threat level to the U.S. military and can take one of five different values. Each of these different values has implications for the way in which a military network is configured. Thus a change in INFOCON level could be used as an external policy trigger (in the form of a policy event) to trigger policy actions that report network parameters in the appropriate manner depending on the INFOCON level.

5.3.2.2 *Adaptive Reporting Based on Network Status.*

The previous section discussed flexible reporting based on external policy triggers; this section discusses adaptive reporting based on network events. Network events are defined as events that are observable by a network management system via monitoring of the network. Examples of such events are congestion and fault events. Network congestion can be detected via monitoring of various network parameters such as the number of dropped packets at network nodes; whereas network faults can be detected by monitoring the network hardware devices, which typically report hardware faults via SNMP traps.

It is often desirable to adapt network management functions such as the reporting of performance statistics based on network events such as network congestion. If a certain network segment is congested, it would make sense to reduce the amount of unessential network management traffic being sent over that network segment so as to avoid aggravating the problem. This can be done by creating policies that automatically trigger adjustment of the frequency and content of such reports when congestion events occur in the network.

Another example of adaptive reporting is the following: Certain types of faults can occur in the network that require diagnostic tests to be run to further diagnose the nature of the fault. Thus a fault event can trigger reporting of the results of the diagnostics.

5.3.2.3 *Adaptive Reporting Based on Threshold Crossings.* It is often desirable to report information over the air only when certain predefined thresholds are crossed. As an example, consider the following policy:

- *Event*: CPU utilization is collected.
- *Condition*: Collected CPU utilization exceeds 70%.
- *Action*: Report CPU utilization to parent node.

If this policy is used in place of the one shown earlier, which was used to unconditionally report CPU utilization every 30 seconds, then nodes will only report their CPU utilization to the GPA node if it exceeds 70%. The assumption here is that if the CPU utilization is lower than 70%, then there is no need to report this to the GPA. This is a very useful capability, as it provides the ability to save network bandwidth by suppressing unnecessary information reporting.

5.3.3 Reporting Aggregated Management Information

Since reduction of bandwidth usage by management applications is an important consideration in MANET environments, it is important to aggregate management information to the extent possible prior to sending it over the air. This capability is illustrated in the following scenario. Here, information about the performance of an application server is collected and aggregated. For this purpose, the following policies are created:

- Monitoring Policy:
  - *Event*: None
  - *Condition*: None
  - *Action*: Monitor application server response time every 30 seconds.
- Aggregation Policy:
  - *Event*: None
  - *Condition*: None

- *Action*: Average application server response times over the last three sampled values.
- Filtering Policy:
  - *Event*: Average application server response time is collected.
  - *Condition*: Average application server response time exceeds 5 seconds.
  - *Action*: Report average application server response time to the parent node.

Following the distribution of these policies to all nodes, all nodes begin to monitor and aggregate the observed application server response time at their local nodes. The idea behind aggregating and filtering reported information is to reduce the management bandwidth overhead in the network by performing aggregation and filtering at the source.

5.3.4   Server Relocation upon Soft Failure

One of the stated benefits of policy-based management is that it increases automation and enables automated network reconfiguration in response to network events. The following scenario demonstrates a thread where such automated reconfiguration is performed to address a performance problem (also known as *soft failure*). First, the following policy is created:

- *Event*: Receipt of average application server response time.
- *Condition*: Average application server response time for the network exceeds 8 seconds.
- *Action*: Determine whether to relocate the application server to a more suitable node.

In order to make the decision to change the location of the application server, a policy action is needed that is responsible for such decision-making running at the GPA (this is the action of the above policy). This action makes decisions about whether to relocate the application server or not and, if so, where it should be relocated. The action runs an algorithm that determines whether to relocate servers or not based on data reported by network nodes about application server response times. This policy is restricted to the GPA, since the decision to relocate the server is a network-wide decision.

Note that this scenario demonstrates the ability to handle *soft* failures in the network, as opposed to a hard failure where the application server fails. The ability to recover from hard failures is often handled automatically by implementing a failover mechanism to a secondary server; however, soft failures such as degraded performance are typically not handled automatically in today's network management systems; they are usually resolved manually by having a human being troubleshoot the problem.

5.3.5 Network-wide Reconfiguration

A final scenario is to automate network-wide reconfiguration of QoS policies in the network in response to a mission change or other external input into the system. Here a mission change could be a change in the goals of a military operation or of a commercial deployment, such as a large trade show. The motivation here is that a change in mission often means that the kinds of communication and their priorities will be different from those in the previous mission, thus necessitating a change in QoS policies. Note that "QoS policies" refer to configuration policies here (see Chapter 2). The initial set of QoS policies configured in the network is shown in Table 5.1. The table shows the amount of bandwidth reserved for the different DiffServ [Blake et al., 1998] traffic classes supported, which include AF1-AF4 (Assured Forwarding) [Heinanen et al., 1999], EF (Expedited Forwarding) [Davie et al., 2002], CTRL for control traffic, and BE (Best Effort). This configuration is the same for all the nodes in the network.

In order to enable automated modification of QoS policies based on mission change, assume that an event MISSION_CHANGE($MissionA$) is generated by a system external to the NMS when the mission changes to a new mission A. The following policy results in the desired reconfiguration on all the network nodes:

- *Event*: MISSION_CHANGE($MissionA$)
- *Condition*: None
- *Action*: Reconfigure QoS policies to the values in Table 5.2.

TABLE 5.1 Original QoS Bandwidth Allocation

DiffServ Class	Bandwidth
AF1	2 Mb/s
AF2	2 Mb/s
AF3	2 Mb/s
AF4	2 Mb/s
EF	1 Mb/s
CTRL	1 Mb/s
BE	5 Mb/s

TABLE 5.2 Reconfigured Bandwidth Allocation

DiffServ Class	Bandwidth
AF1	4 Mb/s
AF2	1 Mb/s
AF3	1 Mb/s
AF4	2 Mb/s
EF	1 Mb/s
CTRL	1 Mb/s
BE	5 Mb/s

The use of the above policy enables automated reconfiguration of QoS policies on all nodes in response to the mission change event above. Although this example is rather simple, it is intended to illustrate the ability to trigger network-wide configuration changes in response to pre-defined events. The alternative to the above policy-triggered reconfiguration is manual configuration of multiple nodes, which is a tedious, slow, and error-prone process. The use of policies can therefore greatly simplify the process of network configuration.

5.4 SUMMARY

The purpose of this chapter was to (i) provide a reference architecture for MANETs and (ii) provide motivation for the use of policy-based management for MANETs. The real power of using policies for managing networks is the ability to automatically adapt to changing network conditions via policy specifications. This ability becomes all the more critical in MANETs, because such networks are extremely dynamic and changing network conditions are the rule rather than the exception. As an example, consider the ability to adapt reporting based on network conditions; such an ability is difficult to achieve without the use of policies, unless the adaptive behavior is hard-coded into the management system. Such hard-coding, however, would make it very difficult to modify the behavior of the management system in any way.

The power of the use of policies for managing networks, then, is the ability to have the network management adapt to both network as well as non-network events. In particular, it is very difficult to incorporate external events into network management without using policies; enabling external events to be taken into account for network management allows business goals or mission objectives to influence, or even drive, the operation of the network. Finally, the use of policies allows network operators to easily modify the way a network is managed by editing policies, rather than having to write new code or create scripts that incorporate the requisite changes.

Having explored the subject of policy-based network management for MANETs in depth, the next four chapters provide a deep dive into each of the FCAPS areas (with the exception of accounting management) using the principles that were described in the past few chapters. These topics are addressed in the following order: configuration management, fault management, performance management, and security management.

6

CONFIGURATION MANAGEMENT

This chapter discusses configuration management functions for MANETs. Section 6.1 begins with a high-level overview of configuration management and the importance of configuration management operations for any communications network, including both wireline networks and MANETs. Section 6.2 discusses configuration management requirements and operations process models for MANETs. Sections 6.3 and 6.4 provide an overview of the relevant configuration management standards and network services standards, respectively. Section 6.5 discusses some of the key MANET configuration tasks based on the OSI layering structure and expands upon the configuration tasks in the lowest three layers of the 7-layer OSI model, namely the physical, data link, and network layers. Interdomain policy management poses a special challenge when configuring a network that has to communicate with networks in other administrative domains; relevant aspects of interdomain policy management are discussed in Section 6.5.4. In Section 6.6 an architecture for configuration management of ad hoc networks is described that showcases a novel, forward-looking control and optimization function. Security considerations for MANET configuration management are described in Section 6.7. Section 6.8 provides a brief summary.

6.1 OVERVIEW

Configuration management is perhaps the most complex of the FCAPS (Fault, Configuration, Accounting, Performance, and Security management) functions. Unlike Performance, Fault, and Security management, where read-only operations are performed on network elements to collect status information, configuration management makes changes to the configuration of network elements in order to enable network services. In fact, configuration

Policy-Driven Mobile Ad hoc Network Management, by Ritu Chadha and Latha Kant
Copyright © 2008 John Wiley & Sons, Inc.

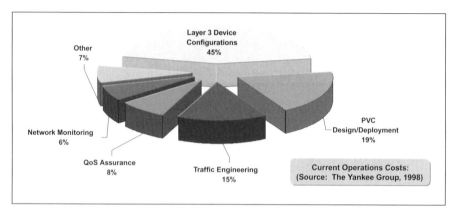

Figure 6.1. Network operations cost.

management is the first step toward obtaining network services; if network elements are not properly configured, there is no network! Furthermore, all of the other network management functions (namely, Fault, Performance, and Security) rely on configuration management functions to help accomplish their output actions. Examples include reconfiguring the network in response to failure events, performance degradations and/or security violations. It is therefore not surprising that configuration management involves a bulk of network operating costs, as illustrated by a 1998 study by the Yankee group [Yankee Group, 1998] which showed that 45% of the cost of operating an IP network is attributable to configuration management (see Figure 6.1).

The predominance of configuration management-related operating costs in Figure 6.1 is a direct consequence of the fact that despite the Internet paradigm of keeping things as simple as possible at the network core, there are a large number of protocols and services that must be configured in order to keep a network up and running. A more recent Yankee group report discusses the need for configuration management systems for managing enterprise networks [Kerravala, 2004]. In this report, a Yankee study is cited that indicates that 62% of network downtime is due to human error in multi-vendor networks with equipment from three or more vendors (see Figure 6.2). This is due to the fact that IP network configuration is extremely complex and is closely related to the previous statistic about the predominance of configuration management cost as a percentage of network operations cost. In a MANET, these problems are greatly magnified due to the dynamic nature of the network. Unlike a wireline network, where assets and network topology are relatively static and where therefore the configuration of network elements rarely needs to change, MANETs are extremely dynamic and require continuous reconfiguration in order to maintain connectivity and high performance. Additionally, the problem is further exacerbated in MANETs since MANETs do not have the concept of edge-only and core-only nodes. Recall

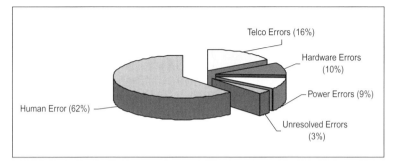

Figure 6.2. Human error as a leading cause of network downtime.

that every MANET node can potentially be an edge node or a core node, in terms of both the applications that the node hosts and the networking functionalities that need to be provided by the MANET nodes.

Although much of the reconfiguration can be handled automatically by network protocols (i.e., at the control plane—e.g., TCP's automatic adjustment of its window size parameters), large improvements in performance can be obtained by judiciously modifying network configuration based on network and other events. Examples include reconfiguring the transmit power at the OSI physical layer to achieve better performance due to temporal fading (as discussed in Section 6.5.1), or reconfiguring the scheduling and queuing parameters at the OSI Layer 2 (MAC/DLL) based on the dynamics of the network to provide appropriate quality of service (as discussed in Section 6.5.2), or partitioning the network to create different routing domains in response to security threats or soft failures (as discussed in Chapters 7 and 8), to name a few. Such reconfiguration actions must be performed by the network management system (i.e., at the management plane). While the above are just a few illustrative example tasks that need to be performed by the Configuration Management function, note that these actions place a great burden on the configuration management component of a network management system for MANETs.

6.2 CONFIGURATION MANAGEMENT FUNCTIONS AND OPERATIONS PROCESS MODELS

This section describes the operations process models for the configuration management task in MANETs. To better understand the operations processes related to configuration management, it is useful to first take a look at the key requirements for this task. Broadly speaking, and as also outlined in the TMN models, configuration management systems are required to perform configuration of network elements and services, configuration database management, configuration auditing, and configuration rollback in case of error. At a high level, these functions can be described as follows:

- **Configuration of network elements and services** deals with sending configuration directives to network elements and software providing network services. This involves interfacing with devices and software using standard or proprietary interfaces. The current configuration view must be stored in a configuration database, so that the network management system always has a current view of the network configuration. Previous versions of the network configuration are also stored in case there is a need to roll back to a prior configuration (see below); the number of versions stored should be configurable and is usually a small number due to storage constraints.

- **Configuration database management** deals with persistent storage of inventory information about all the network assets, including network equipment, interface cards, and relevant networking software such as routing protocols, networking protocols, network services, and so on. This information is stored in a configuration management database that is regularly backed up for survivability.

- **Configuration auditing** is required to verify that the actual network configuration matches the current view of the configuration that is stored in the configuration database. This check needs to be performed periodically to ensure that no tampering or accidental misconfiguration occurs without the knowledge of the network management system. This calls for a *close-knit* set of *seamless* operations coordinated between the configuration and security management components. Note the emphasis on the italicized words, and contrast this with the traditional stovepiped manner of operations in wireline networks, where any coordination between configuration and security management operations is usually achieved by having a human in the loop.

- **Configuration rollback** is a capability that is required in case there is a need to revert to a previous version of the network configuration, for all or for a subset of the network elements. This function may be required in case of error and could also be used for testing purposes.

Having taken a look at the configuration functions that need to be performed for a MANET, the next step is to look at the operations process models for the configuration management task and see how it fits within the overall network management architecture. Figure 6.3 shows the configuration management component and its relationship to other components within the architecture. As discussed earlier, note that the configuration management task is closely coupled with the other key management tasks, namely, Fault, Performance, and Security Management, via a policy infrastructure. In particular, Figure 6.3 illustrates the following. There are several possible sources of input

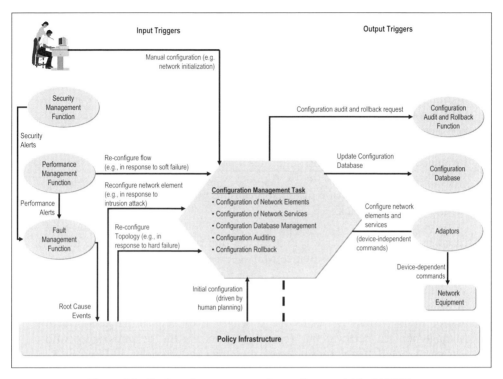

Figure 6.3. Configuration management operations model for MANETs.

for configuration management; these inputs either can be manual or can be obtained from other management components. As an example, at system initialization time, manual input may be provided to configure the network prior to deployment; manual inputs can also be used for reconfiguring the network as needed. Reconfiguration requests are also triggered by other components, via the policy management component, which in turn is triggered directly or indirectly by the other network management functions (namely, the fault, performance, and security management functions). More specifically, note that the "input triggers" to the configuration management task in Figure 6.3 include the outcome of a set of actions that are carried out within the Fault, Performance, and Security Management tasks. As will be discussed in the next three chapters, alerts from Performance and Security Management are sent to the root cause analysis component within Fault Management, where they are correlated and where root causes for the failures are determined. These root cause events are fed to the policy engine, where they trigger ECA (Event–Condition–Action) policies, where the actions are typically reconfiguration actions. These reconfiguration actions are shown in the arrows from the Policy Infrastructure to the Configuration Management hexagon in the center of the figure. In addition, Performance Management sends certain reconfiguration requests directly to Configuration Management for implementation. The reconfiguration action is shown as an output trigger labeled "Configure network elements and services" in Figure 6.3.

As an example, a denial of service attack on a DNS server may be detected by the intrusion detection component within Security Management; this information is sent to Fault Management, where the intrusion alerts are correlated with associated network failures. In this example, there may be associated performance alerts because the denial of service attack would result in abnormally high DNS query response times. Fault Management correlates the performance security alerts, and generates a problem root cause, which is sent to Policy Management. Here, an ECA policy would be defined that would trigger automated reconfiguration of the appropriate network elements to cut off the denial of service attack.

Another example of reconfiguration is the following. A soft failure indicated by network congestion that affects one or more flows in the network may be detected by Performance Management. The appropriate response to this congestion is to downgrade one or more lower priority flows on this path, so that appropriate QoS can be maintained for the remaining, higher priority flows (see Chapter 8 for a detailed discussion of QoS management). Performance Management performs the downgrade by requesting Configuration Management to perform this reconfiguration, via the output trigger whose arrow is labeled "Configure network elements and services" in Figure 6.3.

On its output side, the Configuration Management task essentially performs the functions mentioned earlier in this section, namely, configuration of network elements and services (both initially during network deployment and via reconfigurations in response to Fault, Performance, and Security Management tasks), periodic updates to the configuration management database, and configuration auditing and rollback, as illustrated in Figure 6.3. To accomplish its functions, the Configuration Management task interfaces with a policy infrastructure. More specifically, as discussed in the Policy Management section, policies are defined with actions that perform configuration functions. This enables triggering different configuration actions at different times or in response to different events, such as network or external events. When configuration actions are triggered, the Configuration Management component, which implements the code for these actions, in turn generates device-independent configuration commands. Here "device-independent" means that these configuration commands are not formulated in a syntax that is understood by a particular device from a particular vendor; rather, these commands are formulated in a way that (i) encompasses all the information to perform that particular configuration directive and (ii) is easily translated into the syntax that is required by different vendors implementing the specified functionality.

The translation of device-independent commands into device-dependent commands is performed by *adaptors*. A device-specific adaptor must be developed for each vendor device that presents a different configuration interface.

As mentioned earlier, there is no widely implemented standard configuration interface (although efforts such as *netconf* are attempting to create such standards), and therefore a vendor-specific adaptor will likely have to be developed for each type of vendor device. In order to maintain a modular architecture, it is important to keep these adaptors separate from the Configuration Management component, so that as and when new devices are added or old ones are removed, new adaptors can be added and old ones can be removed without having to modify the Configuration Management component. Furthermore, the Configuration Management component can be developed in a device-independent fashion, without knowledge of specific devices and command syntaxes for these devices.

6.3 IETF CONFIGURATION MANAGEMENT STANDARDS

Over the years, the IETF has made several attempts at standardizing network management protocols. SNMP (Simple Network Management Protocol) [Case et al., 1990, 1999] is a tremendously successful network management protocol defined by the IETF, but its use has been largely restricted to network monitoring, as opposed to configuration. This section provides an overview of some standards that have been developed for configuration management. The standards reviewed here include COPS-PR, SNMP for configuration, including a look at the SNMPconf work in the IETF, and *netconf*.

6.3.1 COPS-PR

The COPS-PR protocol was already described in detail in Chapter 3. Both the COPS-PR and the SNMPconf efforts (see next section) in the IETF were targeted at developing information models (PIBs and MIBs, respectively) that would be implemented by device agents (COPS client for COPS-PR and SNMP agent for SNMP). PIBs were supposed to enable the provisioning of policies within a managed device, as discussed earlier in Chapter 3. However, what COPS-PR and PIBs were really designed for was to provide a protocol (COPS-PR) and information models (PIBs) for configuring devices; the policies of which they spoke were really just configuration policies, as discussed in Chapter 2.

If the vendor community had adopted COPS-PR, it would be a viable approach for configuring network devices. An NMS could use COPS-PR as its mode of communication with the device, via an adaptor. However, as mentioned in an earlier chapter, COPS-PR is not widely implemented and therefore is not currently an option for configuring most types of devices.

6.3.2 Configuration Management Using SNMP

Configuration management is a particularly complex problem because standard configuration interfaces have yet to be adopted by the industry. Even though there is no inherent technical reason why standard protocols such as SNMP cannot be used for configuration, the use of SNMP in the past has mainly been in the area of fault and performance reporting. It is easy to understand why; the trivial security mechanisms supported by SNMPv1 [Case et al., 1990] meant that SNMP could not be used for network configuration, since the prospect of unwanted intruders reconfiguring a network was something that no network operator would want to risk. Fault and performance monitoring is fundamentally different from configuration because the former two activities (fault and performance monitoring) involve *read operations* on the concerned network elements, whereas the latter (configuration) involves *writing* to network elements. Even though it is certainly undesirable to have intruders monitor the status of an operator's network, the effects of such intrusion is generally nowhere as disastrous as an intruder breaking into a network and reconfiguring it at will. Thus network equipment vendors have by and large implemented proprietary configuration interfaces to their equipment— for example, proprietary CLI (Command-Line Interface), proprietary CORBA interfaces, and so on.

With the standardization of robust security mechanisms in SNMPv3 [Case et al., 1999], however, SNMP was viewed as being suitable for use as a configuration protocol. With this in mind, the "Configuration management with SNMP" (snmpconf) working group in the IETF was chartered to document the most effective methods of using SNMP for configuration management and is described next.

6.3.3 SNMPconf

The IETF SNMPconf working group was started in 2000 as a reaction to the work that was underway in the Resource Allocation Protocol (RAP) and Policy Framework working groups at the time (see Chapter 3 for an overview of the RAP working group). As mentioned in Chapter 3, the RAP working group developed COPS-PR as a protocol designed for provisioning, or configuring network elements. The motivation provided by the RAP working group for the development of COPS-PR was that there was no existing standard protocol that was suitable for the configuration of network elements. The SNMP community strongly disagreed with this stance, and it launched a movement to demonstrate that SNMP was well-suited for configuration as well as monitoring of network elements. In addition, since COPS-PR was developed for provisioning *policies* in network elements, the SNMPconf working group went to great lengths to demonstrate that SNMP could easily be used for policy-based network management. The SNMP community believed that SNMP is the protocol of choice for configuration management—largely because SNMP is a well-known, mature industry standard that is widely deployed—and advocated the use of SNMP for configuration as well as policy-based management. One of the tasks of this workgroup was to develop a MIB for representing desired network-wide DiffServ-based QoS behavior. The success of this work thus depended a lot on whether vendors would start to implement these MIBs in their equipment.

In order to understand what the SNMPconf group was trying to achieve, it is necessary to step back and take a look at how they defined policy-based management. In RFC 3512 [McFaden et al., 2003], policy-based management was defined as the ability to:

- Specify "default" configuration data for a number of instances of managed elements, where those instances can be correlated in some data-driven or algorithmic way. The engine that performs this correlation and activates instances based on defaults typically resides in the SNMP agent (although it is allowed to be external to the SNMP agent), and the representation of these defaults therefore reside in SNMP MIB definitions. The object types supporting the notion of defaults are referred to as "template objects."
- Activate instance data derived from template object types via minimal activation directives from the management application.

In other words, they viewed policy-based management as a way of configuring multiple managed objects (e.g., interfaces on a device) at the same time without having to issue individual configuration commands for each object being configured to the device on which the objects resided. In McFaden et al. [2003], they explained how SNMP MIBs should be constructed to facilitate policy-based management; and in Waldbusser et al. [2005], they detailed the contents of a policy-based management MIB, which again facilitates policy-based management. The benefits promised by this approach were:

1. First, the number of SNMP SET commands that need to be sent from an SNMP manager to the SNMP agent residing on a managed device can be greatly reduced if the MIB is structured as specified.
2. Second (and more importantly), the complexity of the processing that needs to be performed by the SNMP manager is greatly reduced, as the burden of the complexity is shifted to the managed device.

In order to explain benefit 1 further, Figures 6.4 and 6.5 illustrate SNMP management without and with a MIB structured for policy-based management. The first figure shows an NMS (Network Management System) managing a device with an SNMP agent resident on it; here it is assumed that the MIBs implemented by the SNMP agent are not structured for policy-based management. As a result, the SNMP manager within the NMS must issue individual SET commands for every managed object on the device that it needs to configure. The SNMP agent receives these SET commands and configures the corresponding managed objects (e.g., interfaces). Thus there is a one-to-one

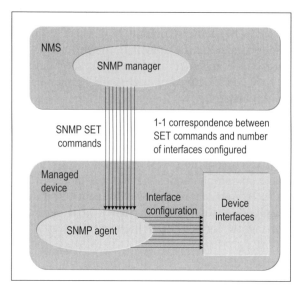

Figure 6.4. SNMP configuration management without Policy-Based Management MIB structure.

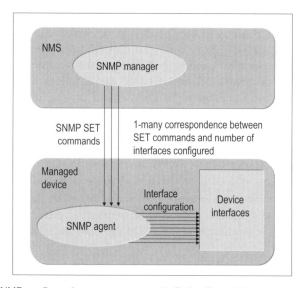

Figure 6.5. SNMP configuration management with Policy-Based Management MIB structure.

correspondence between the SET commands issued by the SNMP manager and the managed objects configured by the SNMP agent.

Figure 6.5, however, shows an NMS managing a device where the MIBs implemented by the SNMP agent *are* structured for policy-based management. As a result, the SNMP manager within the NMS can issue a small number of

SET commands, which the SNMP agent takes and converts into a potentially large number of SET commands for managed objects on the managed device. Here there is a many-to-one correspondence between the SET commands issued by the SNMP manager and the managed objects configured by the SNMP agent.

Let us now look at the advantages, if any, of the above reduction in SET commands in a MANET environment. As already noted in Chapter 5, any NMS for a MANET must be physically distributed across the network, due to scarce over the air bandwidth and intermittent connectivity between network nodes. This means that any over-the-air communication is NMS-to-NMS communication, which is not conducted via SNMP; rather, the information sent between NMS instances consists of ECA policies with configuration information in the action parameters, as discussed in the Policy-Based Management System Architecture section in Chapter 5 (Section 5.2.3.1). Thus a reduction in the number of SET commands does not imply a reduction in over the air bandwidth usage in a MANET. Contrast this with a wireline environment with one NMS instance, where managed devices need to be configured from this remote, centralized management station. Here there is a need to send SNMP commands to remote devices, and therefore the bandwidth savings achieved by the SNMPconf approach may provide some very real benefits.

Within a MANET node, the NMS needs to configure multiple devices. A reduction in the number of SET commands on a MANET node does not confer much of an advantage, since presumably there is a high-speed LAN on the node and bandwidth is not an issue.

Let us now look at the second point above, namely that the complexity of the processing that needs to be performed by the SNMP manager is greatly reduced with the SNMPconf approach as the burden of the complexity is shifted to the managed device. Again, this does not provide much of a benefit if the NMS performs the processing in place of the SNMP agent. The complexity of using configuration defaults, associating managed objects with roles, and determining which and how managed objects need to be configured merely shifts from the NMS to the SNMP agent if the SNMPconf approach is used. It can be argued that this complexity is actually much better suited for implementation within the NMS, for the following reasons:

- *Maintainability*: Logic within an NMS is easier to modify than an SNMP agent provided by a vendor, who may or may not want to make modifications to accommodate NMS requirements.
- *Current Vendor Support*: SNMPconf-style MIBs have not been widely adopted, and therefore equipment vendors do not typically support such MIBs in their products.
- *Ease of Use*: By and large, it is easier to achieve the required configuration objectives by putting the complexity within the NMS rather than within the SNMP agent, since the MIB constructs supported by the latter

may not completely support the NMS requirements. In such cases, the NMS may have to "make do" with the provided SNMP MIBs. A better alternative is to construct the required functionality within the NMS.

At this point, the observant reader may be thinking that the above argument is counterintuitive because it appears to argue against the benefits of policy-based management. After all, didn't the previous chapters explain that it is essential to automate network management for MANETs by using policies? So why make the argument that the policy-based management provided by SNMPconf is not needed for MANETs? The answer is that *there is no argument about the need for policy-based management*; rather, the debate is about *where this functionality should be implemented*. The claim being made here is that this functionality should be implemented within the NMS rather than within the SNMP agent on a device. At the user level, however, this implementation detail is not visible, because the user still specifies network management goals in terms of policies.

6.3.4   netconf

The *netconf* standard [Enns, 2006] was recently published by the IETF *netconf* working group [netconf, 2007]. *netconf* defines a standard for manipulating the configuration of a network device. The standard relies on XML-encapsulated messages that are transmitted using an RPC (Remote Procedure Call) mechanism on top of a suitable transport layer that provides a reliable, persistent connection between the manager and the device being configured. The configuration data for a device are also encoded in XML, and they are manipulated using XML filters to select subsets of the configuration data (see Section 6.3.4.4). In addition, *netconf* connections must provide authentication, data integrity, and confidentiality. Figure 6.6 gives a high-level overview of the different layers in the *netconf* conceptual model.

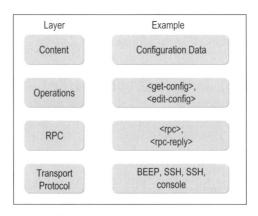

Figure 6.6. *netconf* architecture.

The *netconf* standard uses a client–server model where the entity that is configuring a network device (e.g., a network management system) acts as a software client, and the network device being configured acts as a software server. The transport layer shown in Figure 6.6 provides transport for an RPC mechanism that uses a simple, XML-encoded protocol. The operations layer defines a base set of operations for configuration; these operations are again XML-encoded and encapsulate content that is beyond the scope of the *netconf* standard. In addition to this base set of operations, devices can optionally support additional operations. These additional operations are termed "capabilities" and must be communicated from a server to a client when a session is established between the two. Additional capabilities beyond those listed in the standard may be defined in future IETF documents, thus providing a path for extensibility.

The following sections describe in some more detail the RPC and the operations layers shown in Figure 6.6.

6.3.4.1 RPC Layer. An RPC message uses the <rpc> element to enclose a *netconf* operation. This element must include an attribute that contains a message identifier, which is an arbitrary string (e.g., an integer) that identifies the RPC message. The message identifier is chosen by the client and is used by the server in any responses to the request, in order to identify the request to which the response corresponds. A response uses the <rpc-reply> element. An example from [Enns, 2006] is

```
<rpc message-id="101"
     xmlns="urn:ietf:params:xml:ns:netconf:base:1.0"
     xmlns:ex="http://example.net/content/1.0"
     ex:user-id="fred">
  <get/>
  </rpc>
```

where "get" is an operation that the client wants to perform on the server (see the next section for a description of the "get" operation). An example of a response is

```
<rpc-reply message-id="101"
     xmlns="urn:ietf:params:xml:ns:netconf:base:1.0"
     xmlns:ex="http://example.net/content/1.0"
     ex:user-id="fred">
  <data>
    <!-- contents here... -->
  </data>
  </rpc-reply>
```

The <rpc-error> element is sent as part of an <rpc-reply> message if an error occurs during the processing of the RPC request message. The <rpc-error> element includes information about the nature of the error, which may be one of the following: transport, rpc, protocol, or application, indicating an error in one of these four areas. Additional information about the error is included in an <error-tag> element, along with an <error-severity> element that indicates the severity of the error. An <error-app-tag> specifies implementation-specific errors, if applicable; and an <error-path> element contains the absolute Xpath expression containing the element path to the item for which an error is being reported. Additional tags (<error-message>, <error-info>) contain human-readable and protocol or data model-specific error information.

6.3.4.2 Operations Layer. The base set of operations that a manager can perform on a network device are defined in the *netconf* standard as follows:

- **get-config:** Retrieve all or part of a specified configuration of a device. The parameters for this operation are (a) the datastore from which to retrieve the configuration data (see next section) and (b) an optional filter that specifies which portions of the configuration should be retrieved. The *netconf* specification makes use of "subtree filtering," described in Section 6.3.4.4.
- **edit-config:** Modify the configuration of a device. This operation contains the attribute *operation*, which specifies whether the configuration should be merged with the current one, whether the configuration should replace the current one, whether the configuration should be created from scratch, or whether it should be deleted. The parameters for this operation are (a) the target datastore (see next section), (b) a default operation to be performed (the default is *merge*), (c) the configuration data itself, and (d) testing and error options.
- **copy-config:** Copy an entirely new set of configurations to a datastore, replacing all of its previous contents. The parameters of this operation are the source and destination datastores.
- **delete-config:** Delete the content of a specified datastore. The parameter of this operation is the datastore to be deleted.
- **get:** Retrieve current configuration and device state information. This operation contains a filter as an optional parameter that specifies the portion of the configuration to be retrieved. If it is not specified, all of the configuration is retrieved. The difference between this operation and the *get-config* operation is that the *get* operation retrieves the value of the current or *running* configuration (see next section on datastores for more details), whereas the *get-config* operation can retrieve the value of other configurations as well, such as a stored configuration.

- **lock:** Lock a specified datastore. The parameter of this operation is the datastore to be locked. The purpose of this operation is to allow a client to lock a configuration so that no other client can perform updates to the configuration while it holds this lock. The device must also ensure that any non-*netconf* entities cannot update the configuration while the lock is held.
- **unlock:** Release a previously held lock. The parameter of this operation is the datastore to be unlocked.
- **close-session:** Close an existing *netconf* session. This operation also releases any locks held by the client.
- **kill-session:** Kill an existing *netconf* session. The parameter of this operation is the ID of the session to be killed. Any operations currently in progress are aborted.

6.3.4.3 Datastores. The *netconf* standard describes the concept of "datastores" that a network device uses to store information about the configuration of the device. The terminology used is the following: Every network device must support at least a "running" datastore, which means a store containing the current configuration of the device. The protocol refers to this datastore using the <running> element. In addition to this required datastore, two optional datastores are defined that a device may support; these are a "candidate" datastore and a "startup" datastore. A "candidate" datastore can be viewed as a scratchpad that allows a manager to create and edit configurations without committing them (i.e., without actually making the configuration change in the device). This provides a very convenient way to make a collection of configuration changes, which can then be transferred to the "running" datastore in one atomic operation. A "startup" datastore simply contains the configuration information that the network device will put in place when it powers up. If a network device supports these additional datastores, it must indicate this to the manager by way of the capabilities element at session establishment time.

6.3.4.4 Subtree Filtering. Subtree filtering refers to the specification of XML filters in order to select a subset of the device configuration, which are XML subtrees, for *get* and *get-config* operations. Subtree filters consist of XML elements and their XML attributes. The following components may be part of a subtree filter:

- *Namespace Selection*: Namespace selection allows specification of a namespace, whose contents are returned.
- *Attribute Match Expressions*: Attribute match expressions are used to select the nodes that have attributes that match specified criteria.
- *Containment Nodes*: Containment nodes are those that contain child elements in the XML subtree.

- *Selection Nodes*: Selection nodes in the filter allow specification of criteria that constrain the list of nodes returned. These selection nodes are included in the filter by specifying an empty leaf node in the filter.
- *Content Match Nodes*: Content match nodes in the filter are used to specify information that must be present in nodes returned by the search. An exact match is performed on leaf node elements; and if their content matches that specified in the filter, those nodes are returned.

6.3.5 Summary of Configuration Standards

Among the set of configuration standards described above, the most promising one—in terms of potential vendor support—is *netconf*. Prominent vendors of networking equipment are already starting to implement the *netconf* standard. In contrast, neither COPS-PR nor SNMP have had much success in terms of vendor support for configuration via either of these protocols.

6.4 NETWORK SERVICES: RELEVANT STANDARDS

This section provides an overview of DHCP (Dynamic Host Configuration Protocol) and DNS (Domain Name Service), which are important standards required for configuring and maintaining communications in any network. The overview provided in this section is restricted to the details of the standards; configuration and use of these standards for MANETs will be discussed in Section 6.5.3 later in this chapter.

6.4.1 DHCP

DHCP (Dynamic Host Configuration Protocol) [Droms, 1997] is used for providing IP addresses and other configuration parameters to IP hosts. The protocol is based on a client–server paradigm, where the IP host that needs configuration parameters is a DHCP client and the entity that provides the configuration parameters is the DHCP server. DHCP can be used in one of three modes: automatic allocation, where a permanent IP address is allocated to a host; dynamic allocation, where an IP address is allocated to a host for a fixed period of time; and manual allocation, where a specific IP address is assigned to a specific host based on some host identification provided by the DHCP client. Dynamic allocation is the only scheme that allows addresses that have been allocated to be reclaimed for allocation to other hosts. In an ad hoc network, dynamic allocation is especially useful, since nodes may dynamically enter or exit a network and require new IP address allocations based on the subnet that they are joining.

A DHCP server has two basic functions:

1. *Managing IP Addresses*: A DHCP server controls a range of IP addresses and allocates them to clients, either permanently or for a defined period

of time. The DHCP server uses a lease mechanism to determine how long a client can use a nonpermanent address. When the address is no longer in use, it is returned to the pool and can be reassigned. The server maintains information about the binding of IP addresses to clients in its DHCP network tables, ensuring that no address is used by more than one client.

2. *Providing Network Configuration for Clients*: A DHCP server assigns an IP address and provides other information for network configuration, such as a hostname, broadcast address, network subnet mask, default gateway, name service, and potentially much more information.

The DHCP server can also be configured to update DNS information for clients that supply a host name. The DHCP server populates DNS with the supplied host name and the corresponding IP address that it assigned to the host.

The use of DHCP considerably eases the process of configuring the IP hosts in a network; instead of having to configure each host individually, only DHCP servers have to be configured. In terms of configuration requirements, some of the most common configuration parameters required to configure a DHCP server are:

- Range of IP addresses that the DHCP server should allocate to the clients
- Subnet masks
- Duration of IP address leases
- IP address of the default gateway (router)
- DNS domain name to assign to clients
- IP address of DNS server for clients

6.4.1.1 DHCP Protocol Overview. The DHCP protocol enables host systems in a TCP/IP network to be configured automatically as they boot. DHCP is based on a client–server mechanism, where servers store and manage configuration information for clients and provide that information upon a client's request. The information includes the client's IP address and information about network services available to the client. DHCP evolved from an earlier protocol, BOOTP, which was designed for booting over a TCP/IP network. DHCP uses the same format as BOOTP for messages between client and server, but includes additional network configuration data in protocol messages. A primary benefit of DHCP is its ability to manage IP address assignments by *leasing* IP addresses, which allows IP addresses to be reclaimed when not in use and reassigned to other clients. This enables the use of a smaller pool of IP addresses than would be needed if all clients were assigned a permanent address. DHCP relieves the system or network administrator of some of the time-consuming tasks involved in setting up a TCP/IP network and the daily management of that network.

DHCP offers the following advantages:

- *IP Address Management*: A primary advantage of DHCP is easier management of IP addresses. In a network without DHCP, an administrator must manually assign IP addresses, being careful to assign unique IP addresses to each client and configure each client individually. If a client moves to a different network, the administrator must make manual modifications for that client. When DHCP is enabled, the DHCP server manages and assigns IP addresses without administrator intervention. Clients can move to other subnets without manual reconfiguration because they can obtain new configuration information appropriate for the new network from the DHCP server for that subnet.
- *Support of BOOTP Clients*: Both BOOTP servers and DHCP servers listen and respond to broadcasts from clients. The DHCP server can respond to requests from BOOTP clients as well as DHCP clients. BOOTP clients receive an IP address and the information needed to boot from a server.
- *Support of Local and Remote Clients*: BOOTP provides relaying of messages from one network to another. DHCP takes advantage of the BOOTP relay feature in several ways. Most network routers can be configured to act as BOOTP relay agents to pass BOOTP requests to a server that is not on the client's network. DHCP requests can be relayed in the same manner because, to the router, they are indistinguishable from BOOTP requests.

6.4.1.2 *How DHCP Works.* A DHCP server must first be installed and configured by a network management system (NMS). During configuration, the NMS configures the DHCP server with information about the network that will be handed out to DHCP clients by the DHCP server. After this information has been configured, clients are able to request and receive configuration information.

The following describes the sequence of messaging between a DHCP client and server:

1. The client discovers a DHCP server by broadcasting a *DHCPDIS-COVER* message to the limited broadcast address (255.255.255.255) on the local subnet. If a router is present and configured to behave as a BOOTP relay agent, the request is passed to other DHCP servers on different subnets. The client's broadcast includes its unique ID, which is generally derived from the client's Media Access Control (MAC) address (on an Ethernet network, the MAC address is the same as the Ethernet address). DHCP servers that receive the discover message can determine the client's network by looking at the following information:

- Which network interface did the request come in on? This tells the server that the client is either on the network to which the interface is connected, or that the client is using a BOOTP relay agent connected to that network.
- Does the request include the IP address of a BOOTP relay agent? When a request passes through a relay agent, the relay agent inserts its address in the request header. When the server detects a relay agent address, it knows that the network portion of the address indicates the client's network address because the relay agent must be connected to the client's network.
- Is the client's network subnetted? The server determines the subnet mask used on the network indicated by the relay agent's address or the address of the network interface that received the request. Once the server knows the subnet mask used, it can determine which portion of the network address is the host portion, and then select an IP address appropriate for the client.

2. After determining the client's network, the DHCP server selects an appropriate IP address and verifies that the address is not already in use. It then responds to the client by broadcasting a *DHCPOFFER* message that includes the selected IP address and information about services that can be configured for the client. Each server temporarily reserves the offered IP address until it can determine whether the client will use it.

3. The client selects the best offer (based on the number and type of services offered) and broadcasts a *DHCPREQUEST* message that specifies the IP address of the server that made the best offer. The broadcast ensures that all the responding DHCP servers know the client has chosen a server, and those servers not chosen can cancel the reservations for the IP addresses they had offered.

4. The selected server allocates the IP address for the client, stores the information in the DHCP data store, and sends a *DHCPACK* (acknowledgment) message to the client. The acknowledgment message contains the network configuration parameters for the client.

5. The client monitors the lease time, and when a configurable period of time (corresponding to the lease period) has elapsed, the client sends a new message to the chosen server to increase its lease time.

6. The DHCP server that receives the request extends the lease time; if it does not respond within a certain time interval, the client broadcasts a request so that another DHCP server (if present) can extend the lease.

7. When the client no longer needs the IP address, it notifies the server by sending a *DHCPRELEASE* message that it is releasing the IP address.

6.4.1.3 DHCP Failover. At the outset, it should be noted that DHCP failover has not been standardized, although a number of IETF drafts were written in an attempt to standardize a failover protocol for communication between a primary and a backup DHCP server. This section discusses mechanisms that can be used for DHCP failover in a MANET, as discussed in Droms et al. [2003], an expired Internet draft. This draft describes mechanisms whereby two or more servers can act as backups for each other. The mechanism relies on the use of a *lazy update* scheme, which is a requirement placed on a server implementing a failover protocol to update its failover partner whenever the binding database changes. A failover protocol that does not support lazy update would require the failover partner update to be complete before a DHCP server could respond to a DHCP client request with a DHCPACK. A failover protocol that does support lazy update places no such restriction on the update of the failover partner server, and so a server can allocate an IP address or extend a lease on an IP address and then update its failover partner as time permits. A failover protocol that supports lazy update not only removes the requirement to update the failover partner prior to responding to a DHCP client with a DHCPACK, but also allows gathering up batches of updates from one failover server to its partner for greater bandwidth efficiency, an important consideration in ad hoc networks.

There are two aspects of the broadcast behavior of the DHCP protocol that are key to DHCP failover. The first is that the DHCP protocol requires a DHCP client to broadcast all DHCPDISCOVER and DHCPREQUEST messages when attempting to get a new IP address. Because of this requirement, a DHCP client who was communicating with one server will automatically be able to communicate with another server if one is available. The second aspect of broadcast behavior is similar to the first, but involves the distinction between (a) a DHCPREQUEST for the purpose of renewing an IP address lease (RENEW) and (b) a DHCPREQUEST/REBINDING (explained below). A DHCPREQUEST/RENEW is the message that a DHCP client uses to extend its lease. It is unicast to the DHCP server from which it acquired the lease. However, the DHCP protocol was explicitly designed so that in the event that a DHCP client cannot contact the server from which it received a lease on an IP address using a DHCPREQUEST/RENEW, the client is required to broadcast its renewal using a DHCPREQUEST/REBINDING to any available DHCP server. Since all DHCP clients are required to implement this algorithm, the client can use a different server than the one that initially granted a lease to renew a lease. Thus, one server can take over for another with no interruption in the service as experienced by the DHCP client or its associated applications software.

The basis of DHCP failover involves a set of messages sent between DHCP servers that are providing backup service for each other. The most important one of these is a binding update (BNDUPD) message that is used to communicate to another DHCP server changes in the IP address bindings offered by a DHCP server to its clients. The server receiving the BNDUPD message responds with a binding acknowledgment (BNDACK) message when it has successfully committed those changes to its own stable storage. Other messages include an update request (UPDREQ) message for one server to request binding database information that it has not yet seen from another server; an update request all (UPDREQALL) message to request all binding database information from another server; an update done (UPDDONE) message used by a server to respond to the requesting server to indicate that all the requested updates have been sent and acknowledged; a connect (CONNECT) message to establish a high-level connection with the other server; a connect acknowledgment (CONNECTACK) message to respond to a CONNECT message; a disconnect (DISCONNECT) message to close a connection; a state change (STATE) message to inform the other server of a change of failover state; and a contact (CONTACT) message sent between servers to ensure that the other server continues to see the connection as operational. Such a contact message must be transmitted periodically over every established connection if other message traffic is not flowing.

There exist two failure scenarios that provide particular challenges to the correctness guarantees of a failover protocol. The first is if the primary server crashes after assigning an IP address to a DHCP client, before it can propagate this address assignment information to the other backup servers; in this case, it is possible that a backup server may assign this IP address to another client. Another scenario is when the network partitions so that the primary and backup DHCP servers cannot communicate updates to each other, but each can communicate with one or more clients for address assignment. The solution to both of these scenarios is to assign disjoint address pools for allocation to the primary and backup DHCP servers.

Another problem can occur when the primary server extends the lease time for a DHCP client and then crashes before the backup servers can be updated with this information. Again, in this case, the IP address for which the lease time was extended may get allocated to another client by one of the backup servers. This scenario is handled by the failover protocol through control of the lease time and the use of the maximum client lead time (MCLT). This time is configured on the primary server and transmitted from the primary to the secondary server in the CONNECT message. It is the maximum amount of time that one server can extend a lease for a client's binding beyond the time known by the partner server. See Droms et al. [2003] for further details on this topic.

6.4.2 DNS

The Domain Name Service (DNS) [Mockapetris, 1987] is used to resolve host names to their IP addresses and is a critical component of any IP network. DNS provides a directory lookup service for the Internet analogous to the telephone "411" directory assistance, and it maps user-friendly names (e.g., "www.yahoo.com") to IP addresses. The process of mapping a name to an IP address is called *name resolution*. DNS is implemented using a distributed database of *Resource Records* stored in *Name Servers*. In addition to the standard mapping of names to IP address, DNS provides many other lookup capabilities, such as mail server names, names of other name servers, reverse lookups, public key lookups, server name mappings, and so on.

Names are typically assigned to hosts via DHCP or statically. If dynamic allocation is used via a DHCP server, dynamic DNS can be used to register a newly obtained IP address with DNS so that the host name to IP address mapping is kept up to date in DNS.

In order to provide scalable name resolution, the namespace is divided into a hierarchy. Most of the DNS records are indexed using DNS names based on the well-known domain name hierarchy. Every node in the DNS domain tree can be uniquely identified by a *fully qualified domain name* (FQDN), which is written as the names of each node in the tree from the named entity to the root (using only the characters a–z, A–Z, 0–9, and the dash). A sample naming tree is shown in Figure 6.7.

In general, there are three types of top-level domains below the root:

- *Organizational Domains*: These are named by using a three-character code that indicates the primary function or activity of the organizations contained within the DNS domain.
- *Geographical Domains*: These are named by using the two-character country/region codes established by the International Standards Organization (ISO).
- *Reverse Domains*: This is a special domain used for IP address-to-name mappings (*reverse lookups*), called in-addr.arpa in IPv4 and ip6.int, in IP version 6 [Thomson and Huitema, 1995].

The name hierarchy is divided into different logical domains and zones, where:

- *Domains* are a complete sub-tree that cut the domain name tree in one place. Domains are named based on the non-leaf nodes at the root of their sub-tree (the leaf nodes are the nodes, services, and users).

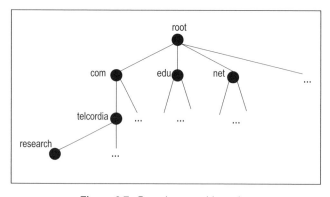

Figure 6.7. Domain name hierarchy.

- *Zones* are any contiguous part of the domain name tree that can cut the tree in one or more places and are under a particular authority.

The DNS standard defines the following:

- Mechanism for querying and updating the DNS database.
- Mechanism for replicating the information in the DNS database among servers.
- Schema for the database.

Applications access DNS through a local client Resolver process. This Resolver sends a UDP packet (on port 53) to the configured IP address of a DNS Name Server (NS). The NS attempts to get the appropriate Resource Record (RR) locally by checking its own authoritative zones (if any) and checking its cache. If it does not have this RR, it can either return an error message or:

- *Provide Proxy Agent Service*: The name server can recursively ask other Name Servers to resolve the submitted request; or
- *Provide Redirection Agent Service*: Here the name server iteratively refers the client to other Name Servers for an answer.

A Name Server can either be a *Master* (or *Primary*) Name Server, or a *Slave*, or *Secondary* Name Server. A Primary Name Server manages any changes in zone for which it is *authoritative* (i.e., for which it is the final authority), and it performs both read and write operations on the database. A Slave Name Server obtains zone transfers from either the master or another slave of the zone. It performs read-only functions on the database. Name Servers can also be *Forwarders*; such servers simply send all requests to another Name Server, and they cache the returned results.

To provide scalability, each Name Server typically has knowledge of a limited number of names and of a small number of other Name Servers. As an example, one of the root Name Servers (".") knows the names and associated IP address of each top-level domain (e.g., ".com" or ".us"). The top-level domain knows where to find its children (e.g., ".com" knows how to find "yahoo.com"). To go up the tree, each Name Server must at least know the IP address of one or more of the root Name Servers (of which there are currently only 13 in the world) and may also know other Name Servers.

6.5 MANET CONFIGURATION FUNCTIONS

This section provides an overview of the configuration functions that need to be performed to configure a MANET. The discussion is structured based on the specific OSI layer being configured, and it describes the various aspects of each layer that require configuration by the Configuration Management component. The focus is on the lower three OSI layers, namely, the physical layer (layer 1), data link layer (layer 2), and network layer (layer 3), since these layers are the ones that need to be tuned by the network management system to create a functional network. While the higher OSI layers (namely the transport layer, or layer 4, through the application layer, or layer 7) also have a wide spectrum of configurable parameters, the configurable parameters for these higher OSI layers are usually not considered to be within the scope of a network management system. Rather, these parameters are typically within the realm of user application management. For example, adjusting video transmission parameters at layer 7 is not a network management function, whereas tuning the MAC protocol parameters at layer 2 falls within the purview of the network management system.

6.5.1 Configuration of Layer 1

This section provides a brief overview of OSI layer 1 (i.e., the physical layer) and discusses the configurable knobs that it provides in a typical mobile ad hoc networking environment. Two key "configurable knobs" at the physical layer are identified, namely, (a) transmission power and (b) rate control (via modulation schemes). Each of them are discussed below, along with an explanation of how they can be manipulated to improve network performance.

6.5.1.1 Transmission Power. A fundamental quantity that belongs to the physical layer but permeates all layers above it is the transmission power. Clearly, the choice of a power level directly impacts the range of a transmission link. That is, for given and fixed coding, waveform, modulation, rate, path loss, and other parameters as well as antenna, amplifier, and embedded processor, the transmission power will determine whether a desired bit-error rate can be achieved at a given distance. In other words, it determines the "reach" of the transmitted signal and hence the existence of a link. Consequently, power control at the physical layer has been widely studied in the literature in the area of wireless network topology control. Network topology or connectivity in turn strongly influence the dynamics of the network in terms of reachability, number of hops between nodes, and so on.

In addition to determining the presence or absence of a wireless link, the transmission power at a node directly impacts the interference level at other nodes, in turn impacting the set of possible simultaneous transmissions that can take place for a given network connectivity. This may impact MAC protocols as well as other neighbor discovery schemes—for example, those employed by routing. In other words, transmission power is a quantity that is controlled at the physical layer level but interacts strongly with all the layers of the protocol stack. It is therefore an important "knob" at the physical layer that must be tuned judiciously by network management systems to achieve "good" (if not "optimal") overall network performance. One way of creating a robust network is to ensure that it is "*k-connected*," where *k*-connectedness is defined as follows: A network topology graph is called *k*-connected or *k*-vertex-connected if its vertex connectivity is *k* or greater.

It is therefore not surprising that power control has been used in wireless networks in the past to improve system performance. Most power control work in the past has concentrated on controlling the power of mobile sets in cellular systems. This type of power control computes the channel loss $h(i, j)$ at the receiver *j* from the transmitter *i* and then computes the required received power at *j* to ensure that the signal-to-interference-plus-noise ratio (SINR) is above a required threshold. The value of the SINR is a function of not only the channel loss and the power at the transmitter, but also the received powers from other interfering nodes. Thus the needed power $P(i)$ at transmitter *i* depends on the powers of all nodes that have measurable effect on receiver *j*. The determination of which nodes belong to this set is in itself a difficult question. All powers for this set of nodes must therefore be selected simultaneously. There are several centralized algorithms that iteratively compute power levels. At iteration *m*, the power levels are a simple function of the power levels used at iteration $m - 1$, where this function is based on measured values of the SINR observed in the $(m - 1)^{th}$ iteration. Many researchers [Chen et al., 1994; Yates, 1995] have proposed distributed methods for computing the power levels so that they converge to the same solution as the centralized one.

6.5.1.2 *Transmission Rate.*

It should be noted that transmission rate control is another important way of controlling connectivity and quality of service by reducing the chance of transmission errors [Li and Ephremides, 2005]. Rate control is not intrusive, since it only affects one link without altering the interference effects at other links. However, rate control requires fine and coordinated adjustments at both transmitter and receiver and, possibly, even tight and adjustable synchronization between them. Rate control can be configured so that, when necessary, the receiver decides to drop bits when stressed in order to maintain a temporarily faltering link for the sake of avoiding costly readjustments to the MAC or the routes. This is the flip side of the power control method and also resides at the physical layer control interface.

6.5.1.3 *MANET Configuration at Layer 1.*

As explained above, power and rate control provide two important "configuration knobs" at the physical layer. The manipulation of these knobs can take place at different time scales. Ideally, it should take place at the packet level—that is, on a per-packet time basis; however, this may be both difficult to achieve and very expensive. It can easily work at the frame level (e.g., a timescale that corresponds to many packets), or even less frequently (e.g., a timescale that spans a few minutes). Of course, canonical trades exist with regard to the resulting responsiveness and overheads when power and rate control are performed at the above-mentioned different time scales. One possible solution is a hybrid scheme that performs both limited control at a small timescale (e.g., at the frame level) and more deliberate control at a longer timescale. The longer timescale control can be provided by a control and optimization function (COF), in a way that optimizes the COF objective function, as will be described in Section 6.6.

A general framework for distributed stochastic power and rate control is described in Luo et al. [2005], where algorithms are outlined that update the transmit powers based on stochastic approximations. The algorithms are distributed in the sense that no global information is needed for the power updates. Interference to each user is estimated locally via noisy observations. The algorithms are shown to converge to the unique optimal solution under certain situations. Here the desired links are used as input to the power control algorithms; these desired links are computed in a cross-layer fashion based on information about (i) existing flows and their destinations, (ii) the priorities of these flows, (iii) existing routes, (iv) slot schedules, (v) channel conditions, (vi) energy efficiency, and so on. Thus, rather than simply trying to maintain an existing network topology, power control algorithms that use cross-layer inputs to create a desired topology are much more effective in the dynamic MANET environment in terms of meeting network objectives and optimizing network effectiveness. One such dynamic cross-layer control mechanism will be described in Section 6.6.

6.5.2   Configuration of Layer 2

This section presents a brief overview of the OSI layer 2 functionality, and it discusses Configuration Management operations that are relevant at this layer. The main purpose of OSI layer 2, the media access control (MAC) or data link layer (DLL), as its name indicates, is to provide access to the transmission medium (i.e., the physical layer) so that application traffic can be transported over the network to its destination. In any network that does not have dedicated point-to-point links, multiple nodes must share the same transport medium (such as a wireless link or an Ethernet cable). The main problem to be solved at the MAC layer, therefore, is to provide access to this shared physical medium in a way that allows every node to transmit information in a "fair" manner. Given that the problem of resolving media access contention is nontrivial even for wireline networks, it becomes even more difficult in MANETs due to the dynamics of the wireless medium.

Broadly speaking, there are three methodologies that can be used to coordinate the transmission of information from the network layer (OSI layer 3) to the physical layer (OSI layer 1). These are: (a) time-based methods, (b) frequency-based methods, and (c) code-based methods. An information packet at OSI layer 3 that needs to be transported through the network can contend for use of the physical medium using any of the above-mentioned three methods (a) through (c). Each of these three methods provides control knobs that can be used for efficient access to the medium. Each of the above three methods is briefly explained below, followed by an identification of appropriate knobs at the MAC/DLL layer that can be tuned by the Configuration Management function to produce an efficient network.

6.5.2.1 Time-Based MAC Schemes. Time-based access schemes, in essence, require nodes to contend for a "time slice" to gain access to the underlying physical medium, which, in the case of MANETs, is the wireless link. Thus the basic approach of a time-based MAC scheme is to allow transmission from only one node at a time over a shared physical medium. No special frequencies or codes are used by the information packets in this case, but rather, all of the nodes use a common frequency and encoding. Time-based schemes can be classified into two broad categories: random-access-based schemes and deterministic-access-based schemes. Each of these is described next.

6.5.2.1.1 Random-Access-Based Schemes. The popular Carrier Sense Multiple Access (CSMA)-based access scheme is an example of a random access scheme. In a nutshell, CSMA-based schemes work as follows. A node that contains a packet for transmission places the packet on the wireless link and listens for (or *senses*) a collision. If no collision is sensed, the node continues to place other packets on the medium. If, on the other hand, a collision is sensed, the node decides to back off and re-send at a later time. The scheme is called a *random-access*-based scheme because the back-off time is determined randomly, based on some algorithm. Several different types of algorithms can be used to decide the back-off time, with the most popular one being based on an exponentially distributed back-off time.

Several versions of CSMA have now come into being since the original CSMA scheme (IEEE 802.3) was first used in Ethernet LANs. Variants of the original IEEE 802.3 scheme include the IEEE 802.11 and IEEE 802.16 family of MAC protocols. Industry and the vast majority of the academic community have focused on modifications and improvements to the IEEE 802.11 DCF, which is a contention-based scheme intended for wireless LANs, and has been shown to render maximum throughput smaller than 18% of the channel capacity due to hidden terminals [Fullmer and Garcia-Luna-Aceves, 1997] and smaller than 5% when multi-hop transmissions are considered. Static bandwidth allocation through manual network planning has been utilized to improve performance. With static bandwidth allocation in a single-channel MANET of N nodes, the utilization of the channel in the neighborhood of a node is of order $O(n/N)$, where n is the number of nodes "competing" for the bandwidth in the neighborhood.

6.5.2.1.2 Deterministic-Access-Based Schemes. The popular Time Division Multiplexed Access (TDMA)-based scheme is an example of a deterministic access scheme. The basic approach of the TDMA-based scheme is to schedule transmissions from different nodes so that each node is assigned specific times at which it is allowed to transmit data (called *timeslots*). If each of the nodes sharing access to the physical medium are assigned distinct timeslots, they are guaranteed that there will be only one node transmitting at any given point in time. Thus, unlike the CSMA scheme, nodes in a TDMA-based scheme transmit packets during timeslots that have been assigned *a priori*. Thus these schemes are also referred to as conflict-free schemes, since nodes only transmit during a preassigned timeslot if they have a packet to transmit. If they do not have a packet to transmit during their preassigned timeslot, they remain silent. Variants of this scheme involve "loaning" of timeslots by nodes that do not need to transmit to other nodes that need to transmit information. If, on the other hand, a node has a packet to transmit but does not have a timeslot, it refrains from any transmissions and waits until its assigned timeslot, and only then does it place the packet on the medium.

Thus TDMA schemes are conflict-free access schemes and are an alternative to the "conflict-based" access scheme (CSMA). The IEEE 802.4 and 802.5 (token bus and token ring) are variants of the conflict-free time-division-based access schemes. In essence, the conflict-free schemes schedule a set of timetables for individual nodes or links such that transmissions from the nodes or over the links are conflict-free. In traditional wireline networks, a token bus or token ring manager (i.e., a specialized node) arbitrates the token sharing and setting up of the timetables. The overall goal is fairness; that is, no node should be deprived or starved of timeslots.

While the problem of timeslot arbitration in wireline networks is in itself a challenging problem, it is intensified in MANETs due to (a) the dynamic nature of the network and (b) the fact that bandwidth is usually in short supply. Thus the goal of a timeslot arbitrator is to schedule nodes in a way that maximizes the amount of time nodes are able to transmit useful information in their neighborhoods, while the overhead used to attain such a schedule is minimized. Within this framework, the ideal channel access scheme would be one in which nodes simply know the times when they can transmit unicast, multicast, or broadcast packets in a way that the channel use is optimal and the intended receivers are able to decode the packets, taking into account all the traffic that must be offered to the shared channels as well as the conditions of the channels around senders and receivers. However, in practice, such instantaneous knowledge is impossible to obtain at each node in real time, and scheduling is an NP-complete problem even with complete system information.

Several practical schemes have been proposed for scheduling conflict-free transmissions that are typically either topology-dependent or topology-independent. Topology-independent transmission scheduling schemes [Chlamtac et al., 1997; Ju and Li, 1998; Rentel and Kunz, 2005b] have been shown to provide average throughput that is at best similar to that of slotted ALOHA (see Rentel and Kunz [2005a]). Clearly, in MANETs, a transmission scheme that combines knowledge of the topology as well as the information exchange requirements would be very helpful in realizing an efficient network. Since bandwidth is scarce in MANETs, scheduling schemes that maximize the use of available bandwidth are essential in MANETs. The area of topology-dependent and information-aware transmission scheduling is an active area of ongoing research. In this book, the use of a COF as described in Section 6.6 is described to help with this problem. More specifically, by programming the COF to use multiple constraints including topology information and a traffic exchange requirements matrix, the COF can help produce an adaptive scheduling mechanism that maximizes a certain objective function (e.g., overall network bandwidth utilization, maximum slot allocation, and so on).

6.5.2.2 *Frequency-Based Access Schemes.*

In a frequency-based access scheme, a node that has a packet to transfer does not contend for a "time" to transmit. Rather, it uses a specific frequency, which is usually assigned ahead of time, to transmit packets whenever it needs to. Thus in this case, there may be several simultaneous transmissions going on over a shared medium. However, the information packets are distinguished by the frequency that they use, resulting in a scheme also referred to as Frequency Division Multiple Access (FDMA).

In FDMA-based schemes, each node is assigned a specific frequency for transmission purposes. This implies that both the transmitter and receiver have agreed to "tune" in to a certain frequency decided either *a priori*, or dynamically, in case an adaptive controller such the COF discussed in Section 6.6 is employed. It also implies the presence of a controlling node/entity, which takes a certain frequency spectrum allocated for the mission/operation and creates frequency slots/bands that are then allocated amongst the network nodes. Of course, since the number of nodes in a practical ad hoc network will likely far exceed the available frequency slots/bands, the problem now boils down to the following question: How should the controller node judiciously allocate the frequencies amongst the network nodes?

The answer to the above question is challenging even in a static wireline environment, and typical solutions require the use of complex optimization techniques. The challenge becomes even greater in the MANET environment due to mobility and unpredictable jamming/interference. Ideally, the controller node will allocate distinct frequencies to nodes within a certain geographic neighborhood and will reuse the frequencies in another geographic neighborhood. However, as mentioned earlier, the fact that there is random node mobility in MANETs complicates such an ideal assignment. Thus, an entity such as the COF as discussed in Section 6.6 holds potential in terms of optimizing frequency allocations by considering cross-layer information, mobility patterns, traffic exchange matrix, and so on.

6.5.2.3 *Code-Based Access Schemes.*

Code-based access schemes for accessing the underlying physical medium are also referred to as Code Division Multiple Access (CDMA). As its name indicates, in this case the nodes in a network are assigned special *codes* that they use whenever they have a packet to transmit. Thus there can exist several simultaneous transmissions that all have the same frequency, but are distinguished by the specific code that they employ. CDMA schemes are popular in commercial cellular (wireless) networks, and several commercial vendors currently offer CDMA-based cellular telephone services.

The operation of CDMA is similar to that of FDMA, with the exception that the use of a special frequency is replaced with the use of a special code. Therefore, similar issues and challenges exist in terms of code allocation and coordination, just as was the case in FDMA with frequency allocation and coordination; these challenges were discussed in the previous section. Additionally, just as in the case of FDMA and time-based access schemes, a COF can help in realizing efficient code allocation and coordination in MANETs.

6.5.2.4 Configuring MAC Parameters. The last few sections provided a brief overview of different approaches to controlling medium access, the key OSI layer 2 function, and described the configuration management operations that need to be applied at this layer (i.e., allocation of timeslots, frequencies, codes, etc., depending on the MAC scheme in use). The layer 2 entities that need to be dynamically and adaptively configured in MANETs include the following.

- Timeslot allocations and back-off algorithm parameters for time-based MAC schemes: Based on a combination of the routing dynamics, traffic paths and demands, and bandwidth fluctuations, the configuration management function needs to perform the following actions:
  - *In the Case of Deterministic Access*: Dynamically reassign timeslots by loaning slots to bottleneck node(s) (i.e., nodes that have become a multiplexing point for many different applications and routing paths) from other nodes. The selection of "other" nodes from which to borrow the slots can be performed based on a combination of (i) traffic priority and (ii) "role" of the node, as two illustrative examples. The "role" of a node describes any special function that the node performs in the network, and it is an indication of its importance in maintaining the network. For example, a gateway node bridges communications between different networks and therefore has higher importance than a "leaf" node in the network. Note that determining which nodes to borrow timeslots from can be a very challenging and complicated action and will require, as inputs, cross-layer information such as routing, traffic exchange matrix, mobility information (trajectory), etc.
  - *In the Case of Random Access*: Dynamically readjust the back-off algorithm parameters so that nodes that are less busy/less important can back off for a longer time than higher priority and/or highly utilized nodes. The overall objective in both cases is to maximize network utilization and to be able to support higher priority services in a timely manner.

- Frequency allocations and reuse coordination in the case of FDMA-based schemes: Examples of specific configuration management actions that must be performed in this case include:
  - Reassignment of frequency bands based on mobility information to minimize cross-talk.
  - Dynamic assignment of multiple and mutually noninterfering frequencies to a high priority node or a bottleneck to alleviate congestion/loading at that node.
- Code allocations and reuse coordination in the case of TDMA-based schemes: Examples of specific configuration-management-related actions that can be performed in this case include:
  - Reassignment of codes based on mobility information to minimize cross-talk.
  - Dynamic assignment of multiple and mutually noninterfering codes to a high-priority node or a bottleneck to alleviate congestion/loading at that node.

Another important feature that can be adapted and dynamically configured to produce improved network performance in MANETs is transmission scheduling at layer 2. For example, a popular queuing and scheduling mechanism at layer 2 is a form of weighted fair queuing (WFQ) which in essence works as follows. In the WFQ paradigm, a real number between 0.0 and 1.0 (also referred to as a *weight*) is assigned to each of several packet queues. This *weight*, in principle, captures the fraction of time that a scheduler services that particular queue. The scheduler essentially performs the function of removing packets from the layer 2 queues and placing them on the physical medium for transmission through the underlying network. Thus the scheduler visits each queue and continues to service that particular queue for a proportion of time that corresponds to the "weight" associated with that queue. During the time that the scheduler is servicing the queue, packets are removed from the queue and made available for transmission on the wireless medium. Thus, for example, a queue with a higher weight will have more packets scheduled for transmission per unit time than a queue with a lower weight. This in turn will result in better service (i.e., lower latency) for the queue with a higher weight. Now, due to the unpredictable dynamics in ad hoc networks, it will be very difficult to decide *a priori* which specific set of nodes/queues will be bottlenecks. Furthermore, the bottleneck nodes themselves will vary, due to a combination of mobility and path loss fluctuations. Thus the queues in MANETs are generally initialized based on very general guidelines. For example, queues associated with high-priority traffic classes (e.g., the Expedited Forwarding class) will have a higher weight assigned to them than will queues associated with the lower classes (e.g., best effort). Once the system is operational, however, due to the dynamics mentioned above, certain nodes can become bottlenecks and hence may need to have their queue weights readjusted. This is precisely where an adaptive configuration management function in conjunction with a COF can help in improving system performance in MANETs.

6.5.3 Configuration of Layer 3

This section describes the major configuration tasks for the network layer, or layer 3, of the OSI protocol stack. More specifically, the configuration tasks associated with routing, addressing, and naming services are discussed in detail.

6.5.3.1 Configuring Routing. A considerable amount of work has been done in the area of routing for MANETs. The IETF has chartered a working group (see manet [2007]) to standardize IP routing protocol functionality for MANETs within both static and dynamic topologies. This working group is standardizing a reactive as well as a proactive routing protocol suitable for use in MANETs. Besides the IETF work, there is a large body of research dealing with MANET routing [Toh, 2002]. Since this topic could easily provide material for an entire textbook, and since the focus of this book is ad hoc network management (rather than MANET routing), details of different MANET routing protocols are not provided in this book. For further details on MANET routing, the reader is referred to the vast literature available on this subject. The focus of this section is to look at the configuration management tasks required to correctly configure routing for MANETs. Although the detailed configuration is necessarily dependent on the details of the deployed routing protocols, the high-level configuration and planning tasks can be discussed in a general manner; this discussion is relevant to any routing protocol.

In small MANETs, routing can be done using a flat address space, with no aggregation or summarization of routes. However, in any large IP network, nodes must be grouped into routing domains to allow routing scalability. It has been shown that MANETs with flat address spaces cannot scale beyond a small number of nodes [Eriksson et al., 2004], and therefore some form of hierarchical routing is required [Manousakis et al., 2002]. When routing domains are used, nodes are assigned IP addresses from one or more IP subnets that are associated with the routing domain to which they belong. This allows routing summarization to take place. One node per domain is selected as the domain gateway and advertises itself as the gateway to all addresses in that domain.

Furthermore, all nodes within a domain route IP packets to nodes in other domains via the border node. Thus only domain border nodes are involved in routing information exchanges with other border nodes, and route summaries are exchanged between these nodes. In other words, these domain border nodes function as aggregation points for aggregating routing information from their domain for distribution to other routing domains. Within a routing domain, nodes only need to know how to reach other nodes within that same domain; to send data outside that domain, they send data to their border node and the border node takes care of routing information outside of the domain to the intended destination. The implication of such a scheme is that when a node moves from one IP addressing domain into another, it must receive a new black IP address associated with that domain (refer to Chapter 5 for a discussion of "red" and "black" communications and addressing).

Dynamic IP address reconfiguration due to mobility is discussed in Section 6.5.3.3.2. In Eriksson et al. [2004], it is assumed that every node has a unique identifier; and the paper discusses node lookup, which is a distributed lookup that maps every node identifier to its current IP address. In the MANET reference architecture used in this book, however, red-side IP addresses are static and are used to identify the nodes; thus there is no need for a node lookup facility as described in Eriksson et al. [2004]. However, in a red–black network, when black IP addresses change, there is a need to update red-to-black IP address mappings throughout the network, so that red-side flows can continue to be routed to their destinations over the black network. The mapping of red to black IP addresses is discussed in further detail in the Security Management chapter (Chapter 9).

The above paragraphs discussed routing for the wireless (or black) portion of the network. In addition to routing in the black wireless network, note that routing must also be configured on the red side when there is more than one radio on the platform, connected to different wireless networks. In this case, a red-side routing protocol is required so that the router on the platform knows which wireless interface should be used for sending out IP packets for a given destination.

Figure 6.8 shows three red (unencrypted) routing domains labeled AD_x, AD_y, and AD_z, where AD stands for administrative domain. Due to the small size of the red networks in Figure 6.8, each red network segment is within one routing domain—that is, within one single administrative domain. The nodes labeled Nx, Ny, and Nz denote border routers that interconnect each of the three red networks with the black wireless network. The three red network segments are interconnected by a black (encrypted) wireless network.

Due to the scale of the wireless network, which could have a large number of nodes, the black network is typically composed of more than one routing domain. Figure 6.9 shows an example of a black wireless network composed of three routing domains.

The nodes on the black wireless network are labeled as EN (Encrypted network Nodes). The superscripts on these nodes refer to the domain to which they belong. For example, EN_a^1 denotes a black network node "a" in routing domain *1*. Likewise, EN_b^1 denotes a node "b" in routing domain *1*, and EN_a^2 denotes a node "a" in routing domain *2*. Since multiple routing domains are depicted in the black network in Figure 6.8, border routers are required that allow interdomain communications between these domains. The border nodes are denoted as BEN with a superscript that indicates the two routing domains

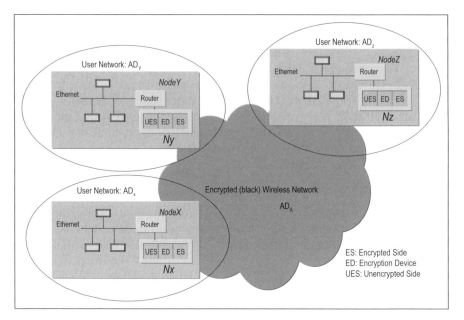

Figure 6.8. Red-side routing domains in a MANET.

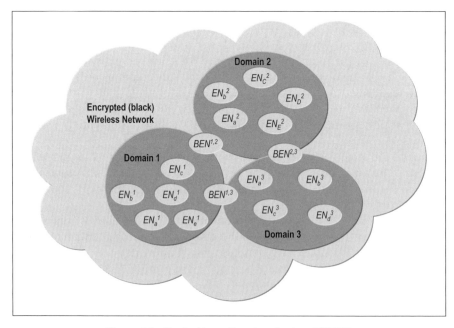

Figure 6.9. Black-side routing domains in a MANET.

that the given BEN connects. For example, $BEN^{1,2}$ represents the border node between routing domains 1 and 2, $BEN^{2,3}$ represents the border node between routing domains 2 and 3, and $BEN^{3,1}$ represents the border node between routing domains 3 and 1.

Note that although the nodes in the figure above are labeled as *encrypted* nodes, they also have unencrypted (red) sides, just like the nodes in Figure 6.8; however, the details of the red side of each node were omitted for simplicity.

The Configuration Management function needs to perform the following routing-related configuration operations on the various network elements represented in Figure 6.8.

- Planning tasks: Plan the following:
  - Number and membership of wireless routing domains: Discussed in Section 6.5.3.2.
  - Border nodes and backup border nodes for each wireless routing domain: Discussed in Section 6.5.3.2.
  - Routing protocol to be used within each wireless domain (intra-domain routing protocol): The selection of which protocol is used for different routing domains will be a policy-driven selection, based on network dynamics. Examples of intradomain routing in MANETs include members of a proactive family of routing protocols (e.g., OLSR) as well as reactive protocols (e.g., AODV). Different network conditions dictate the use of different routing protocols. For example, if there is a lot of mobility within a certain domain, leading to rapidly changing routes, then a reactive routing protocol may be indicated. A reactive routing protocol calculates routes only when traffic needs to be sent; thus, if nodes are relatively silent (i.e., there is very little application traffic), there is very little routing messaging overhead during that time. On the other hand, if the nodes in a routing domain are relatively static, or are moving together so that routes do not change much, then a proactive routing protocol may provide better performance. The amount of application traffic in the network plays an important part in this decision process too.
  - Routing protocol to be used across wireless domains (interdomain routing protocol): The interdomain routing protocol is usually determined based on advance manual planning; an example of a commonly used interdomain routing protocol in IP-based networks is BGP.
  - Routing protocol to be used on the red side: The protocol used on the red side is usually chosen statically, based on advance manual planning, as the red-side network is wired and static. An example of a protocol that is suitable here is OSPF.
- Configuration tasks: Configure the above information in the MANET nodes. This includes configuration of routing protocol parameters within each node.

6.5.3.2 Configuring the Routing Hierarchy. The previous section discussed the fact that hierarchical routing is required for scalability purposes for any nontrivial MANET, even though hierarchical routing may result in suboptimal routing paths [Rastogi et al., 2003] (the suboptimal factor is also called the *stretch* factor). This implies that the routing hierarchy must be (i) formed and (ii) maintained in the face of mobility. Algorithms have been developed for the initial hierarchy formation [Manousakis et al. 2005] that attempt to minimize stretch. The approach described in Manousakis et al. [2005] is not dependent on which hierarchical routing protocol is used. The described approach makes use of a modification of Simulated Annealing [Kirkpatrick et al., 1983], which is a global optimization algorithm. The approach relies on the definition of an appropriate cost function, which the Simulated Annealing algorithm attempts to minimize. The cost functions described in Manousakis et al. [2005] can be used to generate a routing hierarchy that minimizes the routing path suboptimality and also satisfies the hierarchy generation objectives, which could include considerations such as the generation of clusters of balanced sizes and/or diameters, the minimization of border routers, the grouping of nodes with similar mobility characteristics, or combinations of all of these.

One aspect that is not discussed in Manousakis et al. [2005] is how the hierarchy is recomputed after the existing hierarchy becomes suboptimal. This could happen due to a variety of reasons, such as node mobility, elimination of nodes (e.g., due to nodes losing power and shutting down), and so on. Manousakis and McAuley [2006] describes *active hierarchy maintenance* techniques that operate locally and use the same cost functions used for hierarchy creation. The approach is generic and does not depend on the routing protocol used.

Section 6.6 provides a generalized approach to the problem of configuration decision-making. This approach also makes use of a utility function to drive network configuration and reconfiguration. Further discussion of the approach will be provided in Section 6.6.

6.5.3.3 Configuring Addressing. In any IP network, IP addresses are used to uniquely identify nodes and also to encode routing information. The IP address of a node provides an indication of the routing domain to which it belongs.

Configuration of IP addressing for a MANET such as the one described in Chapter 5 can be divided into two categories:

1. Configuring the IP addresses on the wired MANET platform LAN(s)
2. Configuring the IP addresses on the wireless radio interfaces

The first task can be handled by simply configuring a DHCP server on each platform LAN to hand out IP addresses to the hosts on the LAN. Note that these are "red" IP addresses (refer to Chapter 5 for a description of red and black communications) that are assigned to hosts on the wired LAN. These IP addresses are not visible on the "black" side (i.e., on the wireless ad hoc network) because they are encrypted and encapsulated within a new IP packet with a new black IP address as destination. The encrypted packets with black source and destination IP addresses travel over the air to the destination MANET node and are decrypted upon arrival. This is illustrated in Figure 6.10.

The second task is more complex. IP addresses can be handed out to the wireless interfaces on a MANET via DHCP servers on the black side, but the selection of a node to serve as the DHCP server, along with the maintenance of DHCP server data, adds additional complexity to the task. On a wired LAN, connectivity can be assumed to be more or less permanent, so that all nodes are connected at all times to the machine serving as DHCP server. However, on a wireless ad hoc network, intermittent connectivity may result in the node serving as DHCP server being unreachable, which will result in problems for other nodes attempting to acquire IP addresses. This means that a mechanism must be in place for other nodes to take over the DHCP server function when the assigned DHCP server for a set of nodes is no longer reachable. A DHCP failover mechanism such as the one described earlier can be used for this purpose.

The second complexity is the need to handle IP address changes due to mobility. As mentioned in Section 6.5.3.1, when a MANET node moves from one IP addressing domain into another, it must receive a new IP address associated with that domain. Note that this complexity only applies to black-side IP addresses, since these are the ones that are used to route information over the wireless network. Red-side addresses, on the other hand, do not need to change as a result of mobility; only the red-to-black address mappings need

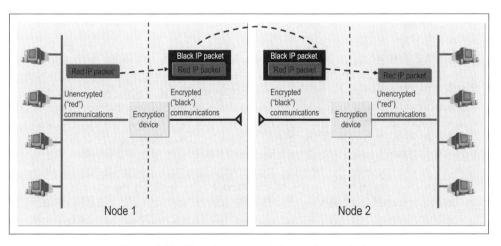

Figure 6.10. IP packet encapsulation and transmission.

to change. The addressing requirements for MANETs are discussed next, followed by appropriate solutions.

6.5.3.3.1 Addressing Requirements for MANETs. This section discusses the requirements for dynamic address assignment for the wireless interfaces of a MANET (as opposed to the wired interfaces on a platform LAN in a MANET). The following are the requirements of any solution for configuring IP addresses on such interfaces.

1. *Automation*: IP address assignment/reconfiguration must be automated and should not require any manual intervention.
2. *No IP Addressing Overlaps*: The IP address assigned to a node must be unique.
3. *Efficient Use of Address Space*: It should be possible to reclaim IP addresses which are not in use. Reclaimed addresses can be assigned to other client nodes.
4. *Low IP Address Assignment Delay*: Address assignment delay is defined as the difference between the time at which the client node requests an IP address and the time at which it receives an IP address. The address assignment delay must be within reasonable bounds, as determined by the functional requirements of the network.
5. *Robustness*: The IP address assignment solution should be robust and resilient to node failures.
6. *High Availability*: The IP address assignment solution should be highly available, so that IP addresses are available to any node entering the network.
7. *Mobility Handling*: When a node moves from one IP subnet to another, it must automatically be assigned a new IP address for the new IP subnet.

6.5.3.3.2 IP Address Assignment for MANETs. The previous section outlined the requirements for an address assignment solution for MANETs. Note that DHCP satisfies the first four requirements listed above. Regarding the robustness and high availability requirements, as mentioned earlier, nodes in a MANET are characterized by intermittent network connectivity. Thus, in order to provide a robust and highly available address assignment solution, it is not sufficient to have just one backup DHCP server per IP subnet. Since there is a nontrivial probability that multiple nodes in an IP subnet may become disconnected, in order to use DHCP, any node in the subnet would need to be capable of serving as a DHCP server, so that even if the currently configured primary DHCP server as well as the backup DHCP server become disconnected from the network, another node will be able to take over the DHCP server function for the subnet.

Let us now analyze the feasibility of using DHCP for black-side IP address assignment. First, if every node in the network must be capable of acting as a DHCP server for its subnet, then it must be configured with an address pool for that subnet, so that it can hand out addresses to nodes in that subnet. Second, any node can move to any subnet, so, taking the above observation one step further, *every node in the network would need to be able to act as a DHCP server for every subnet in the network!* This implies that every node would have to be configured with an address pool for every subnet in the network. Given this observation, it clearly does not make sense to use a DHCP-based solution for black-side IP address assignment.

In Eriksson et al. [2004], the authors suggest the following scheme for dynamic IP address assignment: When a node joins the network, it listens to the periodic routing updates of its neighboring nodes and uses these to identify an unoccupied address. It also registers this address, along with the node's unique identifier, in a lookup table. Note that this lookup table is needed so that other nodes can find out what the current IP address is for a given node whose identifier is known. This last step is not required for the MANET architecture presented in this book, because red-side addresses remain fixed for every node in the MANET reference architecture used in this book, and applications can continue to send traffic to the same destination (red) IP addresses, regardless of black IP address changes. The problem with this IP address assignment scheme is that it does not take into account a critical fact about MANET nodes—namely, that these nodes may sometimes shut themselves down to conserve battery power or other reasons. Using the suggested scheme, whenever a node is not actively participating in routing updates, its address is up for grabs by another node joining the network or moving into that segment of the network. This can cause address collisions if a node goes to sleep, a second node joins the network and assigns itself the first node's address, and then the first node wakes up and continues to use its original address.

An alternate solution is for every node to be preconfigured with one IP address per subnet, for its own use. Rather than have nodes hand out addresses to other nodes via DHCP, or deduce what addresses are available by listening to advertised routes, each node is made responsible for handing out addresses *only to itself*. This guarantees a robust, survivable solution to IP address assignment.

Let us now look more closely at how this solution would work in practice. The main question is the handling of mobility. When a node moves into a new subnet, how does it know to change its IP address? To understand this, let us step back and recall why different IP subnets are used in the first place. As explained earlier in Section 6.5.3.1, in order to scale to large numbers of nodes in a MANET, routes are summarized by routing domain for distribution throughout the network. The node functioning as the border router for the domain can therefore broadcast a periodic beacon announcing the routing domain that it controls. When a MANET node moves into the vicinity of this domain, it hears this announcement beacon. It may happen that a node hears more than one beacon, in which case it can decide which subnet to join. If it is currently a member of Subnet 1 and hears beacons from Subnets 1 and 2, it can decide to stay in Subnet 1 or to join Subnet 2. If the node decides to change subnets, it then autonomously changes its IP address and joins the new subnet. Its routing protocol implementation then exchanges route information with the border router for its new subnet, and routing stabilizes.

To summarize: In order to implement the above solution, the following steps are needed:

- Each node is preconfigured with an IP address per subnet.
- The border router for every routing domain must broadcast a periodic beacon announcing its domain.
- Every node must listen for this beacon and must be able to change its IP address if it moves into a new routing domain (i.e., if it hears a beacon from a new border router).

There is one final item to take care of: the mapping of red to black IP addresses. When a node changes its (black) IP address, it is necessary to update the red-to-black address mappings throughout the network to enable communications between red networks. This is taken care of by the HAIPE (or equivalent) security mechanism that is used for securing the over-the-air communications. A detailed discussion of the HAIPE mechanisms used for distributing and maintaining accurate red-to-black address mappings throughout the network is provided in Chapter 9.

Looking back at the requirements for IP address configuration, it is clear that all of the requirements are satisfied with the above solution, with the exception of the third requirement, which relates to the efficient use of IP addresses. With the above solution, if the number of nodes is n and the number of subnets is m, then $n*m$ IP addresses are required for this solution, which can be a rather inefficient use of IP addresses if m is large. However, with the use of IPv6, efficient use of IP addresses is not necessarily a high priority, and therefore the above shortcoming may not pose a problem.

6.5.3.4 *Configuring DNS.*
In ad hoc networks, the deployment of DNS is fraught with challenges due to the need to balance the requirement for rapid name resolution with the requirement to minimize over-the-air bandwidth usage by DNS updates. Planning a DNS solution for an ad hoc network requires determining where to place authoritative DNS servers throughout a network. Second, a delegation hierarchy needs to be planned to allocate responsibility for different portions of the DNS naming tree to different servers.

The two major requirements of any DNS solution for MANETs are: First, over-the-air bandwidth usage must be minimized due to scarce wireless bandwidth; and second, high availability must be assured, so that any host that needs to resolve a name is able to do so in a timely manner.

In order to satisfy the above requirements, an appropriate server placement strategy is needed for MANETs. Clearly, a solution with one name server that acts as the authoritative server for the entire network will not satisfy either of the above two requirements, since (i) over-the-air DNS queries may consume considerable bandwidth; and (ii) if the central server becomes unreachable, nodes will not be able to resolve names.

An important feature of the MANET architecture that was described in Chapter 5 is that IP addresses of hosts on the red side of the network need not change once they are configured. Only black addresses are affected by mobility between IP subnets and may change frequently. Thus, for a red-side DNS solution, it can safely be assumed that IP address changes will be infrequent. This has important implications for the server placement architecture. Each platform can be defined to be a zone, and each platform can host an authoritative server for the names on that platform. Furthermore, since addresses are relatively static, all IP addresses for all platforms are cached on all other platforms, and a caching resolver is configured on each platform with cached pre-planned IP addresses for each platform. Thus any name resolution requested by a node can be resolved by a local query (to the authoritative server on that platform or to the caching resolver on that platform); or, if the cache is not valid, the name can be resolved by a direct query to the authoritative server for that name. Since red-side name-to-address mappings are not expected to change, off-platform queries will be relatively rare and will be restricted to the case where there are dynamic hosts that can attach to platforms and receive dynamic IP addresses from the red-side DHCP server.

DNS works hand in hand with DHCP in this type of deployment. Advance planning is used to plan IP addresses for every red-side host. Each red-side host is also assigned a preplanned host name. At configuration time, red-side DHCP servers are configured with a list of red IP addresses associated with host names. When a host asks for an IP address, it provides its host name in the request. The DHCP server responds with the IP address assigned to that name. Since the assignment of IP addresses to names is statically planned, DNS can be preconfigured with red-side address-to-name mappings. A pool of IP addresses can also be configured into each DHCP server for allocation to dynamic hosts (e.g., laptops that may be attached to a platform in an unplanned way). Also, since there are no applications running on the black side, the assumption is that there is no requirement to provide DNS for black IP addresses.

To summarize, the following tasks need to be performed to configure DNS for a MANET:

- *Planning Tasks*: Plan the following:
  - Host names for red-side hosts and the corresponding IP addresses.
  - Pool of addresses for unplanned, dynamic hosts that may attach to platforms in an unplanned way.
  - DNS server locations for each red-side enclave.

- *Configuration Tasks*: Configure the caching resolver on every platform with the preplanned name-to-address mappings for every red-side host on every platform. Authoritative name servers are updated automatically by DHCP when the DHCP server hands out IP addresses to hosts as planned.

6.5.3.5 Configuring QoS. The DiffServ standards were reviewed in Chapter 3. In order to implement DiffServ, the network must be configured with the appropriate traffic classification, policing, and scheduling mechanisms. As described earlier, traffic classification is accomplished by marking every IP packet header with an appropriate DSCP (DiffServ Code Point). This DSCP is the label carried by a packet that is examined by every router that forwards this packet in order to determine the treatment to be given to this packet. IP packets are policed at the ingress into the network based on information in the IP header, such as source address and port, destination address and port, protocol, and DSCP. Finally, at every router that forwards an IP packet, scheduling mechanisms are used to determine which packet should be transmitted next. If there is no congestion in the network, then the rate at which packets are received by a router and the rate at which they are relayed to the next hop toward their destinations are equal. If there is congestion, however, the rate at which packets arrive into a router may exceed the rate at which they can be transmitted. It is then that the scheduling mechanisms come into play. Scheduling mechanisms are used to schedule transmissions of packets based on various criteria. The simplest form of scheduling is the "first come, first served" or "FIFO" (first in, first out) paradigm, where packets are scheduled for transmission based on their arrival time at the router; the packet that has been waiting the longest gets scheduled for transmission next. However, the implementation of DiffServ requires that differentiated services be provided for IP packets; in other words, it should be possible to provide "better" service to certain IP packets based on their DSCP markings. This is accomplished by configuring queuing structures and scheduling mechanisms on the routers in the network.

Queuing structures refer to the number of queues at each router and the assignment of packets to queues. Queuing structures may exist in individual routers, switches, processors, and hosts. The purpose of these queuing structures is to prioritize packets and to respond to congestion at the node either by buffering packets for later delivery, or by dropping packets in the case of severe congestion for that particular traffic class. Upon arrival at a router, a packet's DSCP is examined and the packet is placed into the queue corresponding to its DSCP. There may be a one-to-one mapping of DSCPs to queues, in the case where each DSCP gets its own queue; or there may be a many-to-one mapping, in the case where multiple DSCPs share the same queue. In order to configure the queuing structures in a router, manual engineering must be carried out prior to network deployment to determine (a) how many DSCPs need to be supported and (b) the corresponding queuing structure.

Scheduling mechanisms are used to determine how to *service* queues, or how to schedule packets for transmission from the different queues in a router. A wide range of scheduling mechanisms have been defined and are implemented in commercially available equipment. Two of the most popular available mechanisms are:

- *Priority Queuing*: Each queue is serviced based on its priority. Thus if there are four queues, ordered by priority, then packets in the first queue will be serviced first, followed by packets in the second queue, and so on. The implication of priority queuing is that if there is congestion, then it possible that packets in the lower-priority queues may never get transmitted, because preference for transmission is given to the higher-priority queues. This is referred to as *starvation*. Priority queuing is appropriate for situations where certain types of traffic need absolute priority over any other type of traffic. In MANETs, it is sometimes necessary to use priority queuing due to the scarcity of MANET bandwidth, because it may be the only way to ensure that certain high-priority messages get transmitted to their destinations. In wireline networks, priority queuing is less popular because bandwidth is typically plentiful and it makes more sense to share bandwidth more equitably among different traffic classes.
- *Weighted Fair Queuing (WFQ)*: Each queue is serviced in a round-robin fashion in accordance with a set of weights. A weight is assigned to each queue to indicate the proportion of the total bandwidth capacity that will be made available to the queue. For example, assume that there are four queues on a router that are assigned the following weights: 0.1, 0.2, 0.3, and 0.4. Also, assume that the outgoing link data rate for the router is R. Then transmission rates for each of the four queues will be, on an average, $0.1R$, $0.2R$, $0.3R$, and $0.4R$, respectively. The advantage of WFQ is that no queue is starved of bandwidth, since every queue gets serviced in turn based on its weight.

Some commercially available implementations of these queuing mechanisms allow nesting of different types of queues within other queues. This will be illustrated below via an example.

The configuration of priority queuing requires specifying the number of queues, their priorities, and their respective data rates, or rate limits. The configuration of WFQ requires specifying the number of queues and their weights. While defining the number of queues and their priorities is relatively straightforward, based on an analysis of the traffic requirements of the MANET, the specification of the rate limits or weights is very challenging. This is because the rate limits or weights for the queues should reflect the actual needs of the network applications as accurately as possible, which is difficult to do in a MANET. This is illustrated by an example below.

Example. Suppose that a MANET supports four different traffic types, classified as voice, video, data, and critical messaging. Furthermore, assume that within each of these traffic types, there are low- and high-priority flows. In order to differentiate all of these different types of traffic, eight DSCPs are defined as shown in Table 6.1.

The following queuing structure is set up for the above and is illustrated in Figure 6.11.

- Two priority queues P_1 and P_2 are set up for critical messaging, one for traffic with DSCP = 11 and the other for traffic with DSCP = 12. Their priorities are 1 and 2, respectively.

TABLE 6.1 Example DSCP Assignments for Different Traffic Types

Traffic Type	Priority	DSCP
Critical messaging	High	11
Critical messaging	Low	12
Voice	High	13
Voice	Low	14
Video	High	15
Video	Low	16
Data	High	17
Data	Low	18

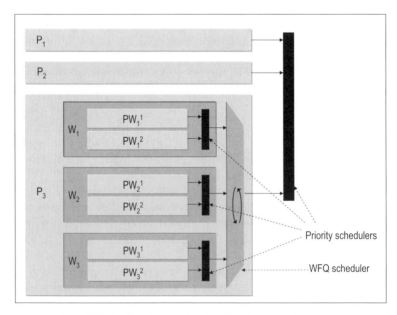

Figure 6.11. Sample queuing structure for four traffic types.

- A third priority queue P_3 with priority 3 is set up for the remainder of the traffic. Within this queue, three embedded weighted fair queues W_1, W_2, and W_3 are set up. The first queue is for traffic with DSCP = 13 or 14 (voice traffic), the second queue is for traffic with DSCP = 15 or 16 (video traffic), and the third queue is for traffic with DSCP = 17 or 18 (data traffic).
- Within each of the three weighted fair queues defined above, two priority queues are embedded with priorities 1 and 2. This defines six new queues: priority queues PW_1^1 and PW_1^2 within queue W_1; priority queues PW_2^1 and PW_2^2 within queue W_2; and priority queues PW_3^1 and PW_3^2 within queue W_3. The higher-priority traffic within each weighted fair queue is assigned to the higher-priority queue, and the lower-priority traffic is assigned to the lower priority queue.

Although the above queuing structure appears rather complicated, it can be explained as follows. Critical messages are the most important type of traffic for this network, and they are therefore allowed to starve the other classes if bandwidth becomes scarce. Thus packets in queue P_1 are transmitted in preference to any other packets, followed by packets in queue P_2. If there is still bandwidth available, queue P_3 is serviced. Now, within queue P_3, bandwidth is shared between voice, video, and data in accordance with the weights allocated to each of these classes. Finally, when bandwidth is allocated to one of these types of traffic, the higher-priority traffic is transmitted in preference to the lower-priority traffic.

Now consider the problem of allocating rate limits and weights to the priority and weighted fair queues, respectively. The rate limit assigned to a priority queue should reflect the rate at which traffic in that queue is expected to be sent through a particular router. Similarly, the weights assigned to a set of weighted fair queues should reflect the relative volume of traffic in each queue. If the rate limit assigned to a priority queue is too small, then some traffic will not be transmitted; similarly, if a weight assigned to a given queue is too small, traffic that should have been transmitted may be dropped. On the other hand, if a large rate limit is defined for a priority queue such as P_1, then all other traffic may be permanently starved, depending on the total bandwidth available in the network. What makes the selection of appropriate rate limits and weights so challenging is the fact that the amount of bandwidth in the network varies dynamically in MANETs.

Furthermore, traffic paths vary dynamically due to mobility, and therefore traffic patterns also change in an unpredictable fashion. For example, if voice traffic is mostly restricted to a certain region in the network, then it would make sense to configure queuing in that region to allocate more bandwidth to voice; and similarly, a smaller amount of bandwidth should be allocated to voice in network regions where voice traffic is not being sent.

An approach to generating the appropriate configurations for queuing mechanisms using an optimization approach is described in Chiang et al. [2007]. The basic idea is to take a simulation model of the network and performance objectives (represented by a utility function) as input and to generate the required set of configurations that generate the optimal utility value as output. Starting with a simulation model of the network, any single simulation run produces one evaluation of the utility function to be optimized. Adaptive simulated annealing [Ingber, 1989] is used as the optimization heuristic here. Multiple sets of parameter values are selected, and multiple utility function evaluations are performed by running simulations in parallel. This allows utility values to be generated in parallel for each selected set of parameters. Once a set of parallel tasks has completed, the algorithm again generates multiple sets of parameters for evaluation. This process repeats until the result has converged. This approach has been shown to be effective for generating queue configurations for MANETs.

6.5.4 Interdomain Policy Management

The subject of interdomain policy management is one that has generated much interest in the network management community. When two or more networks that need to communicate with each other are controlled by different administrative entities, the question of how to reconcile policies defined in each of these domains arises. This section addresses the issue of interdomain policy management. Recall that policies were defined to be of three types in Chapter 2—namely, event–condition–action (ECA) policies, access control policies, and configuration policies. Interdomain policy management is a term that is typically used to refer to the third category of policies—that is, configuration policies. Consider two networks in different administrative domains. At a high level, in order for these two networks to be able to communicate, they must be configured in a manner that allows them to:

- Form network links at the physical and data link layers.
- Exchange IP routing information across network boundaries.
- Provide end-to-end QoS for application traffic flows.
- Provide security services such as end-to-end encryption of application traffic flows.

Each of these topics is discussed in the next four sub-sections.

6.5.4.1 *Interdomain Policies at Physical and Data Link Layers.* At the lowest layers of the protocol stack, some minimum requirements need to be met to allow two networks to communicate. The radio waveforms used by the border nodes for these networks must be compatible to allow links to be formed between them. This implies that the configuration parameters (or policies) at these layers must be chosen in a way that allows interoperability. The configurable parameters for the physical and data link layers were discussed earlier in Sections 6.5.1 and 6.5.2.

6.5.4.2 Interdomain Policies for IP Routing. In order to be able to exchange routing information across administrative boundaries, the network administrators for each domain need to run an external gateway protocol such as BGP to exchange routing information. Each network contains one or more border nodes that establish BGP sessions with border nodes on the other network. Routes are then exchanged between the border nodes using these BGP sessions. The border nodes are responsible for distributing the relevant routing information to other nodes in their domain using internal BGP (iBGP). An added complexity when dealing with MANETs is that the identity of border nodes may need to change dynamically, based on mobility.

BGP is a protocol that allows a large number of configuration policies to be defined to control the routes used between domains and the mechanisms used to aggregate routes [Caesar and Rexford, 2005]. While this allows for a great deal of flexibility in optimizing the behavior of routing, the complexity of configuring all of the available parameters increases the likelihood of misconfiguration. Errors in configuration policies in one domain can affect other domains, if routes are not advertised and propagated correctly. One possible approach to reducing the likelihood of configuration errors is to use configuration checking tools that can search for different types of inconsistencies based on predefined rules [Feamster and Balakrishnan, 2005; Qie and Narain, 2003].

6.5.4.3 Interdomain Policies for QoS. QoS configuration was discussed earlier in Section 6.5.3.5. In order to provide end-to-end QoS across two different administrative domains, the domains must implement compatible QoS mechanisms, such as DiffServ. In addition, the traffic classes used in the two domains must be identical or must be mapped to each other via some predefined set of rules. For example, if one domain supports 10 different traffic classes (with 10 different DSCPs) and the other only supports eight classes, then a mapping needs to be defined between these two sets of traffic classes/ DSCPs. In addition, the border nodes between the two domains need to be configured to re-mark the DSCPs for every packet going from one domain to the other in accordance with the defined mapping.

Next, SLAs (Service Level Agreements) need to be defined that specify the amount of traffic per DSCP that will be allowed to cross from one domain into the other. This ensures that the amount of traffic injected from one domain into another remains within certain predefined bounds. This is especially important for MANETs due to the scarcity of bandwidth in such networks. Following this, policers must be configured on all border nodes to ensure that the specified traffic rate limits are observed. Policies for how to deal with excess traffic should also be configured. For example, all traffic in a certain traffic class that exceeds its limit could be dropped, or marked to a different DSCP.

Finally, the question of how to provide end-to-end QoS assurances for traffic originating in one administrative domain and terminating in a different administrative domain needs to be addressed. This will be discussed in more detail in Chapter 8, as part of the discussion on automated end-to-end service quality assurance in MANETs.

6.5.4.4 Interdomain Policies for Security. Configuration policies for security services also need to be coordinated across administrative domains. As an example, the configurations of firewalls in one domain affect what traffic is allowed to flow across domain boundaries. The more restrictive set of configurations will dominate; for example, if the first domain blocks all FTP traffic, even if the second domain does not block FTP traffic, the second domain will not be able to send or receive any FTP traffic to or from the first domain. Also, administrators may want to configure border nodes with a set of firewall rules that apply to all traffic originating from external networks. These rules need to be negotiated between domains ahead of time.

In order to provide end-to-end communications security, traffic must be encrypted prior to transmission over the air. Although encryption and other communications security services will be discussed in more detail in Chapter 9, it is important to note here that communicating endpoints in different domains need to be able to negotiate security parameters prior to communicating. In particular, interdomain key distribution and revocation mechanisms are required to support security services. As will be discussed in Chapter 9, the use of the Public Key Infrastructure is one way to support interdomain key exchanges.

6.6 POLICY-DRIVEN CONFIGURATION MANAGEMENT IN AD HOC NETWORKS

This section takes a look in more detail at how configuration is controlled in a policy-driven network management system. The concepts that will be described here are relatively new and have not been thoroughly tested or deployed in large-scale MANETs. They are included here to give the reader some ideas about interesting directions that are currently being researched in the area of automated configuration for MANETs. Additionally, some of the techniques that are referenced here require a basic understanding of Bayesian networks, decision graphs, and machine learning. Appropriate references are provided in the text below to enable readers who are unfamiliar with these topics to gain an understanding of these subjects.

6.6.1 Configuration Decision-Making

As the reader has probably realized, *the chief challenge in configuration management is determining the best way of configuring the network*. In wireline networks, this can be done manually, since the configuration of the network does not need to change frequently. However, in MANETs, given the dynamic nature of the network, configuration changes need to happen constantly in order to maintain optimal network performance. This rapid rate of change precludes the use of manual analysis to come up with the best possible configuration settings for the network. In this section, a forward-looking *Control and Optimization Function (COF)* architecture is presented that leverages operation-specific or mission-specific knowledge external to the network and applies learning and reasoning algorithms [Michalski and Tecuci, 1994] to predict future network state and correspondingly optimize the behavior of the network. Some preliminary work in this direction [Poylisher et al., 2005] has yielded promising results.

Of course, existing networking protocols have some degree of adaptability built into them; as an example, routing protocols adapt to node failures by computing new routes to other nodes in the network as needed. The difference between such protocols and the COF described here is that the COF is cognitive and deliberative and acts over longer time frames, while network protocols are autonomic and reactive and act in real time within the constraints specified by the COF.

One of the features of the COF is that it performs cross-layer optimizations wherever possible. Cross-layer optimizations are optimizations across different layers of the protocol stack that are performed in concert to obtain the best possible overall network performance. Although cross-layer optimizations go against the principle of separation of concerns on which the Internet protocol suite was built, such optimizations can provide performance improvements that are hard to achieve if each protocol layer is designed and configured without considering what is happening at other layers of the protocol stack (e.g., see Kyasanur et al. [2005]).

It should be noted that cross-layer design is a double-edged sword. Although it can lead to efficiencies that cannot be achieved with a strict layered design, it is also fraught with dangers and can result in deteriorated performance [Kawadia and Kumar, 2005]. For this reason, it is critical that any cross-layer optimizations be carefully coordinated and performed in a systematic way. The COF provides a sound basis for controlling network parameters and thereby dynamically reconfiguring the network during operation. Before getting into the details of COF functioning, the driving requirements for this function are listed below:

- First, any control decisions made by the COF must aim at *maximizing network effectiveness*, which is represented by an objective function (or utility function) that is expressed in terms of application "happiness."
- Second, in order to make adequate control decisions, the COF must have access to up-to-date network measurements. However, to maintain low bandwidth overheads, such measurements should be obtained locally to the extent possible.

- Finally, certain control decisions must be coordinated across more than one node; however, such coordination should also be minimized and incur minimum bandwidth overhead.

6.6.2 How Does This Approach Relate to Policy-Based Network Management?

Having looked at a high-level description of the COF approach, the following question immediately arises: How does this approach relate to Policy-Based Network Management? After all, Policy-Based Network Management (PBNM) was introduced as a way to help manage and reconfigure the network based on policies. So how does a COF that uses reasoning techniques to determine how to reconfigure the network fit into the picture?

Recall that in Chapter 2, three different types of policies were defined: ECA policies, AC policies, and configuration policies. The COF makes control decisions that reconfigure the network; in other words, it generates *configuration policies*.

The next question that arises is, If the COF generates configuration policies, and ECA policies are used to reconfigure the network via policy actions, then how do configuration policies and ECA policies interact, and how can undesirable interactions between them be prevented? (The reader should note that interactions with AC policies are not of concern here, since AC policies do not reconfigure the network and therefore do not interact with configuration policies.)

The answer to the above question is not simple. Policy conflicts are always difficult to deal with, as was discussed in Chapter 4. However, at a high level, it is clear that there must be some way to ensure that the reconfiguration performed by the COF is not at odds with the reconfiguration performed by ECA policies. This is easier said than done; however, there are some strategies that can be used to make it easier to reconcile the actions of ECA policies with the configuration policies generated and enforced by the COF.

The recommended strategy is to have disjoint sets of configuration parameters that ECA and COF-generated policies are allowed to modify. In other words, ECA policies cannot modify configuration parameters that COF-generated configuration policies are allowed to modify, and vice versa. This strategy may appear like oversimplification, but it makes perfect sense to separate the areas of concern for the COF versus ECA policies.

Having said this, how does one determine which variables should be under COF control and which should be under ECA policy control? The answer here is a little more subtle than the previous one. In a nutshell, configuration parameters that require human oversight and control should be controlled via ECA policies, and the remaining configuration parameters can be left to the COF to control. The rationale behind this is the following: If there is a configuration parameter that requires human oversight, then this parameter can be controlled by an operator-generated ECA policy. Whenever the network operator needs to modify the way that this parameter is set or controlled, he or she can modify the corresponding ECA policies. Another category of configuration parameters that should be controlled by ECA policies are those that cannot be incorporated into the COF algorithms for any reason, such as complexity, risk, and so on. In other words, there might be some configuration parameters that operators do not want to allow the COF to control, either because it is too difficult to automate decisions about the appropriate values for these parameters or because there is a risk that the COF may choose inappropriate values for these parameters that would result in unacceptable network performance. Finally, parameters that do not need to change based on network performance do not need to be controlled by the COF.

Below are some examples that illustrate the above guidelines.

- *Example 1*: A category of configuration parameters that is often appropriate for configuration via ECA policies is DiffServ bandwidth allocation. When DiffServ is configured on network nodes, bandwidth must be allocated for each supported class of service, so that appropriate QoS treatment can be provided to differently marked packets. The amount of bandwidth allocated for different classes of service depends on many factors, such as the amount of traffic that needs to be supported for each class, the amount of bandwidth that is expected to be available, the relative importance of different types of traffic, service level agreements (SLAs) with customers or other users of the network, and so on. Since SLAs are influenced by these decisions, network operators typically prefer to retain control over parameters that control DiffServ bandwidth allocations. Thus ECA policies should be used to configure these parameters; the events that cause modifications of these parameters could be based on time triggers specified by the network operators, or specific events such as a change in mission state (from surveillance mode to active battle mode, etc.).

- *Example 2*: Another category of configuration parameters that should not be configured via a COF is IP address information. Such information is typically planned by the operator using tools that facilitate the task; a pool of IP addresses is supplied as input to such tools, and the tools help to generate an IP addressing plan for the network. This includes allocation of IP addresses to specific network interfaces on network nodes, as well as allocation of IP address pools to DHCP servers. This is a function that would not make sense to assign to the COF, because there is typically no need to dynamically modify IP address assignments based on network performance. Even though black IP addresses need to change due to mobility, these changes can be handled as described in Section 6.5.3.3.2.

- *Example 3*: Let us now look at an example of configuration parameters that would be best controlled by a COF. Configuration parameters for routing protocols include link weight assignments that are used by the routing protocol to compute costs of different routes and select the lowest-cost route. A very simple way of assigning link weights is to assign a weight of 1 to every link; this results in shortest-path routes (i.e., routes with the smallest number of hops) being selected by the routing protocol. However, it is often desirable to use other criteria to select the "best" route to a destination, including dynamic criteria that take into account link utilization, link error rate, and so on. Such criteria need to be measured in real time and used to make dynamic decisions about link weights. Thus the configuration of link weights is an ideal candidate for COF control.

• *Example 4*: Another example of configuration parameters that would be best controlled by a COF relates to the building and maintenance of the routing hierarchy (refer to Section 6.5.3.2). An initial routing hierarchy can be manually planned and configured, especially in military deployment scenarios. However, the routing hierarchy will have to be maintained by making alterations as needed, due to mobility and network dynamicity. The COF can make such decisions, by using an objective function that incorporates the desired metrics, such as minimization of route stretch factor, minimization of cluster diameter, and so on. Again, these decisions need to take into account current network status, node locations, routing information, and so on, thus making such decisions appropriate for the COF.

The next section discusses the details of a COF within the network management system.

6.6.3   COF-Based Solution Architecture

Figure 6.12 depicts a high-level view of a solution architecture that includes a COF. As discussed earlier, every MANET node contains a set of protocol components, along with an instance of a network management system. Within the configuration Management component, a Control and Optimization Function is used to generate configuration policies, as discussed above. Decisions about how to configure an appropriate set of network parameters are provided by the COF.

It is important to distinguish between the nature of decisions being made by network protocols and the COF. The protocol stack makes real-time reactive decisions about power control, transmission scheduling, routing, and forwarding, based on its configuration, locally available information, and information obtained via signaling messages exchanged with peer nodes. This information is therefore directly or indirectly observable from the network itself. The COF, on the other hand, makes proactive decisions based on additional information that is *not* directly observable from the network, but rather is obtained from external sources and from learning and model-predictive algorithms. In a military setting, for example, a critical external source of

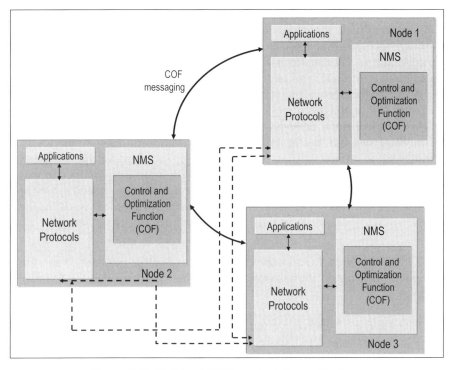

Figure 6.12. High-level COF-based solution architecture.

information is a *mission plan*, which contains information about network assets, planned mobility patterns for different assets, information exchange requirements (IERs), QoS policies, and so on. Such information is usually at a high level and is an approximation of what will actually happen during a mission; for this reason, it must be combined in an intelligent way with actual network events to constantly update and maintain an accurate view of the network and predict its near-term behavior. Such predictions are extremely valuable in guiding the behavior of the protocol components. As an example, if the mission plan dictates that a certain node must enter a cave at a certain time during the mission, where it is known that network connectivity will not be available, then the scheduling and routing components can be directed by the COF to accordingly optimize their behavior—for example, by (i) proactively discovering new routes that do not traverse this node; (ii) not scheduling any timeslots for this node's transmissions; and so on. As another example, if the mission plan calls for an initial reconnaissance phase where enemy surveillance will be conducted via data reported from deployed sensors, followed by an attack phase where precision attack missiles will be launched, the COF can instruct the MAC layer to allocate more timeslots for sensor data in the first phase of the mission and for network fire commands in the second phase of the mission.

The network protocols and the COF exchange information with other nodes to coordinate their actions. The COF exchanges information with other COF instances on other nodes in order to share its set of beliefs about the network and to coordinate its actions as required. As an example of such coordination, the COF could decide to change the percentage of slots that are reservable for transmission; such a change would have to be coordinated across a two-hop neighborhood.

The protocol stack provides functions including QoS, transport, routing, forwarding, and medium access control. In addition to these functions, the protocol stack provides (a) an application layer interface for applications to request and use network services and (b) a COF interface. The COF creates and maintains a view of the current and predicted network state in a *Network Context Manager* and generates actions that tune the protocol stack through a *Behavior Manager*. Both of these components are described below.

6.6.3.1 *Control and Optimization Function (COF).* In order to meet the requirements listed in the previous section, the COF provides the following features:

- The COF is a *supervisory controller* for the network, and it functions at a larger timescale (in the order of seconds/minutes based on the network situation) than do protocol decisions. This in turn implies that measurements from and coordination messages with other nodes only need to be exchanged infrequently, thereby imposing an insignificant network overhead (as discussed in more detail later in this section).
- The COF optimizes an *objective function* which is expressed in terms of application requirements that are assumed to be provided by applications via an application layer interface (to be discussed further in Chapter 8). The objective function is a priority weighted sum of separate objective functions, each associated with an application stream. The global objective function is decomposed so that its expected value can be computed in terms of *mostly locally available* network measurements.
- The COF is a *model predictive controller* in that it looks ahead in time and picks a move or a sequence of moves which has the best expected value, with respect to the objective function, over all future move sequences considered. In order to assess the future, it uses (a) a distributed Bayesian network [Gamez et al., 2005] to determine current state and (b) learned policies to evaluate the effect of possible actions.
- Bandwidth overhead is minimized by (1) locally summarizing any measurements that need to be transmitted over the air before sending them and (2) sending COF decisions to other nodes only if and when necessary and only to the set of nodes that need to know about them (e.g., all nodes in a one-hop neighborhood).

As described earlier, the COF works in conjunction with network protocols by fine-tuning its operation based on COF analysis. The COF makes decisions based on fusing network information with information that is not directly observable from the network, but obtainable from external sources (e.g., a mission plan, terrain information) and from learning and reasoning on long-term past experience. It is important to draw two distinctions between (a) actions that the COF performs to manipulate network configuration parameters and (b) actions that network protocols perform autonomously: (i) the COF functions on a larger timescale than network protocols and (ii) the set of knobs that the COF tunes are **constraints** on network behavior and are not adjusted by protocols autonomously.

At a high level, the COF functions as shown in Figure 6.13. It uses mission knowledge, situation models, and network context information to *predict* network state; *analyzes* this information to determine appropriate actions; performs these actions by *tuning* configuration parameters; and then *senses* the actual network state. This cycle is performed periodically and on receipt of significant events such as mission changes as described below.

Sense: The COF gets real-time data on network state. This data is used both to verify expected behavior and as input to the prediction phase, to help predict the near future network state.

Predict: The COF uses knowledge about the mission, expert knowledge stored in situation models, and real-time information gathered from the network to predict the state of the network in the near future. The following kinds of information are predicted:

- *Mobility Patterns*: Information about predicted mobility is typically available in mission plans and can be continuously adjusted based on

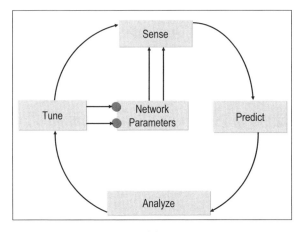

Figure 6.13. COF functioning.

observations. Knowledge about mobility is critical in MANETs because it affects the network topology.

- *Link Failures, Interference, and Channel Characteristics*: Real-time measurements obtained from the network as well as mission knowledge are combined to predict link quality and imminent failures. Knowledge about mobility patterns can also be used to estimate interference and channel quality and predict links that are about to disappear.
- *Traffic Flows and Their Characteristics, Such as Application Types/ Priorities*: A mission plan typically contains information about the expected traffic flows in different mission phases. This knowledge can be combined with actual observed flows to predict expected flows.

Analyze: Based on the predicted state, the COF performs an analysis that uses a Bayesian network-based representation [Heckerman and Wellman, 1995] in combination with Factored Markov Decision Processes, guided by an objective function. The outcome of this analysis is a set of *actions* that are used to tune network parameters. Some examples of the kinds of actions performed are:

- *Assign weights* to quantities that act as protocol constraints when the protocols compute schedules, routes, and so on—for example, battery power; connectivity; energy consumption; whether a node should forward traffic or not; and so on. For example, if many nodes are low on battery power and a certain mission phase is being conducted in an environment where batteries are not going to be recharged soon, then battery power may be assigned a relatively high importance; this in turn constrains the choice of a node as a forwarder (a node with very low battery power should not forward packets) and therefore influences routing.
- *Configure* relevant configuration parameters, such as the weights assigned to different nodes within the election algorithm (such weights can bias the election in favor of certain nodes that are sending high-priority traffic), the percentage of slots assigned to different traffic classes, the amount of bandwidth to allocate to different types of traffic based on destination, and so on.

Tune: The set of actions chosen in the above analysis are executed by *tuning* the appropriate network parameters.

6.6.3.2 Illustration of COF Decision-Making Process. Before delving into the architecture and technical approach for the COF, this section provides a high-level example to illustrate the models and decision-making processes within the COF. The following are the steps toward building the COF and using it for control:

Build Situation Models: The first step toward creating the COF is to build models that capture the relationships between *observable network measure-*

ments, the *knobs* that can be tuned by the COF, and the *objective function*. This step is performed before deployment of the COF. Such a model is initially created by capturing expert domain knowledge about these relationships and adjusted with statistical learning as needed. Bayesian networks to represent situation models and their decision-theoretic extension, Decision Graphs [Jensen and Nielsen, 2001], are used to reason about actions. An example of a Decision Graph created by capturing such relationships is shown in Figure 6.14. Here, rectangular boxes represent *actions* or *knobs* that can be tuned by the COF; ovals represent *variables* that can be directly measured or indirectly inferred (e.g., from measurements, from mission knowledge, or from past experience); arrows represent *relationships* or *dependencies*; and the objective functions are represented by hexagons. The figure shows two linked Decision Graph fragments, each residing in a different MANET node. Sharing of information across node boundaries is represented by arrows across nodes. Note that this figure is purely an illustration and is not complete in any sense. The types of expert knowledge captured in this model are:

- Admission or preemption of a traffic flow (*action* represented by the "Admit_flow" box) affects whether the corresponding application is running (variable represented by the "App_running" oval), which in turn affects the objective function.

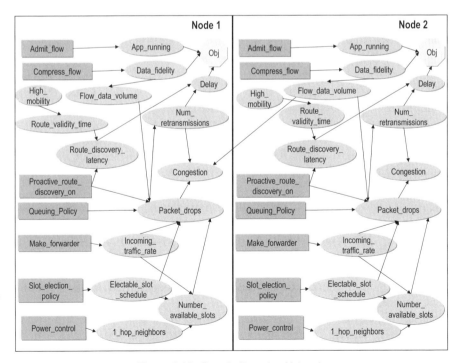

Figure 6.14. Sample Bayesian Network.

- Compression of a traffic flow ("Compress_flow" box) affects data fidelity (variable represented by the "Data_fidelity" oval), which in turn affects the objective function.
- Changing the power affects the set of one-hop neighbors, which affects the number of timeslots available to a node for transmission, which affects the number of packet drops at the node, which is an indicator of congestion. Here congestion is a "virtual variable" inferred from measurements.
- Assume that there is a flow that originates at Node 2 and flows through Node 1; hence the "Congestion" variable on Node 1 is affected by "Flow_data_volume" on Node 2.
- An example of a variable whose value could be derived from mission knowledge is "High_mobility"; a particular mission phase may dictate a high degree of movement, which affects how long routes remain valid ("Route_validity_time") and in turn could affect the decision to discover routes proactively or reactively.

Learn Dependency Probabilities: Again, this step is performed before deployment of the COF. Figure 6.14 shows dependencies among actions, variables, and the objective function, but does not quantify the magnitude of the effects. The next step is to provide this quantification by using expert input and statistical learning to discover the appropriate weights to place on each dependency arc in the Bayesian network. This is done during a *training period* (before deployment), where the appropriate weights are provided/learned in a simulated MANET.

Build Behavior Models: Once the above model has been developed, the next step is to use it to generate *Behavior Models*. This step is also performed before COF deployment. Again, expert knowledge is used to constrain the set of appropriate actions in given network states, and a training period is used to learn the weights for the arcs between actions and variables in the above Bayesian network, unless provided by experts.

Generate Network Context: Once all these models have been developed, the COF can be deployed and used to control network configuration. As stated earlier, this is done on a longer timescale than typical protocol decisions. The COF is periodically invoked to determine appropriate control actions for the network. When invoked, it must generate a *network context*, which involves obtaining measurements for all the variables represented in the situation models developed above, instantiating the model, and making it available for determining control actions. Coordination among nodes may be required to generate a consistent context when variables have dependencies across node boundaries. The representation not only provides the information about who and what needs to be coordinated but also can assess the criticality of the coordination. This function is performed by the *Network Context Manager* (see Section 6.6.3.3.1).

Determine Appropriate Control Actions: Given a network context, the final step is to use it to determine the appropriate control actions to apply to the network. The COF determines the effects of a possible set of actions on the objective function, using the network context and the behavior models developed earlier, and chooses a set of actions that results in the highest possible "network effectiveness," as expressed by the objective function. This function is performed by the *Behavior Manager*, described in the next section.

6.6.3.3 COF Architecture. The primary objectives of the COF are to control network parameters with the goal of optimizing the global network *utility*, or effectiveness. Figure 6.15 shows a high-level view of the internals of the COF. The COF has two major components: (1) the *Network Context Manager*, which generates and maintains the *network context* at each node of the MANET, and (2) the *Behavior Manager*, which that determines appropriate control actions and executes them via the COF interface to the network protocols. Each of these components is described in more detail in the following sections, followed by a discussion of the *objective function* that guides the COF's choice of control actions.

6.6.3.3.1 Network Context Manager. The Network Context Manager is responsible for generating the *network context* in a MANET node. The approach is to model the global belief state of the set of MANET nodes as a composition of node belief states from locally observed network context. Generally, it is intractable to compute large situational beliefs in real time, and

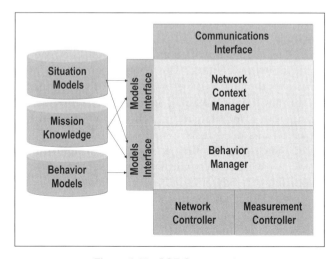

Figure 6.15. COF Components.

limited and incremental communications add significant latency to the propagation of local-to-regional network context information. The technical approach described below addresses these challenges.

Network Context Generation: Network context generation is the process of matching evidence from the measurements to the *situation models* to generate hypotheses of what is observed by the node and the state of the network from the perspective of the node. Initially, the network context is generated based on local information at the node including the measurements, component configuration, and mission knowledge. Situation hypotheses are representations derived from alternative matches of evidence to models. The situation hypotheses for network context are maintained within the Network Context Manager and are updated as new measurements are made available. A representation based on Bayesian Networks is used to capture the uncertainties in the measurements and the matching process. The approach based on probabilistic relational models complements the overall component-based guiding principle of this architecture. Model fragments corresponding to the different network protocols are built and maintained within a knowledge base. At run time the component configurations and measured parameters are used to retrieve the appropriate model fragments and are composed into the network context.

Coordination: Coordination for supervisory control is accomplished by constructing a hierarchical fusion framework. Each individual node's network context is represented as a Bayesian Network, as described earlier, which is a compact representation of the relationships among a set of variables that represent network state. A number of *component level* Bayesian Network fragments (which represent information about individual traffic flows, etc.) are aggregated into *node level* Bayesian Networks, and groups of node-level Bayesian Networks are combined within a *coordination group* (a set of nodes sharing information—e.g., a group of one-hop neighbors) to create a *group* network context. The group network contexts are then combined to create the global aggregate network context. The composition of all node Bayesian Networks constitutes the global Bayesian Network that represents the global belief state. Note that the global belief state is *never represented explicitly* at any single location but is maintained implicitly across the network.

When control actions need to be taken, some coordination across nodes may be required. Such coordination is organized around the concept of a *coordination task*, for which a dynamically formed set of nodes agree to share contexts and coordinate behaviors. Within a coordination task, each node has a role and shares with other session members only those aspects of the context and behavior that are relevant to the node's role within the session so that bandwidth overhead is minimized. For example, a set of nodes in close geographical proximity within a mission phase may coordinate their transmit power and timeslot scheduling.

The key challenges to the Network Context generation and maintenance are (i) to construct a distributed global Network Context dynamically with limited communication support and (ii) to converge quickly. The amount of inter-node communication required can be limited by noting that a message is valuable to the extent that it changes the recipient's beliefs to reflect information available only to the sender. Ordering change messages and sending highest impact messages first helps conserve bandwidth. Paskin and Guestrin [2004] describes a distributed architecture for solving these problems that is robust to unreliable communication and node failures in which session members assemble themselves into a junction tree and exchange messages to solve the inference problem efficiently and exactly. A key part of this is an efficient distributed algorithm for optimizing the choice of junction tree to minimize the required communication and computation.

6.6.3.3.2 Behavior Manager. The function of the Behavior Manager is to determine appropriate control actions to apply to network parameters, while maintaining the highest possible "network effectiveness," as expressed by an *objective function*. Control decisions have long-term effects; for example, powering up a large number of nodes may increase the overall bandwidth right now, but will significantly limit the lifetime of some nodes. A strategy is needed that maximizes the long-term utility of the system. Any current estimate of the network context is necessarily uncertain, and, moreover, predictions of future contexts and application needs are highly stochastic. The challenge is to optimize the *expected long-term utility* given such uncertainty.

The Behavior Manager reasons about the optimal action using a representation that is a decision-theoretic extension of probabilistic relational models [Laskey, 2004; Takikawa et al., 2002]. The control actions that can be taken for each component (the adjustable knobs) are added to the Bayesian Network fragments to construct decision network fragments. At run time, the composed Network Context gets implicitly extended to the decision network for use by the Behavior Manager. To reason over sequences of actions (not just the myopic next action), the decision networks are extended to dynamic decision networks by indexing the time-dependent variables by time [Glesner and Koller, 1995; Ghahramani, 1998]. Like the situation models, the behavior models too can be constructed before the run-time use of the models, using a combination of expert knowledge and training data. Learning the decision theoretic aspects of the behavior models is based on using multiagent reinforcement learning with training data. Optimization of expected long-term utilities is formalized using the *Markov decision processes (MDP)* framework [White, 1993]. In the framework, a long-term objective function is computed to represent the long-term effect of taking a particular action. For each possible joint set of actions for every node, the objective function specifies a long-term utility, and the best joint strategy is the one that maximizes the total value.

The distributed nature of the MANET prohibits a centralized optimization procedure, and a distributed MDP solution requires significant coordination.

Challenges include the lack of a global Network Context; the need to minimize overhead required for coordination across nodes; limited decision times; possible lossy communications; and so on. To address these challenges, an *anytime coordination* approach can be used that builds upon three recent advances: (i) *factored MDPs*, a unique approach for solving large-scale MDPs that exploits problem structure; (ii) *coordination graphs*, a distributed coordination algorithm that allows multiple nodes to efficiently find the optimal action setting through neighborhood communication; and (iii) *distributed inference* for coordinated decision-making. The impact of the communication required for coordination on the available resources is also incorporated. This effect can be evaluated by combining the models with the predictive approach developed for the Network Context Manager. Using estimates, the level of coordination can be traded against the bandwidth required to achieve this coordination.

Some of these ideas have been put to test in a sensor network environment. Robustness to failures is a key issue in sensor networks. To address this issue, Paskin et al. [2005] describes a robust distributed inference architecture that addresses a wide class of important inference tasks. The distributed nature of the approach provides formal guarantees in terms of robustness to unreliable communication and sensor failures. Planning and control techniques using factored MDPs have been used with great success in sensor networks that control actuators and have local parameters that can be optimized.

6.6.3.3.3 Objective Function. Broadly speaking, the goal of the objective function is to enable the COF to estimate the "goodness" of proposed control actions, in order to select the best offered action. This "goodness" is measured in terms of the effect of the action on the perceived *utility* of the resulting network from the point of view of a user, or application. A user-centric view of the network utility is adopted here; network-wide concerns (e.g., maintain low communication profile, conserve power) are represented as *constraints* on resource availability rather than part of the utility itself. The network utility is a function of two factors: **(1) application utility**, given by acceptable latency, loss, jitter, out-of-orderliness specifications and an unhappiness function that assigns a utility value to a particular point in the space of QoS parameters; and **(2) mission priority**, given by a function of the mission situation that assigns relative weights to particular applications at a particular point in time, independently of their type.

Application utilities can be combined to compute the network utility, given mission priorities. The network utility model is based on a weighted sum of such individual utilities. The goal of COF is to *minimize the sum, over the course of the mission, for every application that had any traffic and every measurement period, of average mission priority-weighted application unhappiness values in the measurement period.* In computing the supervisory control actions, node COFs collectively optimize such a multiattribute objective function. Decisions are based on maximum expected utility computations. The proposed form of network utility allows for easy accommodation of new application types and network technologies, because it does not make any assumptions about the network. Modification of the objective function entails modification of application QoS requirements and mission priorities.

Below are some of the challenges in using this approach.

Tractability: Representing the objective function in terms of measurable network parameters is a challenging problem. In general, for each state of the network, there is an associated utility value. Any detailed specification of the objective function will have many dimensions, and the number of possible states grows exponentially with the number of variables in the state description. Thus the challenge is to *make the computation of the objective function tractable.* Domain knowledge comes to the rescue: The proposed type of the objective function is highly structured and the structure can be described in terms of *independence*. If the state of the network is described by a set of variables V, a set of variables X is *utility independent* of everything else if, as $V - X$ is held fixed, the induced preference structure over X does not depend on the values of $V - X$. Judgments of utility independence appear to be natural and common [Keeney and Raiffa, 1976], and domain experts are fairly good at identifying independence. Utility independence has strong implications. For this effort, the most important is that if there are k utility independent partitions X_k of the system variables, then there is a function $f_k(X_k)$ for each partition such that the utility function is a multilinear combination of the f_ks. The powerful idea of *conditional additive (CA) independence* [Bacchus and Grove, 1995, 1996] is used here. CA-independence provides a decomposition of an objective function that is very useful computationally for run-time expected utility computations. It has been shown that every objective function has a perfect CA-independence graph: a graph where vertex separation corresponds exactly to CA-independence. This allows computation of expected utilities by reasoning on such graphs. The CA-independence also allows an additive functional decomposition of the objective function.

Computing Utility Based on Locally Available Information: It is expected that *the global utility will be decomposed along the lines of individual application utilities*. Applications map to flows and since each node participates only in a small subset of all flows, the local behavior computation will depend only on the computation of the factors associated with the flows on the node. This implies that each node only participates in a small number of such factors, and further the computation of the factors requires only short-distance message passing among a small number of neighbor nodes. The multilinear combination form further allows for additional efficiencies, because the computation can be performed as the messages are propagated within a tree that links all the nodes with the variables in the factor. If there is no factored representation of the utility function, each and every state of the system must be considered individually. However, if an additive decomposition exists, only the marginal probabilities of the variables are needed to compute the expected utility (by linearity of expectation). Finding the marginal probabilities can be easy for the singly connected Bayesian Network (linear time). More importantly, the factors require only a subset of variables for the computation. The maximization of the contribution of the factor to the global expected utility will maximize the global utility. Therefore, local decisions can be made by optimizing only the associated factors.

6.6.3.4 COF Implementation: Benefits and Challenges. The preceding sections described the basic ideas underlying the concept of a Control and Optimization Function for determining how configurations should be altered in response to changing network conditions. In addition, information that is not available from the network, such as mobility plans, can be used to make decisions about network reconfiguration. The effectiveness of the COF is largely dependent on (a) the degree to which the situation and behavior models are able to capture the network dynamics and dependencies and (b) the closeness with which the objective function is able to represent the performance of the network.

The COF fits within the larger policy-based network management umbrella within the Configuration Management component. As discussed earlier, the COF generates configuration policies for controlling certain configuration parameters that need continuous adjustment during network operation. The COF operates autonomously and is not controlled by the Policy Management component. At the same time, the Configuration Management component performs network reconfiguration based on policies; these reconfigurations are triggered by the Policy Management component. These interactions are depicted in Figure 6.16.

This figure shows the flow of information between Configuration Management and the other network management components. The COF uses information from the Fault Management, Performance Management, and Security Management components to feed its decision-making process that triggers network reconfiguration. Network reconfiguration is also triggered in parallel based on policies; here, ECA policies are triggered by network events as well as other events, and they result in the invocation of configuration actions implemented by the *Configuration Action Refiner* within Configuration Management. The actions performed by each of these components—that is, the COF and the Configuration Action Refiner—are distinct and non-overlapping, as explained earlier in Section 6.6.2.

6.7 SECURITY CONSIDERATIONS FOR CONFIGURATION MANAGEMENT

The rigid separation between black and red sides of a network that is enforced by encryption devices (see Chapter 5) poses a challenge in the design of any optimization function such as the COF described earlier, due to the need for

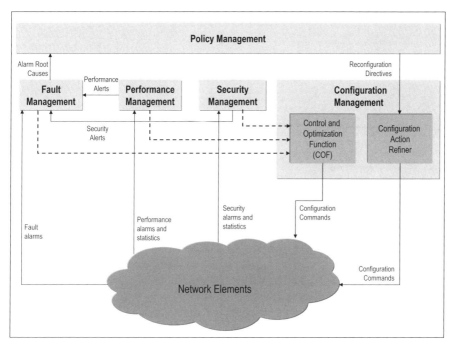

Figure 6.16. COF and policy-triggered reconfiguration.

information exchange between components operating on different sides of the encryption device. There is also a need to be able to send configuration information from the red side (where the NMS resides) to the black side network elements, in order to configure these network elements. Since the encryption device that sits at the red–black boundary operates at the network layer, any protocol function "above" the network layer—along with applications, including the NMS and in particular the COF—must reside on the red side, whereas the protocol functions "below" the network layer, including the monitoring of link status, reside on the black side. The COF needs information about all protocol layers on the black side, such as topology, status, and performance information. In order to address the above problem, there is a need for a Cross-Domain Guard, or CDG (discussed in more detail in Chapter 9) between the black and the red side, as shown in Figure 6.17. The purpose of this guard is to allow limited exchange of monitoring information from the black side to the red side, and configuration information from the red to the black side.

6.8   SUMMARY

This chapter provided an overview of the configuration management functions required for a MANET. An adaptive configuration model for MANETs that

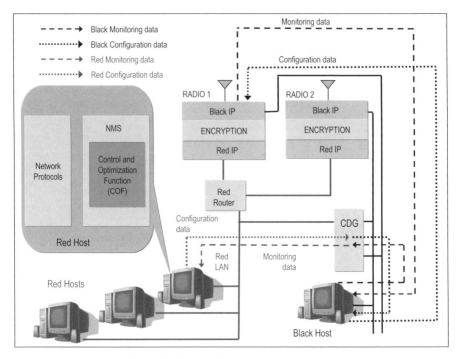

Figure 6.17. Red–black information passing.

seamlessly integrates fault, performance, and security management tasks via a policy infrastructure was described. As noted early on in this chapter, while the other management functions (namely, Fault, Performance, and Security) perform read-only operations on network elements and services, the configuration management function performs both read and write operations on the network elements/services. To accomplish such read/write (i.e., configuration) tasks, appropriate management standards have been defined; the relevant configuration management standards were reviewed in this chapter.

In MANETs, a variety of configuration management-related functions need to be performed at the various layers of the OSI stack. A detailed review of some of the key configuration functions that need to be performed at the various OSI layers was provided. The chapter concluded with a discussion of the critical problem of how to make "good" configuration decisions, and it presented a forward-looking Control and Optimization Function (COF) to assist with "smart" configuration decisions. More specifically, the COF uses an objective function and information about the network and about the mission plan to make configuration decisions to deal with the constant change that characterizes a MANET.

Before concluding this chapter, recall that although the individual FCAPS functions are described in separate chapters in this book, it is important to note that all of the FCAPS tasks function in an integrated manner in MANET environments. That is, unlike their typical wireline counterparts, the FCAPS functionalities in MANETs are not stovepiped. Each of the FCAPS functionalities are presented in separate chapters mainly to facilitate the discussion of each of the FCAPS functions. Thus, while the next three chapters focus on the details for MANET Fault, Performance, and Security Management, respectively, all of these MANET management tasks work in an integrated manner and are knit together via a policy framework, as explained in this and each of the following three chapters.

7

FAULT MANAGEMENT

Fault management deals with monitoring, diagnosing, and recommending solutions to network failures and is thus vital to maintaining the "health" of the underlying network. Once a network has been configured, it must be monitored to ensure that network elements and services are functioning correctly and are providing an acceptable level of performance. When there is a malfunction in the network, it is the fault management function that must collect all the evidence of this malfunction and diagnose the root cause of the problem. This is one of the key functions performed by a fault management system.

This chapter presents issues and solutions for fault management in MANETs. The contents of this chapter are organized as follows. Section 7.1 begins with a high-level overview of fault management for MANETs. Next, in Section 7.2, fault management operations process models are presented, along with a description of what aspects of the TMN fault operations process models can be leveraged for MANETs. This is followed by a brief overview of monitoring in MANETs, with a particular focus on red–black separation issues in Section 7.3. Section 7.4 discusses root cause analysis for MANETs, and it points out the shortcomings of existing approaches for static wireline networks when applied to MANETs. Section 7.5 discusses self-healing for MANETs using the policy-based management approach. Detailed fault scenarios for MANETs are provided in Section 7.6 to illustrate the issues that arise in MANETs and the way in which such issues can be handled. Section 7.7 concludes with a brief summary.

7.1 OVERVIEW

Before delving into the details of fault management operations and models for MANETs, note that failures in MANETs can be classified as belonging to

Policy-Driven Mobile Ad hoc Network Management, by Ritu Chadha and Latha Kant
Copyright © 2008 John Wiley & Sons, Inc.

two broad families or classes: (a) the class of "*hard*" failures and (b) the class of "*soft*" failures. A *hard failure* is defined as a failure that is associated with equipment—typically an equipment failure or malfunction. The root cause analyses of hard failures are typically deterministic in nature; that is, the fact that equipment is broken or not is diagnosed with probability 1. A *soft failure* is defined as a failure that is associated with performance degradation. Examples of soft failures include service degradation due to excessive loss and/or delay. Soft failures are typically stochastic in nature; that is, the "root cause" of any particular soft failure is difficult to diagnose with absolute certainty, and therefore the probability that a given root cause diagnosis for a soft failure is correct lies somewhere between 0 and 1. For example, the root cause explaining why packets are being dropped in large numbers could be (a) excess application traffic, (b) a denial of service attack that floods the network with traffic, (c) equipment malfunction that causes excessive retransmissions, and so on.

MANETs by their very nature are stochastic—in part because of the unpredictable environmental conditions and mobility of the MANET nodes, and in part because of the scarcity and variability of bandwidth resources. Thus a majority of failures in MANETs are of the "soft failure" type. This is in contrast to wireline networks, which are rarely subject to soft failures, due to significant overprovisioning of network capacity. Also, wireline networks typically function under benign environmental conditions, because they are typically not subjected to adverse environmental effects such as temporal fading that produce fluctuating bandwidth, or mobility changes that result in variable network topology. Soft failures are therefore less frequent in wireline networks, as compared to the dynamic and unpredictable wireless mobile ad hoc networking environments.

7.2   FAULT MANAGEMENT FUNCTIONS AND OPERATIONS PROCESS MODELS

This section discusses the details of fault management operations in MANETs. As mentioned in the introductory chapter, a significant amount of work exists in the TMN standards with regard to defining network management operations and models. However, as also mentioned earlier, these models and definitions were developed for wireline networks. Due to the fundamental nature of the differences in the operations and characteristics of MANETs, the models cannot be applied unmodified to MANETs. Nevertheless, much can be learned from the TMN operation models, and the lessons learned in turn can serve as a constructive guide to designing fault management operations processes and models for mobile ad hoc networking environments. Thus, rather than reinvent the basic model definitions and operations, the TMN models are leveraged here and are enhanced for the MANET environment.

In the following paragraphs, a brief review of the TMN operations process models and definitions for fault management operations are provided. The aspects of these models that can or cannot be applied to MANETs are described. Armed with this knowledge, the fault management operations models for MANETs are presented.

7.2.1   TMN Fault Management Operations Process Models

Figure 7.1 provides a schematic representation of the fault management model based on TMN guidelines. The left side of Figure 7.1 represents "inputs" or "triggers" to the Fault Management task, while the right side of the figure represents "outputs" or "actions" taken by the fault management task. The circles on the left and right represent, respectively, the "sources" for the input triggers and the "destinations" for the output actions. The core responsibilities of the fault management task are summarized in the central hexagonal box. In essence, the fault management operations correspond to functionalities that enable detection, logging, isolation and correction of abnormal operations in a network. The acronyms EML, NML and SML, denote the Element Management Layer, Network Management Layer, and Service Management Layer, respectively.

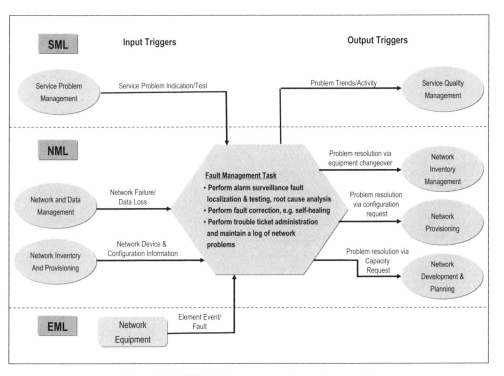

Figure 7.1. TMN fault management operations model.

While the general definitions of fault management functionalities in Figure 7.1 are broad enough to apply to either wireline or wireless networks, the operations model in Figure 7.1 is more suited to a wireline environment than a MANET environment for several reasons. These reasons are discussed in the remainder of this section. It should be noted that the TMN models do provide several useful lessons, which are elaborated upon in Section 7.2.2.

7.2.1.1 Nature of the TMN Layer Classification.

The "layer" classification in Figure 7.1 includes a Service Management Layer (SML), a Network Management Layer (NML), and an Element Management Layer (EML). In wireline networks it is typical to have service personnel operate the SML, where they perform a variety of manual operations, such as checking to see if the services are up and running in accordance with the negotiated service parameters. If deviations are observed, a "service problem indication" is issued, as shown in Figure 7.1. Additionally, the operators at the SML may issue periodic "test" triggers (as also shown in the figure) to proactively check network status.

However, in the case of mobile ad hoc networking environments, given that MANETs often have to function in remote terrains, the concept of an SML with a large army of operators that can periodically test the network, as in the wireline networks, typically does not exist. However, the philosophy of having an SML can still be applied to MANETs, albeit without the requirement for a large number of operators. More specifically, by specifying policies that automatically check for quality of service (QoS) degradations—which in the wireline world would be the equivalent of an operator checking for service violations—the philosophy of an SML can be effectively captured.

7.2.1.2 Stovepiping of Network Management Functions.

Another aspect of typical implementations of the TMN model is that the different FCAPS management functions are stovepiped and do not exchange information. Thus there is little or no explicit dependency between the fault management task and other management tasks such as configuration management or security management. Such a stovepiped mode of operation in turn involves expensive and delay-ridden human intervention to coordinate the individual FCAPS functions, whenever interactions are required. In wireline networks, which typically do not require frequent reconfiguration due to the static nature of the network, such human intervention does not pose a significant problem. However, such a stovepiped manner of operation is definitely unacceptable in wireless mobile ad hoc networking environments. The reason for this is that the dynamic nature of MANETs results in a need for frequent reconfiguration due to soft and hard faults, and therefore the amount of human intervention that would be required to deal with this high volume of manual network operations functions would be unacceptable. For this reason, as mentioned in the introduction in Chapter 1, it is imperative that fault management operations be closely coupled with configuration, performance, and security management operations in MANETs.

As an illustrative example, one possible outcome of fault management involves configuration management operations to reconfigure the network to "heal" around the failure. Two examples will be presented in later sections: one for the wireline case as illustrated in Figure 7.4 in Section 7.4.1.3, where the network reconfiguration is not dynamic and requires manual intervention; and another for the wireless case that is dynamic and seamless as illustrated in Figure 7.8, Section 7.6.2. Nevertheless, such a configuration operation— whether it be automatic or via manual intervention—will result in a new set of network relationships between network elements. These relationships must be maintained by the fault management system in the form of network element dependency models (discussed in Section 7.4.1.2 for wireline networks and in Section 7.4.2.2 for MANETs) in order to perform fault diagnosis and root cause analysis. In turn, this implies that the new dependency information needs to be updated as quickly as possible, so that subsequent root cause analyses and self-healing actions initiated by fault management can be consistent with the underlying network state and topology. This can only be achieved via a closely coupled set of FCAPS operations models. Heavyweight message exchanges that involve human intervention between stovepiped FCAPS operations, as is typical in wireline operations models, will therefore be ineffective in MANET environments, as will also be explained during the discussions on self-healing actions in Section 7.5.

7.2.1.3 *Response to Network Faults.* As a final remark on the need to enhance the TMN operations models for MANETs, the process for responding to network faults needs to be examined more closely. As shown in Figure 7.1 (see the last bullet item in the fault management task description in the hexagonal box), this function is handled via trouble ticket administration (TTA). Trouble ticket administration, as defined by the TMN models, typically involves an operator and works as follows. Upon receipt of a fault indication, a trouble ticket is issued by the system. While the issuing of a trouble ticket is an automated process in most of the wireline FCAPS operations models, the actions taken upon receipt of a trouble ticket typically involve human intervention in most wireline systems. Example corrective actions include reconfiguring network elements or renegotiating service level agreements (SLAs), for example, which again require manual intervention. For example, reconfiguring network elements in wireline networks require the SML operators to manually initiate a modification of network configurations on their edge nodes; reconfiguration of core nodes and customer premise nodes typically occurs by scheduling a workforce task to travel to the equipment site to perform the reconfiguration. Another alternative is to assist users in reconfiguring their equipment (such as home networks) with the help of network call centers. Although such an approach eliminates the need for on-site presence of a workforce task, it still involves human intervention via call centers. Response times

are typically high when there is manual intervention required (in the order of hours or sometimes even days—for example, when new equipment has to be added). Similar observations apply if SLAs have to be renegotiated with customers or among service providers.

While manual intervention and high response times may be the norm in wireline environments, they are unacceptable in MANETs. This is partly because MANETs are expected to be deployed and operational in remote terrains. Recall that MANETs are used in areas either where establishing infrastructure networks is infeasible (e.g., in remote terrains) or where networks must be deployed on-the-fly, as is the case for battlefield operations. This precludes substantial manual intervention during network operations. More importantly, as mentioned earlier, the stochastic nature of MANETs results in a need for frequent network reconfiguration, which cannot be handled manually. In many circumstances, network dynamics arising from environmental changes as well as mobility in MANETs may necessitate automated responses in the order of seconds or at most a few minutes. Such low response times are obviously unachievable without a highly automated mode of operation.

7.2.2   Fault Management Operations Model for MANETs

As pointed out in the previous section, although they require several enhancements, the TMN models provide a good starting point for designing a policy-based network management system for MANETs. This section first summarizes the aspects of TMN models that will be leveraged for fault management, followed by a description of the policy-driven networking management approach to managing MANET faults.

To begin with, a relevant concept that can be retained from the TMN operations models is the need for "intelligence" in the network management system (NMS) to either reactively (e.g., via the trouble ticket administration, workforce tasking, and SML operators) or proactively (e.g., via the SML operators) reconfigure the network. Additionally, the broad definitions of fault management responsibilities in terms of event logging, alarm correlation, root cause analysis, and problem correction [Hasan et al., 1999] are also aspects that are useful and applicable to MANETs. However, as mentioned earlier, while the above concepts still apply to MANETs, the subsequent operations models and process definitions require some enhancements.

In particular, the detailed operations processes and models that form the core of the fault management operations for use in MANETs need to be enhanced so that

- FCAPS operations are tied together via policies.
- Responses to network faults are automatically handled via policies to the extent possible.

It should be noted that manual intervention cannot be completely eliminated in MANETs, since hard equipment failures may ultimately require human intervention to replace malfunctioning hardware; however, the need for immediate manual intervention can be *minimized* by providing for a self-healing network that attempts to keep the remaining nodes in the network fully functional in the face of failure of other nodes. The fault isolation and self-healing activities are automated via root cause analysis and self-healing, as described in Sections 7.4 and 7.5, respectively. As will be explained in Section 7.4.2, novel root cause analysis techniques are required for the challenging MANET environment that can handle both soft and hard failures and are resilient to imperfect information resulting from the stochastic MANET environment. Sections 7.5 and 7.6 describe self-healing using policies, which help to overcome the shortcomings of manually driven trouble ticket administration and workforce tasking described in the previous section.

Figure 7.2 provides a schematic of the operations model for the fault management process within a MANET management system. The following are its salient features:

- The fault management operations process model has strong dependencies and is integrated with the operations process models for configuration,

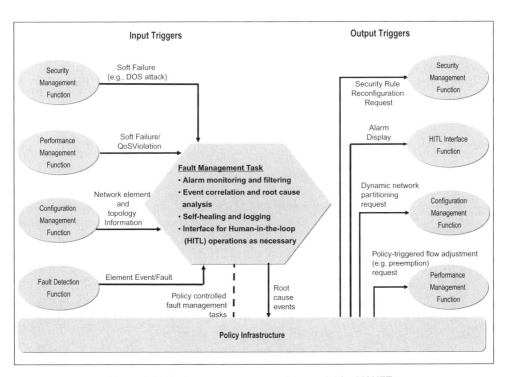

Figure 7.2. Fault management operations model for MANETs.

performance, and security via policy management, as captured via the input and output triggers.

- A policy infrastructure ties together the various management functionalities, namely, fault, performance, configuration, and security management functions, resulting in an integrated ad hoc network management system that can take timely and seamless decisions in a consistent manner, as directed by policies.

- The self-healing operations that correct problems diagnosed by root cause analysis that form a major portion of the fault management process model are automated via policies, to enable the critically needed rapid response to problems in the dynamic MANET environment. Bearing in mind the potential need for human intervention, this policy-based management system provides knobs that can be used by a network operator to override automated actions, should the need arise, as shown via the "HITL (Human In The Loop) Interface Function" in Figure 7.2.

Finally, note that the text on the output triggers (arrows) in Figure 7.2 are illustrative examples associated with that functionality. For example, the output trigger that says "Dynamic Network Partitioning Request" to the configuration management function is one example output generated by the policy management function in response to a root cause event, to produce self-healing by partitioning the network to "heal" around the failure (hard or soft, as the case may be), based on policies. Likewise, the output trigger labeled "policy-triggered flow adjustment request" to the performance management function is an example of a QoS-related action that the performance management function needs to take. As will be discussed in Chapter 8, managing QoS assurances is a major task of the performance management function in MANETs.

7.3   NETWORK MONITORING

The most basic function of Fault Management is to monitor the network. This involves both (i) periodically checking the operational state of network elements and (ii) listening for fault notifications from network elements. Both of these functions are typically performed via SNMP. Network elements support MIBs (Management Information Bases) that contain a broad array of information about a network element. The network management system can poll the SNMP agent on a device to obtain the values of MIB variables and can also listen for SNMP traps, which are unsolicited notifications that are sent by devices to notify a management system about problems.

An important issue to consider when discussing monitoring of a MANET is the red–black separation discussed in Chapter 5. Typically, information flow between the red and black sides of the network is prohibited, or at the very least severely restricted. Note that information flow between red and black sides of the network here refers to information flows that originate on the black side of the network and terminate on the red side, or vice versa—as opposed to information flows that originate and terminate on the red side of the network and use the black network as a transit path. The latter flow of information is permitted.

Since most of the expected network problems in a MANET originate on the wireless, ad hoc network—that is, the black side of the network—it is imperative that some management information from the black side of the network be allowed to flow from the black network elements to the red network management system. The network management system itself resides on the red side, since it needs to be protected from intrusions and attacks. The flow of monitoring information from the black side to the red side may be permitted via the use of a *cross-domain guard*, which is a device that filters traffic and allows certain types of traffic to cross from the black side to the red side and from the red side to the black side. This security architecture allows gathering of monitoring information from the black side of the network, and also reconfiguration of the black network, initiated by the red-side network management system. This is further discussed in the chapter on Security Management (Chapter 9).

7.4 ROOT CAUSE ANALYSIS

An extremely important task performed by Fault Management is the ability to collect and filter network alarms and perform root cause analysis to pinpoint the root cause of a perceived problem in the network, which in turn is used to trigger self-healing actions. A particular network problem, such as a link failure, may give rise to multiple network alarms, such as "interface down" alarms from nodes using such a link, performance alarms due to the resulting network congestion, service alarms resulting from the ensuing network partition, and so on. Rather than flooding a network operator console with all these alarms, it is important for the Fault Management system to correlate these alarms and find the root cause of these alarms. If the root cause can be fixed, it is expected that all related alarms would be cleared. Furthermore, it is only when the root cause of a problem is successfully located that an appropriate corrective action can be taken in order to provide the critically needed self-healing capability in MANETs. Thus, techniques that can perform efficient and accurate root cause analysis are critical to the successful functioning of fault management operations in MANETs.

While the design of fault localization and root cause analysis mechanisms for communications networks is in general a very challenging task even for wireline networks, the challenges are compounded even further in MANETs due to the following reasons:

- Frequent random and sporadic hard failures arising from the ad hoc nature of the network coupled with random mobility; such failures could be caused by environmental fluctuations (e.g., due to cloud cover, fading), hostile environments resulting from jamming, power failures, and so on.
- Frequent random soft failures, again arising due to the stochastic nature of the underlying network; such failures could be caused by network congestion due to scarce and varying bandwidth, network partitions resulting in service unavailability, and so on.
- Frequent transient failures—for example, caused by nodes going to sleep to conserve power; this results in a need to distinguish between transient and nontransient behavior.
- Rapidly varying network topology, which makes it difficult to construct dependency models (discussed in the next section).

The remainder of this section is organized as follows. First, an overview of root cause analysis techniques for traditional wireline networks is provided, along with a brief explanation of the shortcomings of traditional root cause analysis techniques that are currently in use for these networks. This is followed by a discussion of root cause analysis techniques for MANETs.

7.4.1 Root Cause Analysis for Wireline Networks

This section provides a brief overview of the current state of the art in the area of mechanisms for root cause analysis for wireline networks, and it presents models for deriving network element relationships that are required for fault diagnosis and root cause analysis in such networks.

7.4.1.1 Brief Overview of State of the Art in Wireline Root Cause Analysis. The bulk of existing fault diagnosis methodologies for wireline networks are geared toward diagnosing faults related to network connectivity at the physical layer. Examples include Wang and Schwartz [1993], Nygate [1995], Katzela and Schwarz [1995], and Yemini et al. [1996]. As also noted in follow-on work by Steinder and Sethi [2001, 2002a, 2004b], the fault diagnosis work during the mid- to late 1990s focused mostly on diagnosing problems associated with faulty cables, interfaces, and other problems typically encountered at the physical layer. Additionally, while commercial tools for IP network management have been appearing on the market over the past decade, they do not incorporate "network layer" information while performing fault diagnosis/root cause analysis. Furthermore, most of the techniques used in the traditional wireline fault diagnosis mechanisms use deterministic models that assume that all the relationships in the network element dependency model (as explained in the next section) are: (i) known with 100% certainty and (ii) static—or at least, do not vary much over time, unless manual reconfigurations are performed, which happens relatively infrequently. Assumptions (i) and (ii) are both invalid for MANETs, where network topological changes (and hence relationships between the network elements in the dependency models) are frequent and where connectivity information between network elements may be unknown or uncertain.

7.4.1.2 Network Dependency Models for Fault Diagnosis in Wireline Networks.

In order to perform fault diagnosis and analyze the root cause of an observed symptom, the network management system (NMS) maintains information about the relationships among the various network elements that are interconnected. In particular, the fault management component of the NMS obtains information about the number and type of network elements and the manner in which the network elements are interconnected (i.e., mesh, ring, star topology), and it stores the information in the form of *network dependency models*. These network dependency models, together with a set of fault diagnostic rules, are used by the fault management component to perform root cause analysis and subsequent self-healing. This is because information about the individual elements and relationships between the network elements is required to be able to make inferences both about the types of faults that can occur and in tracing a given type of fault to a root cause.

A network dependency model (NEDM) constructed for fault management purposes can be viewed as a graph where the vertices of the graph represent the network elements (e.g., routers, switches) and the edges of the graph represent interconnectivity (direct links) between the network elements. The amount of information maintained and the complexity of the NEDM essentially determine the granularity of the faults that can be analyzed.

In today's wireline networks, the information in the NEDM model is restricted to the lowest OSI layer—namely, the physical layer—with a limited amount of information about the data link layer. The NEDMs for traditional wireline networks are typically constructed as follows. First, information about all the core network elements (such as routers, switches, etc.) are obtained by the fault management system. In today's systems, the fault management system typically does this by (a) interfacing to the configuration management component of the FCAPS systems, where the network inventory is maintained, or (b) employing special auto-discovery mechanisms. As a specific example, consider the representation of routers in a network. In this case, the vertices of the graph depicting the network dependency model will each represent a router element and will contain information such as type of router, number of active inbound and outbound interfaces, and number of passive inbound and outbound interfaces (to be used in the case of automated protection switching self-healing, as discussed in Section 7.5.1). Note that this list is not intended to serve as an enumeration of all possible information in the network dependency model, but rather provides a sample of the type of information stored.

Next, information about the direct links between two given network elements is obtained by the fault management system as follows: (a) by again interfacing with the configuration management system or (b) via special auto-discovery mechanisms. The information typically includes information about the type of link (fiber, twisted-pair, Ethernet, etc.) and its speed.

The following pseudo-code describes the construction of such a network dependency model by the fault management system for a network with N nodes.

Let n represent a node number, that is, $1 \leq n \leq N$.

Let the tuple $\mathbf{R}_{<T, NAI NPI, NAO, NPO>}$ be used to denote the information associated with a router, where

T = Router type
NAI = Number of active inbound interfaces
NPI = Number of passive inbound interfaces
NAO = Number of active outbound interfaces
NPO = Number of passive outbound interfaces

For each $n \in N$,

- Consult the configuration management system or use auto-discovery to obtain information about the details of $Router_n$.
  - Construct the 5-tuple $\mathbf{R}^n_{<T, NAI, NPI, NAO, NPA>}$, with the superscript n denoting the information associated with $Router_n$. This 5-tuple is used to represent $Router_n$ as a vertex in the dependency graph.
  - Store in system database.
- Based on the information obtained from the previous step, determine the number of active interfaces for $Router_n$.
  - Denote the number of active interfaces by the number K. Let k be an integer such that $1 \leq k \leq K$.
- For $k = 1 \ldots K$ do the following:
  - Get the name of the router and the specific interface number that this link terminates.
  - Store this information in the system database.

Figure 7.3 shows an example wireline network consisting of five routers with six interconnecting links.

The network dependency model for the network in Figure 7.3 can be represented as follows. Following the 5-tuple convention described above for denoting the details of a router, and assuming that the total number of interfaces (including both active and passive interfaces) on each router equals five and the router type is denoted by the string "*rtr-type*," the network dependency model $NEDM$ can be represented as follows: $NEDM = (NE, DL)$, where NE represents the set of network elements (routers, switches) and DL represents the set of direct links between pairs of network elements. Here

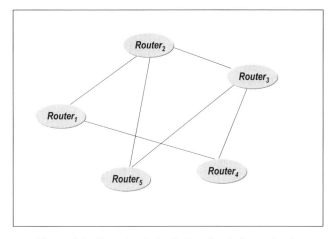

Figure 7.3. Example: A simple 5-node wireline network.

$$NE = \{\mathbf{R}^1_{<\text{rtr\_type}, 2, 3, 2, 3>}, \mathbf{R}^2_{<\text{rtr\_type}, 3, 2, 3, 2>}, \mathbf{R}^3_{<\text{rtr\_type}, 3, 2, 3, 2>}, \mathbf{R}^4_{<\text{rtr\_type}, 2, 3, 2, 3>},$$

$$\mathbf{R}^5_{<\text{rtr\_type}, 2, 3, 2, 3>}\};$$

$$DL = \{L_{R1\_R2}, L_{R1\_R4}, L_{R2\_R1}, L_{R2\_R3}, L_{R2\_R5}, L_{R3\_R5}, L_{R3\_R2}, L_{R3\_R4}, L_{R4\_R1},$$

$$L_{R4\_R3}, L_{R5\_R2}, L_{R5\_R3}\};$$

the superscript on R denotes the router number; and L_{Rx_Ry} represents a link between *Router$_x$* and *Router$_y$*. Note that the link speed is also stored along with the link endpoints as part of the information in an *NEDM*. Also notice that L_{R1_R2} and L_{R2_R1} have been listed explicitly to accommodate asymmetric links (although in most traditional wireline networks, bidirectional links are usually symmetric).

An observant reader will quickly draw the parallel between a *NEDM* and the network topology map. Recall that the network topology map is essentially a map that captures node connectivity in a network. This analogy is indeed valid but with the following caveats. Unlike a network topology map in which no detailed information about the node types is maintained, a *NEDM* constructed for fault diagnosis is designed to maintain details both about the node type and the link type, as mentioned in the preceding paragraph, along with their interconnectivity pattern. For example, while a network topology map may merely store the direct link between *Router$_1$*, and *Router$_2$*, the *NEDM* will store more information pertaining to *Router$_1$*, *Router$_2$*, and their interconnecting link. In particular, as described above, the *NEDM* will store the 5-tuple $\mathbf{R}^i_{<T, NAI, NPI, NAO, NPA>}$ for each of the routers 1 through 5, as well as their interconnecting links.

Thus in essence, the *NEDM* contains information about the network topology (i.e., which node has a direct link with which other node in the network), the capabilities of each of the network nodes (in terms of the types of router and the input/output interfaces), and the link speeds of the direct links that interconnect any two nodes. With the help of such a *NEDM* and fault diagnosis rules, the fault management component can then perform root cause analysis, as explained next in Section 7.4.1.3.

Summarized below are the salient points associated with *NEDMs* for wireline networks:

1. The *NEDM* can be thought of as a graph that captures the physical layer relationships between the various network nodes, by (a) using vertices to represent the network elements (e.g., routers, switches) and (b) using edges to represent the direct communication link (fiber, twisted-pair, etc.) between any two vertices.

2. The connectivity between the vertices (or network elements) is known *a priori* and is relatively static. The reason that connectivity is not completely static is because it may happen that backup equipment comes on-line, as a consequence of self-healing based on automatic protection switching (as also discussed in the context of traditional self-healing mechanisms in wireline networks in Section 7.5 (Step 3.1.1.1) and Section 7.5.1), which causes a change in the associated *NEDM*.

3. The properties of the *NEDM* are known and fixed. For example, the router types do not change, and neither do the wireline link speeds. Furthermore, as noted in the previous paragraph, there is almost static connectivity between network elements. Thus the dependencies in terms of the edge definitions between two vertices (i.e., links between two routers) also do not change frequently. Network links may change when network expansion takes place—for example, when new cables are being laid—but this is a rare event and is typically accompanied by a great deal of manual planning, which includes updates to dependency graphs.

4. *NEDM* construction is a fairly static process for wireline systems. Changes to the *NEDM* are typically performed on rare occasions as a result of planned network expansions.

5. Wireline *NEDMs* are better suited to handle:

   a. Hard failures (e.g., cable disconnect, equipment malfunction) rather than soft failures (performance threshold violations). This is further explained in the concluding portion of the next section.

   b. Single fault types or, at most, multiple instances of the same fault type, rather than multiple heterogeneous network faults.

7.4.1.3 Root Cause Analysis in Wireline Networks—Example. This section is devoted to illustrating via a simple example how the *NEDMs* described in the preceding section can assist with fault diagnosis in wireline networks. More specifically, the topological dependencies captured via the *NEDMs* described in the preceding section, together with a set of *a priori* rules that prescribe a set of actions based on the *NEDM* information, are used in analyzing faults and diagnosing the root cause of an observed problem as described in this section. The contents of this subsection are organized as follows. Bearing in mind the salient properties of the *NEDMs* for wireline systems (summarized in Section 7.4.1.2), the focus will be on the subset of faults that can be handled via these *NEDMs*, namely the class of hard failures (e.g., equipment failures). This is followed by a brief discussion on possible self-healing actions that can be taken. The subsection concludes with a short qualitative discussion of why the *NEDMs* as defined in Section 7.4.1.2 for traditional wireline systems cannot handle a broader spectrum of failures encompassing a wide variety of soft failures—that is, performance-related failures, as well as multiple failures occurring simultaneously, and thus paves the way for discussions on MANET root cause analysis in Section 7.4.2.

Given that routes between sources and destinations in wireline networks are relatively stable, ATM [Wu and Yoshikai, 1997] permanent virtual circuits (PVCs) and switched virtual circuits (SVCs) are widely used in these networks, with these PVCs and SVCs being established *a priori*, during network planning stages. Another technology that is used is MPLS [Rosen et al., 2001], where label switched paths (LSPs) are also defined during network planning. In the example that follows, a wireline network is depicted that is assumed to use either ATM PVCs or SVCs, or MPLS LSPs.

Figure 7.4 illustrates a simple five-node network similar to the network in Figure 7.3. The network figure on top labeled "Before Failure" is an illustration of the network under normal operational conditions, when no failure has occurred. The two figures on the bottom labeled "Restoration via APS-based self-healing" and "Restoration via Reconfiguration" are illustrations of the network after the failure has occurred and two possible corrective actions are taken, as explained in Step 3.1.1.1 and Step 3.1.1.2.2 below. The failure scenario that is being illustrated is failure of the transmitter on $Router_5$ on the link between $Router_5$ and $Router_2$. The following steps will occur to accomplish fault detection, root cause analysis, and self-healing.

1. An alarm is issued by the element management system at $Router_5$, indicating a transmitter failure.
2. At the same time, the host attached to $Router_5$ notices that it is not receiving any packets from the host attached to $Router_1$ and issues a trouble ticket.
3. The fault management system receives the alarm from $Router_5$. Note that the fault management system is not shown in the figure, but it is assumed to reside in some centralized location. Additionally, the trouble ticket also reaches the fault management system. To reduce time lags, most of the current wireline fault management systems do not necessarily wait for trouble tickets to be issued, but rather automatically trigger predefined rules that use the information in a NEDM to attempt self-healing

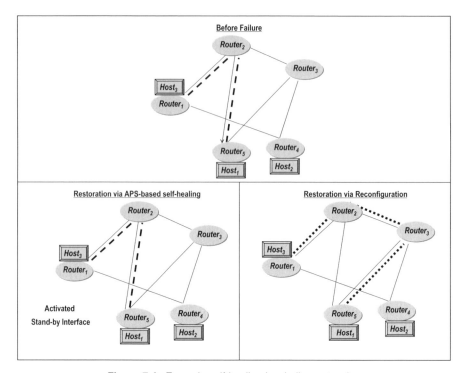

Figure 7.4. Example self-healing in wireline networks.

as described below in Step 3.1. If the automatic recovery action fails, then a ticket requesting manual assistance is opened. There may also be some cases where the end customers notice the problem before the system generates a ticket and call the service provider personnel to open a trouble ticket. In such a case, the fault management subsystem will try to initiate an automatic recovery action, and, should the problem be resolved, it will send a clear notification to the trouble ticket system.

3.1 The fault management system uses the *NEDM* information together with a set of predefined fault diagnosis/self-healing rules to automatically attempt a self-healing action. For the example cited above, the following set of rules (described as a set of steps in the discussion below) will be invoked as part of fault-diagnosis and self-healing:

3.1.1 Check to see if there is a spare transmitter that it can use for the failed interface. This is done by checking the *NEDM* and the associated router details, namely, $\mathbf{R}_{<T,\ NAI\ NPI,\ NAO,\ NPO>}$, as described in Section 7.4.1.2.

3.1.1.1. If so, perform an automatic switch-over (this is an example of APS, as also discussed in Section 7.5.1) and update the *NEDM*. This involves updating the information on the number of active and passive interfaces in the *NEDM*. The network diagram in the lower left-hand corner of Figure 7.4 labeled "Restoration via APS-based self-healing" illustrates this type of self-healing action.

Note: In this case, the communication path through the network between *Host₁* and *Host₃* will remain unchanged before and after the failure and is shown via the dashed lines in Figure 7.4. Furthermore, the switch-over occurs at the element management layer (EML) and seldom has to reach the network management layer (NML) per the TMN nomenclature.

3.1.1.2. If not, then do the following:

3.1.1.2.1. Consult the *NEDM* to see if there exists another active transmitter on another edge (i.e., another active interface on another link). In Figure 7.4, this corresponds to the link between *Router₅* and *Router₃*.

3.1.1.2.2. If yes, query the configuration database in order to determined whether another set of links can be used to recover from the fault. More specifically, based on the identified link from the dependency graph in Step 3.1.1.2.1 together with the current list of PVC, SVC, or LSP information stored in the configuration database, trigger PVC/SVC/LSP configuration, as applicable. In other words, a new PVC, SVC, or LSP, as the case may be, is configured such that the default route between *Host₁* and *Host₃* now uses *Router₃* and *Router₂* as its transit nodes. Additionally, make the following updates:

3.1.1.2.2.1. Update the list of currently active PVCs/SVCs/LSPs to reflect the addition of the new route via *Router₃*, and, to delete the route *Router₅*-*Router₂*-*Router₁*.

3.1.1.2.2.2. Update the *NEDM* to: (a) reflect the new number of active and passive interfaces associated with *Router₅* and (b) remove the link between *Router₅* and *Router₂* to reflect the fact that the link between *Router₅* and

Router$_2$ is unusable since there is no active transmitter connecting the two routers.

Note: In this case, the communication path through the network between *Host$_1$* and *Host$_3$* will change. The changed path through the network is shown via the dotted line in the lower right-hand corner of the network that is labeled "Restoration via Reconfiguration" in Figure 7.4.

3.1.1.2.3. If not, issue a trouble ticket with a requirement for manual intervention.

There are a few points to note in the above fault diagnosis, root cause analysis, and self-healing steps. In Step 3.1.1.1, after the switch-over (via APS) is made to a passive (standby) transmitter, note that there is an action to update the *NEDM*, namely, an action to reflect the latest number of active and passive transmitters within the affected router (*Router$_5$*). This type of update is considered a passive update—that is, an update that has not occurred due to changes in the network topology, but rather, an update to reflect changes within a network element. More precisely, the number of active/passive transmitters within a router is updated in the *NEDM*. Furthermore, the self-healing action that is triggered in this step is very simple: It is essentially an automatic switch-over to another active transmitter within a router, and does not involve any changes to the relationships between routers. Thus such a change does not require communications with other network management functions such as configuration management or performance management. This is in contrast to Step 3.1.1.2.2 discussed next, which requires more complex actions and updates.

Step 3.1.1.2.2, like its counterpart Step 3.1.1.1, is a step that involves updates to the *NEDM*; that is, the number of active and passive transmitters change for *Router$_5$*. However, in addition to this change, another required update to the *NEDM* is the deletion of the link in the *NEDM* between *Router$_5$* and *Router$_2$*. This link has to be deleted to reflect the absence of a functioning link between *Router$_5$* and *Router$_2$*, since there is no transmitter transmitting on the physical link any more, and thus there is no neighboring relationship between *Router$_5$* and *Router$_2$* in the NEDM. Furthermore, the self-healing actions involved in this case are more complex than in the previous case. This is because in Step 3.1.1.2.2, the fault management system essentially has to work with the configuration management system to establish new PVCs/SVCs/LSPs (shown by the dashed lines for *Host$_1$* and *Host$_3$* in Figure 7.4). Recall that *a priori* routes are typically realized via establishment of PVCs, SVCs, or MPLS LSPs during system startup and stored in the configuration database.

Changes to the routes will, however, require a certain amount of time to implement the reconfigurations and may also necessitate a HITL (human in the loop). More specifically, the route reconfigurations will involve a call to the configuration management component. Recall that due to the stovepiped nature of the FCAPS functions in traditional network management systems for wireline networks, such an interaction with the configuration management system will typically take time (i.e., will not occur seamlessly) and possibly involve a HITL. Furthermore, the actual reconfigurations themselves will likely involve human intervention, to ensure that the new PVCs/SVCs/MPLS LSPs created (shown via the dotted lines in the lower right-hand corner of the network diagram labeled "Restoration via Reconfiguration" in Figure 7.4) are indeed available for use and have no "route looping" issues.

Thus, Step 3.1.1.2.2 is a recovery action that will not be frequently performed. It is included in the discussions here to serve the following dual purpose: (a) to highlight the need for an integrated and policy-driven system network management system for MANETs and (b) to point out a potential course of action available to the current network management systems. In particular, such a reconfiguration is an excellent example that underscores the need for a policy-driven management system—in addition to, of course, the need to design an *integrated* (i.e., a non-stovepiped) network management system for MANETs. Additionally, Step 3.1.1.2.2 can be viewed as an enhancement to corrective actions that can be taken based on existing *NEDMs* in current wireline networks. However, even with such enhancements, the current fault diagnosis and root cause analysis techniques cannot be used for MANET environments, since they require fundamental changes (e.g., *NEDMs* built to handle uncertainty, integrated network management systems driven by "smart" policies that can adapt to uncertainties about the underlying networks, etc.), as discussed in Section 7.4.2.

Another observation to be made is that in general the above *NEDM* and recovery steps can also be applied to tackling the problem of a fiber cut between a pair of routers—for example, *Router₅* and *Router₂* in Figure 7.4. The details of some of the steps will obviously differ. For example, instead of an alarm to indicate a failed transmitter, the fault management system will generate a "loss of signal" to indicate that the underlying fiber is not operational. Furthermore, this loss of signal will be generated by the element management systems at both *Router₅* and *Router₂*. The fault management system will then use this information, along with the *NEDMs*, to determine that the link between *Router₅* and *Router₂* is unusable. Using this information, the fault management will then skip Step 3.1.1 (i.e., the step in which it looks for a spare transmitter to use, since in this case the cable itself is broken) and attempt to perform Step 3.1.1.2.1 (i.e., look for an alternate PVC/SVC/LSP to use), unless of course there is a spare cable available—which is not usually the case due to the expenses involved.

Finally, before concluding this section, the following points should be observed. It is easy to see why the *NEDM* as described in Section 7.4.1.2, and its usage as described in this section, cannot handle soft failures. For example, consider the scenario where the link between $Router_5$ and $Router_2$ becomes congested, and Hosts 1 and 2 complain about bad service. When a trouble ticket is issued, there is no information available for the fault management system to perform root cause analysis, other than the trouble ticket indicating that there is a service problem. This is because the *NEDM* as discussed in Section 7.4.1.2 does not store nor convey any problem other than a physical problem (hard failure-related). While this is a very simple example—that is, it represents only a single soft failure, whereas in realistic MANETs, multiple simultaneous failures can occur—it is very easy to see why the techniques in place for wireline systems need significant changes in order to be used in MANETs. This is described in detail in the next section.

7.4.2   Root Cause Analysis for MANETs

7.4.2.1   *Brief Overview of Need for Enhanced Root Cause Analysis in MANETs.*   The approach described in the previous section for root cause analysis for traditional networks is inadequate for MANETs for a number of reasons, which are enumerated here. First, while wireline networks have the concept of "core" and "edge" nodes with core nodes being solely responsible for network services and edge nodes being the interface nodes between end systems (hosts) and the underlying network, in MANETs, every node is both a core node (in terms of the network services it provides) and an edge node (in terms of housing an application/user/host). Consequently, while fault diagnosis is typically restricted to the lowest layers of the protocol stack in wireline networks, in MANETs the fault diagnosis has to extend to all of the layers of the protocol stack. For example, it is very common to have problems at the network layer of the stack in MANETs, while the network layer in wireline networks is relatively stable. Thus while the dependency models for traditional wireline systems are built with a focus on capturing details of only the lower layers of the stack (e.g., only the physical layer), MANETs require a more complex dependency model that keeps track of the entire protocol stack. More specifically, item 1 in the summary portion of Section 7.4.1.2 therefore becomes a significant shortcoming and the NEDMs need to be considerably enhanced for use in fault diagnosis/root cause analysis in MANETs.

Second, whereas existing techniques frequently use a deterministic model and assume that all dependencies and causal relationships are known with 100% certainty, the dynamic and unpredictable nature of MANETs underscores the need for stochastic dependency models. The stochastic models for MANETs, unlike their deterministic counterparts for wireline networks, should not only be resilient to incomplete and imperfect information, but should also be amenable to dynamic changes in the causal relationships, as the MANET nodes move and change positions. Additionally, even if the nodes do not move physically, they can effectively be "hidden" due to environmental conditions and/or hostile jamming. Thus the NEDMs need to be dynamic in nature, to reflect the stochastics of the underlying MANET. Thus items 2 and 3 presented in the summary portion of Section 7.4.1.2 become impediments to successful root cause analysis in MANETs, calling for enhanced models.

Furthermore, due to the unpredictable nature of MANETs and consequently the dynamics in the relationships between the elements in the NEDM, there will be frequent changes to the information captured by (and contained in) the NEDM. This in turn precludes any lengthy interactions between stovepiped configuration management and fault management functions. Rather, what is needed is an integrated network management system whose management functions work in a non-stovepiped manner. As is obvious, item 4 in the summary portion of Section 7.4.1.2 once again becomes a serious impediment and hence renders the described approach inadmissible for use as is in MANET environments.

Last, but certainly not least, while resource (e.g., bandwidth) availability problems are relatively rare in wireline networks, in contrast, MANETs are plagued by resource scarcity. Due to the dynamic nature of MANETs, available bandwidth is a fluctuating entity, with the available bandwidth swinging between extremes (namely, very low bandwidth availability to almost "full" bandwidth availability). Consequently, the assumption of only one fault existing in the system at a given time, which is often made in wireline root cause analysis techniques, is largely invalid in MANETs. More specifically, the wireless root cause analysis techniques for MANETs must have the ability to deal with the existence of *multiple simultaneous faults of different types*. Thus item 5 in the summary portion of Section 7.4.1.2 also becomes a significant impediment when one tries to apply traditional fault diagnosis/root cause analysis techniques for MANETs.

The above problems preclude the use of traditional root cause analysis models and techniques that have been defined and used in the wireline communications networking industry. Thus new root cause analysis techniques and models are required that can cope with the unique challenges listed previously in order to be useful in MANETs.

The remainder of this section provides an overview of work in the area of root cause analysis for MANETs. Section 7.4.2.2 describes a multilayer model to capture the dependencies that may exist between entities in multiple network nodes and in multiple protocol layers at those nodes. Following this, a quick review of some of the popular Bayesian algorithms that can be used to perform fault localization in MANETs is provided in Section 7.4.2.3. In particular, since MANET root cause analysis is a new and exciting field of research with a significant amount of research work that is currently active, Section 7.4.2.3 briefly outlines some of the most promising approaches.

7.4.2.2 Layered Model for Fault Diagnosis in MANETs.

As seen from the discussion in Section 7.4.1.2, fault diagnosis mechanisms utilize dependency information among the various network elements to perform root cause analysis. However, as seen from the earlier discussion, wireline *NEDMs* and traditional fault diagnosis techniques cannot be applied to MANET environments in their current form. In particular, they need to be enhanced to handle the following:

- Network uncertainty; that is, they should not require complete knowledge of network dependencies to be able to function.
- A wide range of faults; that is, they should be able to handle both hard and soft failures that extend across multiple layers of the OSI protocol stack. In other words, they should not be restricted to handling faults only associated with the lowest OSI layer—that is, the physical layer. Note that the term "layer" here refers to the OSI layered stack composed of the seven OSI layers, namely the physical layer (layer 1), data link layer (layer 2), network layer (layer 3), transport layer (layer 4), session layer (layer 5), presentation layer (layer 6), and application layer (layer 7). The reader is cautioned not to confuse this with the TMN layering which includes the Business Management Layer, Service Management Layer (SML), Network Management Layer (NML) Element Management Layer (EML), and Network Element Layer, as was discussed in Chapter 1.

Furthermore, since the upper OSI layers are necessarily dependent on the lower OSI layers to provide services, the *NEDMs* discussed in Section 7.4.1.2 need to incorporate this dependency information and make it available for fault diagnosis in MANETs. Additionally, the *NEDMs* for MANETs have to also be augmented to provide information about relationships that exist *across* multiple MANET nodes.

This section describes a multilayer model that captures dependencies that may exist both across multiple (OSI) protocol layers within a MANET node (referred to as "vertical dependencies") and between entities in different nodes within the underlying MANET (referred to as "horizontal dependencies"). It is based on the models in Steinder and Sethi [2001] which are enhancements to the model in Gopal [2000].

The "vertical dependencies" in the multilayer dependency model constructed for fault diagnosis purposes are very similar in philosophy to the layered OSI model, wherein services offered by a given OSI layer are a combination of the protocol(s) implemented at that layer and the services provided by it and the lower OSI layer(s). To better understand such a vertical dependency model, consider the relationship between the lowest OSI layer, namely the physical layer (also denoted by OSI_L1) and the OSI layer just above it—that is, the data link layer (DLL)/Media Access Control (MAC) layer (also denoted by OSI_L2). As its name indicates, the DLL/MAC layer on a node offers *access to the underlying medium* so that the node can send the information over the physical layer. This channel access service offered by

OSI_L2 is in turn implemented via a specific MAC protocol (e.g., TDMA, CSMA etc.) and the services offered by the underlying physical layer that actually transmits the information bits. The physical layer assumes the form of signal in space in wireless MANETs and the form of a cable, twisted pair, and so on, in wireline networks.

The layered dependency model in OSI is in fact a recursive dependency model. Moving up the protocol stack, for example, consider the services offered by Layer 3, namely the network layer (also denoted by *OSI_L3*). The services offered by *OSI_L3* are typically network connectivity-related services, such as establishing a communications path/routing path between two hosts by using routing protocols. Some examples of MANET routing protocols include the Optimized Link State Routing (OLSR) protocol and the Ad hoc On demand Distance Vector (AODV) protocol, to name just a few. These routing protocols help to establish communications paths by constructing routing tables. Routing protocols typically exchange routing messages to discover neighbors and learn the current network topology, in order to establish communications paths. For routing messages to be transmitted from one node to another, routing packets have to contend for the right to transmit data. It is here that the services of *OSI_L2* come into play. As explained in the preceding paragraph, the *OSI_L2* provides the *OSL_L3* with access to the underlying physical layer (*OSI_L1*), so that routing protocol messages can be exchanged successfully, in turn enabling the routing protocol to compute and establish routing paths between a set of interconnected nodes.

It is easy to follow this recursive relationship all the way up to the application layer (*OSI_L7*), which can be viewed as offering the use of a given network to a certain application. The *OSI_L7* implements such a service via application level protocols that the user can use to invoke information exchange through the underlying network—be it a wireless MANET or a wireline network—and invokes the services of the layer below, namely the presentation layer (*OSI_L6*), to help format the user-generated messages for presentation to the underlying network.

Such a recursive dependency relationship can be harnessed for fault management purposes as well. Figure 7.5 presents one such recursive dependency model based on the work in Steinder and Sethi [2001]. The dependency model in Figure 7.5 is adapted from the work in Steinder and Sethi [2001] to highlight a set of sample horizontal and vertical dependencies in a MANET.

In Figure 7.5, each of the nodes labeled Node(a), Node(b), Node(c), and Node(d) represents a MANET node. This figure shows three OSI layers (for illustration purposes) within Node(a) and Node(d), and one OSI layer within Node(b) and Node(c). The topmost layer in Node(a) and Node(d), labeled Layer (L + 1), is shown to correspond to Layer 4 in the OSI model, namely the Transport Layer. The services offered by this layer typically correspond to either bit-oriented or connection-oriented transport of the application packets. Examples of protocols used at this layer include UDP and TCP. The layer labeled Layer (L) in Node(a) and Node(d) beneath the topmost layer in

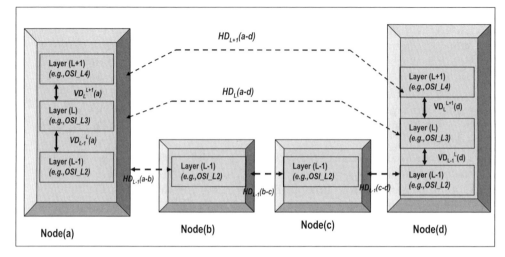

Figure 7.5. Layered dependency model.

Figure 7.5 is the Network Layer (corresponding to OSI Layer 3). As mentioned earlier, this layer offers networking services to the layer above it. The layer at the bottom (labeled (L-1)) in all of the nodes in Figure 7.5 corresponds to the Data Link/MAC layer (OSI Layer 2).

The above vertical dependency model based on the OSI concept of layering captures dependencies *within* a node. While incorporation of vertical dependencies is a much needed enhancement of the *NEDMs* discussed in Section 7.4.1.2, in order for the dependency models to be useful for MANET fault diagnosis, they must also capture relationships and dependencies *across* nodes. This is because the horizontal dependencies provide important "network-wide" information since they capture the relationships between the intermediate nodes that carry an end-to-end service.

Since MANETs are almost never a fully connected mesh, application services originating from a source node will likely use the services of one or more *intermediate* nodes before reaching the intended destination node. The term *transit* nodes will be also be used interchangeably to refer to the intermediate nodes in this book. The horizontal dependencies capture the peering relationships between these transit nodes along an application's end-to-end path. Observe that failures, either soft (i.e., failure to provide an *a priori* "agreed-to" level of service as captured via a QoS requirement) or hard (i.e., element malfunction), can occur anywhere along the path or on an intermediate node. In turn, such a failure will affect the end-to-end service and manifest itself as a fault symptom for further diagnosis by the fault management system.

Figure 7.5 shows how to accommodate both vertical and horizontal dependencies, within and across communication nodes respectively. In particular, in Figure 7.5, solid lines are used to denote vertical dependencies within a node, and dashed lines are used to illustrate the horizontal peering relationships across nodes. With regard to nomenclature, the vertical dependency that exists between Layer (L + 1) and Layer (L) within Node(a) is denoted by VD_L^{L+1} *(a)*. Similarly, the horizontal dependency that exists at layer (L − 1) across two nodes, say Node(a) and Node(b), is denoted by $HD_{L-1}(a − b)$. Now, with regard to the nature of the horizontal dependencies across nodes, observe that the horizontal dependency between two nodes in a MANET can be either logical or physical. For example, Node(a) and Node(d) are considered to be logical neighbors (peers) from the standpoint of the OSI Layer (L + 1) (i.e., Layer 4) due to the fact that the TCP/UDP packets originating from Node(a) terminate in Node(d) and are not inspected in the intermediate nodes Node(b) and Node(c). Thus from the standpoint of the TCP Layer (Layer 4) at Node(a), its peering node is a corresponding TCP entity at Layer 4 in Node(d). This horizontal peering dependency (i.e., logical neighboring relationship) at Layer (L + 1) between Node(a) and Node(d) is denoted by $HD_{L+1}(a − d)$. However, the horizontal dependency at Layer (L − 1), i.e., $HD_{L-1}(a − d)$ between Node(a) and Node(b) can be viewed as a physical peering relationship from the viewpoint of OSI_L2, since the layer 2 frames from Node(a) will be sent to Node(b) before being passed on to their next hop, OSI_L2 physical peer Node(c). Similar recursive observations exist with regard to the fact that the Layer (L + 1) at source node Node(a) will have a logical peer in the horizontal direction with destination node Node(d), where Layer (L + 1) can represent the Session, followed by the Presentation and then by the Application layers (i.e., *OSI_L5* through *OSI_L7*, respectively).

Next, the following provides an example of how such a layered dependency model can be used for fault analysis in MANETs. Consider for example the following problem: "Service between Node(a) and Node(d) experiences unusually large delays but is not completely down." Since the service is not completely down, a hard failure that is typically associated with a network element failure is ruled out. This corresponds to a soft failure.

To diagnose the cause of such a soft failure, one needs to examine the layered dependency model. The following are simple examples that show how fault diagnosis can occur based on knowledge of vertical and horizontal dependencies.

- If it is known that the routers implement a congestion indication mechanism (e.g., Explicit Congestion Notification (ECN)) and the ECN bit has been observed to be set in received packets, then the unacceptable end-to-end delay experienced between Node(a) and Node(d) is probably due to packet drops at the network layer buffers. Examples of corrective actions in this cause could include dynamic processor rescheduling by readjusting the weights for Weighted Fair Queuing (WFQ) at Layer 3 (if WFQ is used), or changing the token rates for the Layer 3 queues if token-based queuing is used, or reconfiguration of routing policies so that less congested paths are used between the source and destination.

- If it is known that the routers implement a congestion indication mechanism (e.g., Explicit Congestion Notification (ECN)) and the ECN bit has not been set, then the unacceptable end-to-end delay experienced between Node(a) and Node(d) is probably due to packet drops due to environmental conditions—for example, high attenuation between two links at the physical layer. In this case, no amount of processor rescheduling will help; however, if the environmental conditions are localized, then a corrective action could involve reconfiguring the network to use alternate paths that minimize the effect of the environmental problems.

Having qualitatively illustrated the use of such a layered dependency model, the next step is to point out the key differentiators with regard to the dependency models for fault diagnoses in MANETs versus their wireline counterparts.

- The dependency models for wireline systems typically only incorporate the details associated with the lowest OSI layer (i.e., the physical layer) and abstract layers 2 and above. Consequently, fault diagnosis in wireline systems is focused on analyzing problems at the physical layer. In contrast, in MANETs, the dependency models and fault diagnosis techniques must extend all the way up to OSI layer 7, as mentioned earlier in this section.
- The horizontal dependencies in wireline systems are relatively static, while the horizontal dependencies in MANETs are dynamic, with the relationships between nodes (i.e., the horizontal dependencies) varying over time. More specifically, while a small subset of the dependencies will be static in MANETs (e.g., the vertical dependency model that corresponds to the lower OSI layers), there is potentially a large set of dependencies that will be dynamic. For example, horizontal dependencies change as nodes move; vertical dependencies change as new services are deployed on demand and/or new transport protocols are used based on the application type; for example, non-real-time video may use streaming protocols like TCP, whereas real-time short-lived video sessions may use RTP/UDP-like protocols. Thus, the fault diagnosis techniques work with "snapshots" of the dependency models that get updated over time.
- Last, and most importantly, while the dependency models themselves are deterministic in the case of wireline networks (i.e., the information linking failures and symptoms is assumed to be known with 100% certainty), in MANETs there is a high degree of uncertainty in linking failures and symptoms, due to the plain fact that the information provided to the fault diagnosis system may itself be uncertain.

Now, once a layered dependency model is constructed—or, more precisely, in the case of MANETs, once a snapshot of a dependency model for the network is constructed—it can be transformed into a *belief network* for use in fault diagnosis as done in Steinder and Sethi [2002a, 2004b]. In particular, as described in Steinder and Sethi [2002a, 2004b], a belief network is a directed acyclic graph (G,P), where

- $G = (V, E)$ is a directed acyclic graph.
- $v_i \in V$ is a binary valued random variable and represents an event in the network.
- $(v_i, v_j) \in E$ represents a causal relationship; that is, v_i *causes* v_j.
- $P = \{P_i\}$, where P_i is a probability associated with variable v_i.

Associated with belief networks are *evidence sets*, where an *evidence set* denotes a partial assignment of values to variables represented in a belief network. With the understanding that a variable v_i in a belief network represents an event, an evidence set essentially denotes assignments of a binary random variable (i.e., 1 (true) or 0 (false)) to the events, based on an observation of events that have occurred. Observe also that this assignment, to begin with, will be partial (i.e., will not cover all of the possible events). Next, given an evidence set, belief networks can be used to make two basic queries related to fault correlation as also described in Steinder and Sethi [2002a, 2004b]:

- Belief assessment—that is, the task of computing the probability that some variable (i.e., events) possesses certain values (i.e., belief that the event is indeed is causing the observed symptom/problem).
- Most probable explanation (MPE)—that is, the task of finding a complete assignment of values (beliefs) to variables (events) in a way that best explains the observed evidence.

7.4.2.3 Quick Overview of Bayesian Approaches to MANET Root Cause Analysis.
Once the belief networks along with evidence sets have been created, they can be used to make queries such as Belief Assessment and MPE (most probable explanation) as mentioned in the preceding section, and thus be used in fault diagnosis. Recall that fault diagnosis is the process of ascertaining with a high degree of confidence (ideally with probability 1, i.e., with absolute certainty) the root cause of an observed symptom, in order to take subsequent corrective (self-healing) actions. The belief assessment and MPE essentially help in diagnosing the root cause by assigning a high degree of confidence (a high probability "value") to an event (i.e., a "variable" in the language of belief networks) and thus hone in on the root cause.

However, both belief assessment and MPE tasks are NP-hard [Garey and Johnson, 1979] in general belief networks. Therefore approximations have to be performed in order to make them feasible for the types of MANETs under consideration. Some of the most actively researched approximations include *Iterative Belief Propagation* and *Iterative MPE in polytrees* [Steinder and Sethi

2002a,b, 2004b] that are based on adaptations of Pearl's iterative algorithms [Pearl, 1988]. While the details of the approximations are outside the focus of this book, these approximations in essence provide for tractability by using simplified belief networks that (a) employ binary-valued random variables and (b) associate an inhibitory factor with every cause of a single effect (noisy-OR model) and assume that they are all independent. The approximations introduced by (a) and (b) essentially help contain the search space (or state space) of the system, in turn paving the path to "tractable" analysis.

For example, use of binary-valued random variables versus a continuum of random variables helps contain the values (0 or 1 in this case) that can be associated with a given event (i.e., a belief network variable). Note that this approximation was introduced while defining belief networks and evidence sets by Steinder and Sethi [2002a, 2004b] in the preceding subsection. Likewise, by introducing the independence approximation in (b), the combinatorics associated with the problem space—namely, the set of possible states that the system can be in due to a given fault and a set of observed symptoms—is reduced significantly. In essence, all of the above approximations help achieve polynomial time complexity and tractability by placing less stringent demands in terms of computation time and memory requirements.

Recent research by Steinder and Sethi [2003, 2004a] has investigated various techniques to deal with the complexity and tractability issues that typically plague the use of Bayesian inferencing techniques for large systems. In particular, in Steinder and Sethi [2003, 2004a], the authors propose a technique called IHU (incremental hypothesis updating) wherein they operate directly on the dependency graphs and propose "smart" heuristics to perform fault diagnosis quickly, thus paving the way for fault diagnosis in large-scale networks. Via popularly used metrics such as detection rates and false positive rates, they illustrate the efficacy of their approach for wireless ad hoc networks. For example, they show that their algorithms achieve high detection rates, where the detection rate metric denotes the percentage of faults that occurred in the network in a given experiment that were detected by an algorithm; and also low false-positive rates, where the false-positive rate metric is defined as the percentage of faults proposed by an algorithm that were not actually occurring in the network in a considered experiment; that is, they were false fault hypotheses. More recently, the authors in Natu and Sethi [2005] provide an extension of the IHU algorithm in the specific context of MANETs to include the concept of temporal correlation which allows the algorithm to account for changes in the dependency graphs over time. For a comprehensive survey of fault localization techniques, the reader is referred to Steinder and Sethi [2004c].

Finally, before concluding this section, one other technique called the *bucket elimination* technique merits mention. The bucket elimination technique has been widely used with belief networks and is one of the most popular algorithmic frameworks for computing queries such as MPE using belief networks and is capable of computing optimal solutions. The optimality, however, comes at the expense of complexity. The bucket elimination technique has exponential complexity in time and space (memory demands). Therefore, despite its potential to produce optimal (high confidence) solutions, the bucket elimination algorithm in its current form is unsuitable for large-scale MANETs.

7.5 SELF-HEALING

The necessity for self-healing mechanisms that provide service survivability (uninterrupted service) amidst random/sporadic failures in MANETs is both critical and obvious. However, as also mentioned in the introduction, the dynamic and unpredictable nature of MANETs, coupled with the fact that mobile ad hoc network resources such as bandwidth are scarce, renders the design of appropriate self-healing mechanisms extremely challenging. This section discusses adaptive policy-driven self-healing mechanisms that can be used to recover automatically from network problems. It begins with a brief overview of existing self-healing mechanisms in communications networks in Section 7.5.1, followed by a description of adaptive policy-driven self-healing mechanisms in Section 7.5.2.

7.5.1 Self-Healing for Traditional Networks

Well-known self-healing systems that are used in practice today exist largely in the context of wireline telecommunications systems. Widespread and commercial deployment of self-healing in wireless networks is still in its early stages [Kant et al., 2002; Kant and Chen, 2005; Sadler et al., 2005]. A majority of the well-known self-healing mechanisms in existence today function within a single layer of the OSI stack, namely the physical layer. In addition, they employ the philosophy of resource redundancy for handling failures; in other words, certain pieces of equipment are dedicated to providing backup functions solely for the purpose of restoration/failure handling. This results in a requirement for standby equipment that remains idle during normal working conditions. In the event of a network equipment failure, an automatic switchover occurs from the malfunctioning or failed equipment to the standby equipment. For this reason, this self-healing mechanism is also referred to as an APS (Automatic Protection Switching) mechanism. Examples of widely used APS self-healing mechanisms in use today are SONET self-healing rings, Bidirectional Line Switched Rings (BLSRs) [Wu, 1992], and so on. Another specific example of APS in traditional wireline networks was illustrated in Section 7.4.1.3 (Step 3.1.1.1); the example there discussed routers that are deliberately not run at full capacity and that have several interfaces that are kept idle intentionally to serve as standby in case of failures.

While the main advantage of an APS philosophy is the excellent response time (with a typical restoration delay of less than 50 msec), it has several disadvantages from the perspective of MANETs:

- First, APS is very resource-intensive, since by definition it works on the principle of resource redundancy. In fact, traditional APS systems not only devote spare interfaces but require idle equipment (e.g., spare routers) to serve as a standby in the event of an automatic switch-over requirement. This is definitely an issue for MANETs, since they are typically resource-constrained.
- Second, APS is limited to handling hard failures (i.e., equipment failures) alone. Due to the stochastic nature of MANETs, it is anticipated that a substantial number of failures will fall under the category of *soft* failures—that is, failures such as excessive performance degradation, poor signal quality, and so on.

APS mechanisms deal with hard failures such as equipment malfunctions; in contrast, *soft* failures (i.e., performance degradation due to lack of network resources) are typically handled in today's telecommunications networks by overprovisioning. Given the falling costs of fiber-optic cables and the increasing processing power of routing equipment, it is more cost-effective for service providers to install excess capacity than it is for them to dedicate human resources to handling soft failures when they occur. However, overprovisioning is not an option for MANETs because wireless bandwidth is typically scarce, and overprovisioning is not cost-effective.

Thus, while APS-based mechanisms and overprovisioning are suitable for the environment that they were tailored for (i.e., telecommunications networks), they are not applicable to the unpredictable and resource-constrained MANET environment. The important point to note here is that the scarcity of MANET resources implies that not all network problems can be fixed; rather, the goal of self-healing mechanisms for MANETs is to maximize the overall network performance, keeping in mind that not all resource demands can always be satisfied. In light of the diverse QoS and survivability requirements of MANET applications, it may even be unnecessary to provide a uniform degree of restoration (e.g., restoration delays of less than 50 msec) to all of the applications. For example, while mission-critical and delay-sensitive applications may require stringent restoration delay guarantees, loss-sensitive applications, such as imagery information transfer, may be satisfied with delayed but guaranteed restoration. Hence, self-healing mechanisms need to take into account the different survivability requirements of MANET applications in order to be effective.

7.5.2   Self-Healing Operations for MANETs

This section describes a policy-driven self-healing strategy for MANETs. The basic idea is simple and has been discussed in earlier parts of this book: Policies are used to define responses that should be taken by the network management system to address a given root cause. Thus policies are used to tie together the monitoring aspect of the network management system (via fault and performance management) and the configuration and security aspects (via configuration and security management, respectively).

Using the terminology defined earlier in Chapter 5, ECA (Event–Condition–Action) policies are used to define responses to network events as follows. In a nutshell:

- The "Event" portion of the policy contains a root cause event.
- The "Condition" portion of the policy contains conditions that need to be checked prior to performing a corrective action, if any.
- The "Action" portion of the policy invokes either (a) the necessary reconfiguration actions for dealing with the identified root cause or (b) additional diagnostic tests to determine what corrective action should be taken.

This is further illustrated in Figure 7.6. Note that the root cause itself could be either a hard failure (such as failed equipment) or a soft failure (such as excessive network congestion). Soft failures can also be handled via judicious Quality of Service (QoS) mechanisms, described in Chapter 8.

Policies are an ideal way to execute self-healing, since they provide a convenient, human-friendly mechanism for controlling the behavior of the network management system. Network operators can specify how they want their

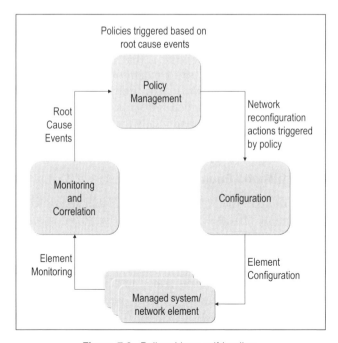

Figure 7.6. Policy-driven self-healing.

network to be reconfigured in response to identified root causes, and this reconfiguration is then automatically implemented as needed by a Policy Decision Point (see Chapter 2 for a definition of Policy Decision Point). The next section provides realistic examples of fault scenarios that occur in MANETs, along with a description of appropriate policies for self-healing.

7.6 FAULT SCENARIOS

This section describes a number of failure scenarios that can potentially occur in MANETs. The scenarios have been carefully selected to provide insights into both (a) the wide variety of failures that can occur and (b) the challenges in locating and fixing the problems associated with the corresponding failures. To this end, the scenarios illustrated below range from *hard failures* (see Scenario 1 in Section 7.6.1) within a single subnet or routing domain, to a complex combination of *hard* and *soft failures* that impact several subnets (see Scenario 3 in Section 7.6.3). Additionally, a scenario to highlight the impact of security-related anomalies (e.g., intrusion attacks that manifest themselves as soft failures) on the fault management system is presented in Section 7.6.5. While the class of "hard" failures is typically thought of as being "easy to fix" and deterministic in nature in wireline systems, the illustrative examples in this section show how even such hard failures transform themselves to becoming more difficult to analyze and not as easy to fix in MANETs. Needless to say, the challenges in diagnosing and fixing *soft failures* in such MANETs are only exacerbated.

The remainder of this section is organized as follows. Each scenario is presented in a separate subsection below. Note that while the size of the networks has been kept small in order to ease explanation, the concepts and steps are applicable regardless of the network size. For each scenario, a high-level description of the scenario is provided first, followed by the processing flow, the steps for detection of the fault, and finally the corrective action(s) that can be performed.

7.6.1 Scenario 1: Radio Fault

In this scenario, a radio in the MANET fails; the other radios learn about this problem via the routing protocol, and reroute traffic. In this scenario, all of the nodes belong to one subnet or routing domain. As more traffic goes through the remaining nodes, congestion is observed.

Figure 7.7 shows the failure scenario described above. Recall that a hierarchical management hierarchy was described in Chapter 5, where every node hosts a network management system (NMS), and nodes are organized into *clusters* for management purposes. In Figure 7.7, Node 3 hosts the NMS that manages the NMSs on the other nodes in this group (labeled as "cluster head" in the figure).

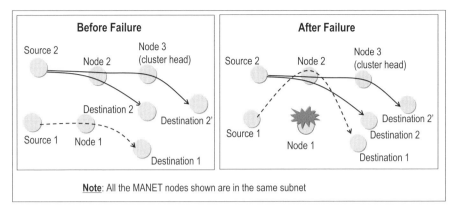

Figure 7.7. Radio failure scenario.

The left-hand side portion of Figure 7.7 illustrates the normal mode of operations, with the right-hand side portion illustrating what happens in the network after a failure has occurred but before any corrective action is actually taken. For the purpose of illustration, three sets of flows are shown, with Source 1 initiating one flow between itself and Destination 1 and Source 2 initiating two flows, one each to Destination 2 and Destination 2′ in Figure 7.7. As will be seen via this example, even with a relatively simple network and flows such as those illustrated, the failure of one network element has the potential to impact all of the network services (despite the mesh nature of the interconnection) and also involves relatively complex fault diagnosis and corrective actions.

In the normal operating mode, the paths taken by these flows are illustrated on the left-hand side of Figure 7.7; namely, Source 1 uses Node 1 as its transit node whereas Source 2 uses Node 2 as its transit node. Note that the actual paths taken between a source and destination can vary over time, depending on the conditions of the underlying network and the metrics used by the routing protocol in computing the reachability information. Furthermore, for purposes of illustration, the intensity of the flows from Source 2 is assumed to be such that it saturates the links that the flows take en route to their destination. The following paragraph discusses what can potentially happen in such a case if the radio in Node 1 fails.

If Node 1 fails, the routing protocol will learn about this failure and adapt by routing the flows from Source 1 around the failed node, which in this case will result in flows being routed via Node 2. However, since the links surrounding Node 2 are already saturated, this additional load will soon lead to a soft failure; in other words, the links will become congested, causing unacceptable loss and delay. Such a situation is captured in the right-hand portion of Figure 7.7. The soft failure will impact *all* of the flows in this scenario, despite the fact that there was a single radio failure.

Sections 7.6.1.1 through 7.6.1.3 discuss the processing flows, fault detection techniques, and corrective actions, respectively, that can be taken by a policy-driven fault management system for this failure scenario.

7.6.1.1 *Processing Flow.* The processing flow for this scenario is as follows:

- *Node 2* detects high packet loss on its wireless interfaces, indicating congestion. Since this congestion could affect other nodes, *Node 2* notifies the NMS on *Node 3* (the cluster head).
- The NMS on *Node 3* realizes that *Node 1* is unreachable. This information is based on the lack of communication from the NMS on *Node 1*; typically, each NMS will regularly communicate a status ("heartbeat") to its cluster head. Alternatively, routing traps from the black routers of the neighbor nodes could also be used to determine that Node 1 is unreachable. Recall that the network management system is housed on the unencrypted (red) side, whereas the failure shown in this scenario corresponds to a radio failure on the encrypted (black) side of the network, with very limited management information exchange between the two. In this scenario, the information exchange includes a selected set of routing traps across the two network segments.
- A QoS problem is detected by the performance management systems on *Source 1* and *Source 2*, and both of these nodes (*Source 1* and *Source 2*) send "QoS failure" notifications to the NMS in Node 3.
- The NMS performs graph analysis via its horizontal dependency model to determine whether any of the probable paths between the troubled node pairs go through *Node 2*, which is known to be congested (see first bullet above). Observe that the horizontal dependency model constructed by the NMS is essentially based on *inferences* that it makes about the network connectivity, since it is housed on the red (unencrypted) side and the failures have occurred in the black (encrypted) side of the network. Assuming that a limited amount of "routing information leakage," as mentioned earlier, is allowed from the black to the red side, the NMS is able to construct the dependency models that it needs to perform fault diagnosis. Of course, it should be noted that since these topology inferences are probabilistic in nature, the resulting outcome (from the root cause analysis) will be a probabilistic one; that is, with some certainty $(0.0 < x \leq 1.0)$, the given solution is a root cause. In this specific example, due to the small size of the network, the fault diagnosis narrows down the root cause and the NMS determines that the soft failures are most likely caused by the congestion (QoS-related problems) on *Node 2*.
- Based on the timing of the radio failure on *Node 1* and the congestion on *Node 2*, and the QoS failure notifications received from *Source 1* and *Source 2*, the fault management system within the NMS on *Node 3* deter-

mines that the congestion on *Node 2* is probably caused by the radio failure on *Node 1*.

In summary, the fault management system interacts with the performance management system—since one of the symptoms here was a Quality of Service problem (i.e., soft failure)—to help in its root cause analysis phase, and it diagnoses the root cause of the congestion problem (soft failure) as being due to a radio (hard) failure.

*7.6.1.2 **Fault Detection.*** The following observable events occur in the network/NMS:

- An actionable alert is generated on the node with the radio failure (*Node 1*).
- Performance alerts are generated by *Source 1* and *Source 2*.
- An alert is generated by the NMS at *Node 3* (the cluster head) upon diagnosing the root cause of the congestion problem, with a level of impact that is indicative of the "criticality" of the problem. The level of impact is a policy-driven input, since it is an artifact that is dependent on the nature of the problem and the nature of the mission that is being supported by the given MANET.

*7.6.1.3 **Fault Correction.*** Fault correction is performed as follows. In the scenario described here there is a *hard fault*; in other words, there is an equipment failure, and the failed radio must be repaired or replaced. A trouble ticket must therefore be generated to trigger this repair/replacement. However, since equipment repair is a manually intensive process and may take a long time, there is a need for alleviating the congestion problem in the meantime. The only way to achieve this is to create additional network capacity. One possible way of dynamically creating additional network capacity in MANETs is to capitalize on the mobile aspect of MANET nodes and to check whether it is possible to re-deploy nodes by moving one or more mobile nodes to the bottleneck region, thereby alleviating the prevailing congestion. In addition, based on the nature of the terrain, there may arise situations that preclude moving ground nodes but may allow bringing in aerial nodes. For example, MANETs deployed in swampy or mountainous terrains often use unmanned aerial nodes (UANs) to provide relay capabilities, wherein the UANs are directed to fly to different locations based on communications requirements. The following flow illustrates the fault correction process carried out at the cluster head (*Node 3*) that can be used to correct the above congestion problem:

- Upon determining the root cause of the problem as described above, the root cause is sent by Fault Management to Policy Management.

- Policy Management receives the root cause event and retrieves the relevant policy. In this case, it is assumed that a policy has been specified that indicates the corrective action to be taken as follows:
  - *Event*: Root Cause indicating (i) location and identity of failed node and (ii) need for immediate capacity replacement to relieve congestion on *Node 2*.
  - *Condition*: UAN asset is available in the vicinity of the failed node.
  - *Action*: Send configuration directive to the UAN asset to move to an appropriate location near the failed element and to configure itself with the appropriate frequencies and subnet parameters to be able to relay traffic for the congested portion of the network.

  Note: Policies such as the one above will typically be derived from *a priori* performance information and/or simulation studies and will undergo analysis in an appropriate testbed prior to deployment.
- The above policy is triggered and executed, resulting in an alleviation of congestion at the site of the failed element.

In the case where there is no UAN available to move into position as a relay for the congested portion of the network, other policies could be in place that attempt different solutions. Examples of alternate mechanisms include (a) moving another ground vehicle closer to the area of congestion so that some flows are routed through this ground vehicle or (b) throttling low-priority flows so that the most critical messages get through. While some of the above alternatives (e.g., dynamic throttling based on priorities) may be automated, some other alternatives, such as moving another ground node, may need to involve human decision-making. This is because it may not be practical to move ground vehicles to different locations, due to a combination of reasons including terrain impediments (such as the presence of swamps/rivers) or simply because of the fact that these other ground nodes may be critical to performing a certain set of functions in their current location.

7.6.2 Scenario 2: Environment-Related Problem

In this scenario, a MANET link becomes unavailable due to environment-related issues; for example, the nodes move into a mountainous terrain whereby two adjacent nodes cannot communicate (see nodes *Source 1* and *Node 1* in Figure 7.8). All of the nodes in this scenario, like the previous case, belong to a single subnet. The nodes that were originally using this link quickly learn about this problem via the routing protocol, and they reestablish new routes over other available links in order to maintain network connectivity and uninterrupted information transfer through the network. However, due to the resultant reduction in network resources because of unavailable links, the remaining network links become congested, causing a soft failure for all of the services using the remainder of the links. Observe that this scenario is

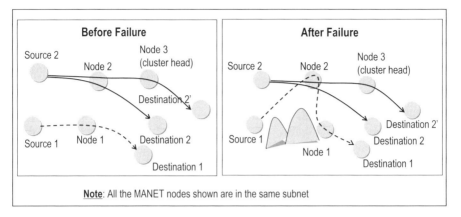

Figure 7.8. Soft failure due to environmental problems.

very similar to the scenario in the previous section, but with a different root cause. Whereas in the previous scenario the root cause was due to a hard radio failure, in this case, the root cause actually happens to be a soft failure; in other words, no network element fails, but a link becomes unavailable due to the nature of the current terrain.

Figure 7.8 illustrates the scenario corresponding to a soft failure due to environmental problems in a MANET. The left portion of Figure 7.8 illustrates the situation under normal operational conditions and is labeled "Before Failure." The right portion of the figure denotes the situation after the failure has occurred, but before any corrective action is taken. As before, the node labeled *Node 3* is the cluster head and contains the network management system (NMS) that manages the set of nodes shown in this example. The explanation for this scenario is largely similar to that in Section 7.6.1, with the caveat that the actual cause in this case is not a hard radio failure, but instead, a soft failure caused by an intervening mountain that effectively disconnects *Source 1* and *Node 1*. However, although the description of the scenario from the perspective of the user is similar to that in Section 7.6.1, the processing flow, fault detection process, and corrective actions differ, as explained below.

7.6.2.1 Processing Flow. The processing flow for the fault scenario in Figure 7.8 is as follows:

- *Node 2* detects high packet loss on the encrypted (black-side) interfaces indicating congestion. Since this congestion could affect other nodes, *Node 2* notifies the NMS on the cluster head node (*Node 3*).
- In the meantime, the fault management function on the NMS on *Node 3* learns about the loss of physical connectivity between *Source 1* and *Node 1*. As mentioned earlier in Section 7.6.1.1, the NMS infers a certain horizontal dependency model (i.e., physical connectivity) via a limited amount

of information obtained from the black network for monitoring purposes.

- A QoS problem is detected by the performance management systems on *Source 1* and *Source 2*, and both of these nodes (*Source 1* and *Source 2*) send "QoS failure" notifications to the NMS in *Node 3*.
- Based on the recent history of physical connectivity and the timing of congestion on *Node 2*, the fault management component determines that the congestion is probably caused by the loss of physical connectivity. This conclusion is arrived at by performing graph analysis to determine whether any of the probable paths between the troubled node pairs go through *Node 2*, which is known to be congested. As before, the inferences obtained via this analysis are probabilistic in nature, due to the stochastic nature of the input (dependency) models. Once again, due to the small size of the network, the root cause analysis narrows down the root cause of the problem, and the NMS determines that the soft failures are most probably caused by the congestion on *Node 2*.

7.6.2.2 Fault Detection. The following observable events occur in the network/NMS:

- Performance alerts are generated by *Source 1* and *Source 2*.
- An alert is generated by the NMS on *Node 3* upon diagnosing the root cause of the congestion problem, with a certain level of impact that is indicative of the "criticality" of the problem.

7.6.2.3 Fault Correction. Since the scenario described here has the same impact as the previous one (in Section 7.6.1), the corrective action processing is performed in the same way as was described in Section 7.6.1.3. The only possible difference here is that if the terrain change is known to be temporary (e.g., if the maneuver plan for the MANET is known and it can be predicted that the terrain blockage is temporary as all nodes will soon have moved past the blockage area), then it may be preferable to take no action to correct the problem. The processing on the cluster head (*Node 3*) is shown below.

- Upon determining the root cause of the problem as described above, the root cause is sent by Fault Management to Policy Management.
- Policy Management receives the root cause event and retrieves the relevant policy. In this case, it is assumed that a policy has been specified that indicates the corrective action to be taken as follows:
  - *Event*: Root Cause indicating (i) loss of physical connectivity between *Source 1* and *Node 1* and (ii) need for immediate capacity replacement to relieve congestion on *Node 2*.
  - *Condition*: Maneuver plan indicates that terrain will not change in the near future.

- *Action*: Send configuration directive to the UAN asset to move to an appropriate location near the two disconnected nodes and to configure itself with the appropriate frequencies and subnet parameters to be able to relay traffic for the congested portion of the network.
- The above policy is triggered and executed, resulting in an alleviation of congestion at *Node 2*.

7.6.3   Scenario 3: Faults with Impacts on Multiple Subnets

The preceding two sections presented a couple of simple failure scenarios with the scope limited to just one subnet, or routing domain. This section presents a slightly more complex example that covers multiple types of failures (i.e., both hard and soft) and spans multiple subnets.

The scenario illustrated in Figures 7.9 and 7.10 captures multiple problems in Subnet 2, including radio failure on one node and loss of physical connectivity due to the presence of a mountain. The network radios learn about these problems via the routing protocol, and reroute traffic. As more traffic goes through *Node 3*, it becomes congested.

As illustrated in Figure 7.9, which shows the "normal" mode of operation, the situation before the failure is as follows. *Source 0* (marked "S0") in Subnet 2 communicates with *Destination 0* (marked "D0") also in Subnet 2 via *Node 1* (marked "1"). *Source 1* (marked "S1") communicates with *Destination 2* (marked "D2") via a set of transit nodes (namely, nodes *Border 1 (B1), Node 3, Node 4* and *Border 2 (B2)*) as shown in Figure 7.9, spanning subnets 1

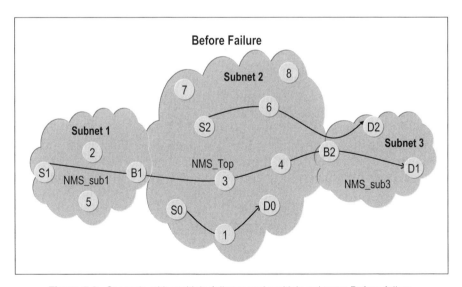

Figure 7.9. Scenario with multiple failures and multiple subnets: Before failure.

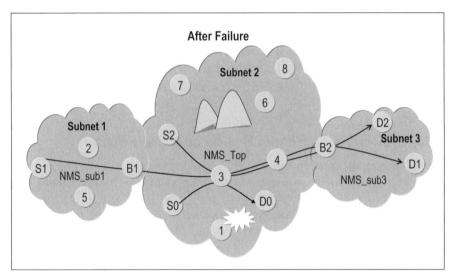

Figure 7.10. Scenario with multiple failures and multiple subnets: After failure.

through 3. Likewise, *Source 2 (S2)* originates flows from Subnet 2 and communicates with *Destination 2 (D2)* in Subnet 3 via nodes *Node 6* and *Border 2 (B2)*. These flows have been selected to illustrate flow origination and termination in different subnets as well as traversal of multiple subnets.

As mentioned earlier, the hierarchical network management system requires that nodes be clustered for management purposes. In this scenario, each subnet forms its own cluster. Thus, each subnet has a cluster head; the NMS on this cluster head manages the NMSs of all the other nodes in the subnet. The NMSs at the cluster heads are marked as *NMS_sub1, NMS_Top* and *NMS_sub3*, respectively, for subnets 1, 2, and 3 in Figure 7.9. Observe that *NMS_1* is located on *Node 5, NMS_Top* is located on *Node 3* and *NMS_sub3* is located on node *Destination 1*. Here *NMS_Top* is the "parent" of *NMS_sub1* and *NMS_sub3* in the network management hierarchy (refer to Chapter 5 for a discussion of the management hierarchy).

Figure 7.10 illustrates the situation when the following multiple failures have occurred but no corrective actions have been taken. First, a radio in *Node 1* in Subnet 2 fails. In addition, the nodes in the upper portion of Subnet 2 are disconnected from the remainder of the network due to mountainous terrain.

Having briefly introduced the scenario, the processing flows, fault detection, and potential corrective actions are discussed next in Sections 7.6.3.1, 7.6.3.2 and 7.6.3.3, respectively.

7.6.3.1 Processing Flow. The processing flow for the fault scenarios in Figures 7.9 and 7.10 are as follows:

- *Source 1* observes service quality degradation. This is a soft failure that is detected by the performance management component at *Source 1* based on QoS degradation. This is due to its traffic being routed through *Node 3*, which is already congested. The NMS on *Source 1* therefore notifies its parent NMS (i.e., *NMS_sub1*) about this soft failure.
- For similar reasons, *Source 0* and *Source 2* observe service quality degradation. The NMSs on *Source 0* and *Source 2* therefore notify their parent NMS (i.e., *NMS_Top*) about these soft failures.
- *NMS_sub1* determines that there is no performance degradation problem for traffic within its own subnet. Consequently, it determines that the soft failure problems are probably caused by problems in other subnets. It therefore notifies its parent, *NMS_Top*, about the soft failure between Subnets 1 and 3.
- In the meantime, *NMS_Top* learns about the loss of physical connectivity between nodes in its subnet (caused by mountainous terrain).
- *NMS_Top* performs graph analysis, as explained in the context of the preceding scenarios, for the probable paths between Subnet 1 and Subnet 3. In this scenario, *NMS_Top* determines that the only path between Subnets 1 and 3 is going through Subnet 2.
- *NMS_Top* investigates the communications between *Border 1 (B1)* and *Border 2 (B2)*. It performs graph analysis to determine if any of the probable paths between *Border 1* and *Border 2* go through *Node 3*. For this example, *NMS_Top* determines that the congestion on *Node 3* would probably impact the communications between *Border 1* and *Border 2*.
- Finally, *NMS_Top* determines that the soft failures reported by Subnet 1 are, with a high degree of certainty, related to the problems in Subnet 2. Note that the probability becomes lower if there are other subnets through which subnet 1 can communicate to subnet 3.

7.6.3.2 Fault Detection. The steps for fault detection are similar as in the earlier examples. Here the following steps occur:

- An alert is issued with regard to the hard (radio) failure on *Node 1* along with an associated policy-driven critical impact level.
- Performance alerts are generated by *Source 0, Source 1* and *Source 2*.
- An alert is generated by *NMS_Top* on *Node 3* upon diagnosing the root cause of the congestion problem, with a certain level of impact that is indicative of the "criticality" of the problem.

7.6.3.3 Fault Correction. Since the faults described in this section are a combination of faults described earlier in Sections 7.6.1 and 7.6.2, the two policies described earlier in Sections 7.6.1.3 and 7.6.2.3 are used for fault correction.

7.6.4   Scenario 4: Soft Failure Due to Unanticipated Overload

In all of the above scenarios, the failures were due to different types of exter-
nally induced anomalies—namely, a radio failure or environment-related
problem that in turn manifested itself as a soft failure in the underlying
MANET. This section provides a MANET failure scenario where there is no
hard failure or environmental problem, but yet there is a soft failure manifested by performance degradation. Such a soft failure can be caused by an
unexpected yet sustained high variance of input traffic. Note that MANETs
often have to be deployed on-the-fly where there exist many unknowns in
terms of the actual amount of load that will be placed on the network. This
can happen, for example, when beginning a new operation, such as deployment
of a network in a disaster area on short notice. In such scenarios, only a limited
amount of *a priori* traffic engineering can be performed before network
deployment.

Furthermore, due to node mobility, the physical locations of the MANET
nodes can vary. This in turn has a myriad of implications. First, the routing
protocols have to find new routes whenever their neighborhood relationships
change as a consequence of node mobility. Next, there is a high probability of
nodes and paths becoming "bottleneck" elements, due to the fact that they
may be the only "closest" neighbor for many other nodes. All of these in turn
would contribute to the unanticipated yet sustained overload condition in the
MANET. Contrast this with wireline networks where there is usually a large
pre-deployment period to estimate projected network traffic. Furthermore,
with static nodes, the routing protocols establish stable neighborhood relationships; that is, the neighbors of a node do not change much over time, unless
there are node failures.

Figure 7.11 illustrates a soft failure caused by unexpected traffic congestion
via a small (in terms of number of nodes) single-domain network. Recall, as
mentioned earlier, that the size of the network used does not matter to illustrate the processing flows, fault detection, and correction explanations in Sections 7.6.4.1 through 7.6.4.3. Of course, the time to perform the root cause
analysis will increase as the problem space—namely the number of nodes and

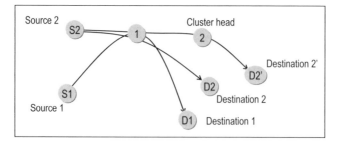

Figure 7.11. Soft failure due to sustained unanticipated congestion.

consequently the number of states that the underlying MANET can be in—becomes large, as is to be expected.

This scenario can be described as follows. Subsequent to deploying a MANET, node mobility can result in a large number of nodes being in close proximity to one particular node. This results in paths through this node becoming the "most favorable routes" through the network. In such a case, the links to and from this node, and the node itself, soon become bottlenecks and start becoming congested. Consequently, the applications that utilize these nodes and links start to receive very poor service quality, eventually resulting in performance violations and soft failures. In Figure 7.11, the node labeled 1 becomes such a bottleneck, subsequently leading to soft failures for all of the nodes that use it as their transit node through the MANET, as shown in the figure.

7.6.4.1 *Processing Flow.* The processing flow for the fault scenario illustrated in Figure 7.11 is as follows:

- *Node 1* detects high packet loss indicating congestion. Since this congestion could affect other nodes, the NMS on *Node 1* notifies the NMS on the cluster head node (*Node 2* in the figure).
- The performance management function on *Source 1* notices numerous performance violations in terms of threshold crossings, for flows from itself to destination node, *Destination 1*.

  Note: The "threshold" values are policy configured variables, to provide for maximum flexibility.
- *Source 1* sends a notification to the NMS on the cluster head (located on the node *Node 1*) indicating a soft failure (QoS problem) between itself and *Destination 1*.
- Similarly, *Source 2* sends a notification to the NMS on the cluster head indicating a soft failure (QoS problem) between itself and *Destinations 2* and *2′*.
- The cluster head NMS on *Node 2* performs graph analysis (as described while explaining the earlier fault scenarios in this chapter) to determine if any of the probable paths between the troubled node pairs go through *Node 1*.
- Once the cluster head NMS determines that the bottleneck is indeed *Node 1* (by applying graph analysis via the horizontal dependency model), it also talks to the NMS in *Node 1* to verify that there is no anomalous behavior in the traffic pattern caused, for example, by security break-ins. It does this by communicating with the security management component of the NMS in *Node 1*. This enables the NMS to distinguish between unexpected but nonmalicious traffic congestion, and a malicious traffic surge, which is the subject of the next scenario in Section 7.6.5. Note that this last step reinforces the need for an integrated NMS for ad hoc

networks, as was discussed in Section 7.2.1.2. Consequently, armed with a positive outcome, the NMS determines that the soft failures are *most likely caused* by the congestion on *Node 1*.

Note: The emphasis is on the italicized phrase "*most likely caused.*" As mentioned earlier, due the stochastic nature of MANETs, the dependency model as well as the inputs (in terms of the dependency relationships) will be probabilistic in nature. Thus the outcome of the root cause analysis will also be probabilistic in nature.

7.6.4.2 Fault Detection. The following action occurs as part of fault detection for the scenario in Figure 7.11:

- If the problem persists—that is, the congestion continues for a certain amount of time exceeding the threshold (with the threshold being a policy-driven parameter)—the NMS on *Node 2* generates an alert to indicate that there is a network-wide service degradation. Recall that as illustrated in Figure 7.11, all of the network nodes use *Node 1* as a transit node, which is why the congestion is a network-wide soft failure.
- Performance alerts are generated by Source 1 and Source 2.

7.6.4.3 Fault Correction. The root cause for the above scenario is the fact that there is too much traffic transiting one particular network node. Even though this root cause is different from those in Scenarios 1 and 2 (in Sections 7.6.1 and 7.6.2, respectively), the possible solution is the same: Create more network capacity by bringing in another node (such as a UAN), or move an existing node to affect routing so that fewer flows get routed through the bottleneck node. Policies similar to those described for Scenarios 1 and 2 can therefore be used to reconfigure the network accordingly.

7.6.5 Scenario 5: Soft Failure Due to DOS Attack

This section is devoted to illustrating the effect of a security-related attack and the consequent service failure on fault diagnosis/root cause analysis and policy-driven self-healing. The specific example considered (for illustration purposes) is a denial-of-service (DOS) attack and its impact on service failures and fault diagnosis followed by self-healing. A DOS attack is, in essence, caused by a malicious host flooding the network with unnecessary packets, so that network resources are wasted in processing useless information and hence become partially unavailable for mission-critical applications. Note that while a DOS attack can begin in a fairly benign manner by being contained to one network segment (domain), it can soon spread, if the border nodes are attacked, to other network segments, and very soon cripple an entire network. Such a DOS attack essentially results in soft failures throughout the network.

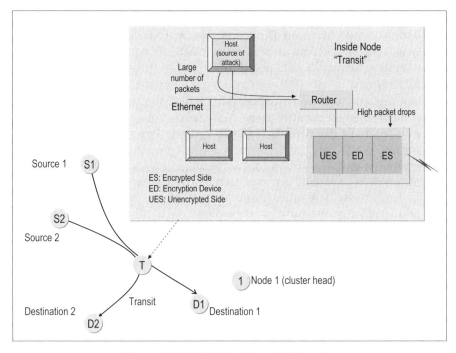

Figure 7.12. Soft failure due to DOS attack.

Figure 7.12 illustrates a DOS attack scenario and the processing performed by fault diagnosis and root cause analysis. As shown in this figure, a DOS attack originates from a host on the node labeled *Transit*. A host on the node *Transit* starts to maliciously inject copious amounts of traffic destined to all of the other nodes in the network. Recall that the hosts within a MANET node are typically connected to a router via a high-speed (usually a Gigabit Ethernet) LAN. The router has one or more wireless interfaces that are connected to the wireless network, and they have speeds that are at least an order of magnitude or more lower than the high-speed LAN.

Thus, while the host may be successful in getting its packets across the inbound interface to its router, the outbound interface from the router will now be overwhelmed, due to the limited speed on its outbound (wireless) link. Such an attack causes the node *Transit* to drop a large number of legitimate packets that are going through it—that is, using it as a transit node to reach their destination nodes. Consequently, legitimate applications are denied service, which is manifested as a soft failure.

7.6.5.1 Processing Flow. The processing flow for this scenario is described below. Before enumerating the steps, the following assumptions have been made for this scenario:

- A host on node *Transit* has been compromised, thus turning malicious.
- The Network Intrusion Detection System (NIDS) within the security management component within the NMS on node *Transit* detects an anomaly—unusually high packets from one of its hosts—and diagnoses a DOS attack.
- Routing traffic is not dropped. This is because network management, routing, and other control-plane messages will, in most practical networks, be allocated a distinct DSCP (DiffServ Code Point) that is not available for user traffic (see the Performance Management chapter for a detailed treatment of QoS). This is done for multiple reasons, the most important of which is that network management and control must be accorded higher priority than other traffic, so as to ensure that critical monitoring and configuration traffic is transmitted in a timely fashion over the network, even in heavy load conditions.

The processing flow for the scenario in Figure 7.12 is as follows:

- The node *Transit* detects high packet loss on its wireless interface(s). Additionally, the NIDS component on *Transit* detects a probable intrusion from a host on its LAN, which is flooding the network with a high volume of traffic. It notifies the fault management component within the local NMS, which correlates these alerts with a low level of certainty, and determines that the root cause of these two problems might be a DOS attack originating from the local platform.
- Since *Transit* cannot be completely certain of the root cause of the problem, it notifies the NMS on *Node 1*, which hosts the cluster head, of the high packet loss and intrusion, as well as the possible root cause.
- As the packets from *Source 1* to *Destination 1* and *Source 1* to *Node 1* are dropped by *Transit*, the performance management component in *Source 1* observes a performance threshold violation, which, if left uncorrected, could potentially lead to a soft failure.
- *Source 1* therefore sends a performance-related alarm to the NMS on *Node 1*.
- Similarly, *Source 2* also observes a similar service performance violation threshold crossing and sends a similar alarm to the NMS on *Node 1*.
- The NMS on *Node 1* determines that the soft failure notifications are probably related to the high packet loss on *Transit*.
- The NMS on *Node 1* correlates all of the available information and determines with a high degree of certainty that the problem is caused by a malicious DOS attack.

7.6.5.2 *Fault Detection.* The following observable events occur in the network/NMS:

- An intrusion alert is generated by node *Transit*. This is correlated with the local high packet loss, and a root cause with low certainty is sent to the NMS on *Node 1*, indicating the possibility of a DOS attack.
- Performance alerts are generated by *Source 1* and *Source 2*.
- An alert is generated by the NMS on *Node 1* upon diagnosing that a DOS attack is the root cause of the congestion problem, with a certain level of impact that is indicative of the "criticality" of the problem.

7.6.5.3   *Fault Correction.* As described earlier, the fault management component on *Node 1* generates a root cause for the DOS attack. The following steps are executed to correct the problem:

- The root cause event triggers a corrective action via a predefined policy that is executed by the policy management component on the *Transit* node. The policy is of the following form:
  - *Event*: Root cause indicating location of DOS attack and identity of malicious attacking node. Note that this event is sent by *Node 1* to node *Transit*.
  - *Condition*: None.
  - *Action*: Send configuration directive to the configuration management component on the *Transit* node, directing it to disable the port on the router to which the malicious host is connected, thereby shutting off the source of the DOS attack.
- The above policy is triggered and executed, resulting in cutting off the source of the DOS attack.

Regarding the condition of the above policy: Even though no condition is included as part of the above policy, in practice it may be necessary to include one or more conditions in the policy. As an example, a network manager could use the concept of a "threat level" for the network, which could indicate various situations such as the existence of a high or low potential for network attack, and so on. The current network threat level could be included as part of the diagnosis process, or as part of the policy to control whether a certain action is very conservative (e.g., err in favor of caution, at the risk of cutting off legitimate network users) or less conservative (e.g., do not cut off all traffic from a given node). Examples of two such policies are given below.

- Conservative policy:
  - *Event*: Root cause indicating location of DOS attack and identity of malicious attacking node.
  - *Condition*: Current network threat level is high.
  - *Action*: Send configuration directive to the configuration management component on the *Transit* node, directing it to disable all traffic entering

the network from the *Transit* node, thereby shutting off any possible source of the DOS attack.

- Less conservative policy:
  - *Event*: Root cause indicating location of DOS attack and identity of malicious attacking node.
  - *Condition*: Current network threat level is low.
  - *Action*: Send configuration directive to the configuration management component on the *Transit* node, directing it to disable the port on the router to which the malicious host is connected, thereby shutting off the source of the DOS attack.

7.7 SUMMARY

This chapter discussed issues, models, and operations related to MANET fault management. Although the TMN fault management operations models provide a good foundation for defining and understanding fault management operations, they need to be extended for MANET environments, which—unlike their wireline counterparts—are more stochastic in nature. This implies that there is a need for an adaptive and integrated (i.e., non-stovepiped) policy-driven MANET fault management model that can cope with the unpredictable dynamics and soft failures that occur in MANETs.

The major fault management tasks in MANETs include network monitoring, root cause analysis, and self-healing. Root cause analysis (RCA) and self-healing tasks become especially complex in MANETs, and traditional RCA and self-healing techniques that have been successfully applied to wireline networks cannot be used unmodified in MANET environments. There is a need for RCA techniques that can handle uncertainties in the inputs to the RCA algorithms themselves. Due to the random and unpredictable nature of the MANETs themselves, probabilistic RCA algorithms—as opposed to the traditionally used static rule-based approaches that assume availability of complete information—hold better potential for assisting with fault diagnosis in MANETs. A multilayer approach that incorporates both vertical and horizontal dependencies, both within a MANET node and across MANET nodes, is a promising approach to assist with the RCA. Note that RCA for MANETs is a new and exciting field and is an active area of ongoing research.

Once the root cause of a problem has been identified, an important function of the fault management task is to provide self-healing capabilities to the extent possible. While self-healing—that is, providing seamless restoration of affected services—is a challenging and complex task in any network, the challenges only become exacerbated in MANETs. To illustrate this, several examples were presented, and the steps involved in fault diagnosis, self-healing, and the use of a policy-driven mechanism to assist with self-healing operations in

MANETs were presented. The example fault scenarios and policies presented did not provide an enumeration of all possible failure and self-healing types, but rather served as illustrative examples to highlight both the complexity of actions as well as the need for a policy-driven integrated set of operations involving performance, configuration, and security management tasks.

8

PERFORMANCE MANAGEMENT

This chapter discusses performance management functions for MANETs. It begins with a high-level overview of performance management for MANETs in Section 8.1. Section 8.2 discusses performance management operations process models, and it explains what aspects of the TMN performance operations process models can be leveraged for MANETs. Two key performance management functions in MANETs include network performance monitoring and providing Quality of Service (QoS) assurances. They are discussed in the remainder of this chapter. In particular, Section 8.3 provides a brief overview of monitoring in MANETs, with a focus on performance management issues, and Section 8.4 discusses approaches for providing end-to-end service quality in MANETs. Section 8.5 concludes with a brief summary.

8.1 OVERVIEW

Performance and fault management operations in MANETs are very intricately tied together, as mentioned in the previous chapter. In particular, recall that due to the stochastic nature of MANETs, a significant percentage of "failures" in MANETs are associated with *soft* failures. This in turn requires deviation from the traditional stovepiped FCAPS implementation of network management operations. More specifically, the performance management process needs to work in close cooperation with the fault management process to accomplish management of MANETs.

However, the complex nature of MANETs requires sophisticated techniques to establish, maintain, and manage these networks. Therefore, just as with the fault management processes and operations models (discussed in the previous chapter), performance management processes and operations models

Policy-Driven Mobile Ad hoc Network Management, by Ritu Chadha and Latha Kant
Copyright © 2008 John Wiley & Sons, Inc.

are more complex in MANETs than in traditional wireline networks. This chapter is devoted to discussing the following:

- The performance management process in MANETs.
- How performance management complements fault management.
- How performance management inter-works in a seamless manner with fault management to deliver uninterrupted end-to-end communications in the complex MANET environment.

The reader should bear in mind the fundamental connection between fault and performance management processes in MANETs. The complexity of MANETs calls for a sophisticated coupling of the fault and performance operations models and processes, in turn underscoring the need for a dynamic policy-based network management paradigm to link them, as discussed in the preceding chapters.

Finally, recall that the complexity of MANETs arises in part due to the inherent randomness of the operational environment in which MANETs are expected to operate, and in part due to the flexibility and dynamism that MANETs offer to network users. This flexibility and dynamism is indeed a double-edged sword because while MANETs offer immense potential via the flexibility of on-the-fly deployment and dynamic reconfiguration to support a wide spectrum of user requirements, these operations not only require a high degree of automation, but also require "intelligence" to make the most appropriate decisions in terms of deployment, reconfigurability, and sustenance. This, in turn, highlights the need for an integrated set of FCAPS operations processes driven by a dynamic policy-based network management paradigm that can take automated decisions based on the dynamics of the underlying network, as opposed to a set of static rules.

8.2   PERFORMANCE MANAGEMENT FUNCTIONS AND OPERATIONS PROCESS MODELS

As in the previous chapter on Fault Management, this section first provides an overview of the TMN operations process model for performance management, following by a model that is specifically designed for MANETs based on the functions that need to be performed by the performance management task.

8.2.1   TMN Performance Management Operations Process Models

The traditional definition of performance management as defined by the TMN includes the process of collecting performance data and performance-related alarms, filtering duplicate alarms, determining current system performance to detect performance deviations or threshold violations, and providing appropri-

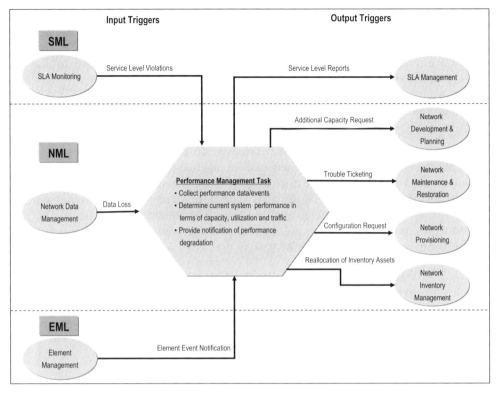

Figure 8.1. TMN performance management operations model.

ate notifications to users. Figure 8.1 provides a schematic diagram for performance management based on the TMN guidelines.

The left side of Figure 8.1 represents inputs, or triggers, to the performance management task, while the right side of the figure represents outputs, or actions taken by the performance management task. The circles on the left and right represent, respectively, (a) the sources for the input triggers and (b) the destinations for the output actions. The core responsibilities of the performance management task are summarized in the central hexagonal box, which in essence are:

- Collecting performance data/events from the network.
- Processing the performance information to check for threshold violations.
- Providing appropriate notifications.

The acronyms EML, NML, and SML denote the Element Management Layer, Network Management Layer, and Service Management Layer respectively, as mentioned earlier.

Just as for the fault management model, observe that while the general performance management definitions and functionalities in Figure 8.1 are broad enough to apply to either wireline or wireless networks, the operations model in Figure 8.1 is formulated with a wireline environment in mind, for the same reasons as outlined in the previous chapter.

8.2.2 Performance Management Operations Models for MANETs

As pointed out previously, the TMN models for performance management provide a good starting point for defining performance management models while designing a policy-based network management system for MANETs. As mentioned in Chapter 7 while discussing Fault Management operations process models, Performance Management is very closely coupled to the remainder of the network management operations models, namely, fault management, configuration management, and security management. The key functions of performance management for MANETs are:

- Performance anomaly detection, via periodic polling of network elements and detection of threshold crossings.
- Generation of performance violation alerts that are used as inputs to Fault Management to perform integrated correlation with fault, configuration and security events, and root cause analysis.
- End-to-end quality of service (QoS) assurance.

Figure 8.2 provides a schematic of the operations model for the performance management process within a policy-based network management system. The figure highlights the following:

- The performance management operations process model has strong dependencies on, and is integrated with, the operations process models for fault, configuration, and security, as captured via the input and output triggers.
- Performance violations are inputs to an integrated root cause analysis engine, implemented within the fault management process (as discussed in Chapter 7 while discussing fault diagnosis).
- A policy management component ties together the FCAPS functions by seamlessly integrating these functions via policies.
- A critical new function for QoS assurance is included that is responsible for providing QoS assurances to applications with differing priorities using the MANET.

8.3 NETWORK MONITORING

Just as the fault management function is responsible for monitoring the network to detect network fault conditions, the performance management

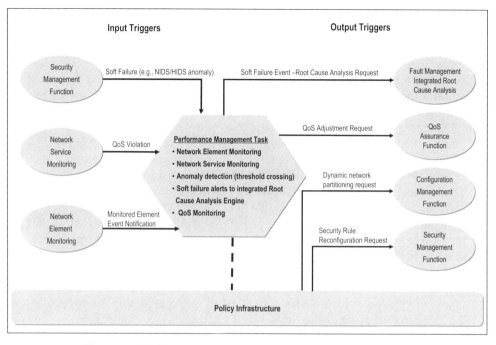

Figure 8.2. Performance management operations model for MANETs.

function is responsible for monitoring the performance of the network on a continuous basis—that is, even when the network is functioning normally. Thus, unlike fault management, which may not have much to do when there are no network faults, performance management must periodically poll network elements and services in order to ensure that key performance parameters are within expected bounds.

8.3.1 Collection of Performance Statistics for Network Elements

The collection of performance statistics from network elements is typically performed via SNMP. For example, the routers that connect a wired LAN and the wireless medium have buffers that can exhibit a sustained overflow, indicating abnormal or unanticipated operational conditions (note that the buffers are normally sized such that they do not overflow under normal operational conditions). An important task of performance management is to monitor the routers to detect such sustained overflows, and report to an integrated fault/performance/security correlation engine, to help with root cause analysis associated with soft failures. Observe that sustained buffer overflows at a router will result in packet loss, in turn leading to a disruption in service.

Another example of a network element in MANETs whose performance needs to be monitored is the device that performs encryption, such as an IPSec or HAIPE device (see Chapter 9 for a discussion of IPSec and HAIPE). A key performance metric associated with encryption devices is latency—that is, the time lag that is introduced due to the encryptions and decryptions performed. In order to provide timely service, it is imperative that these latencies be low. More specifically, the performance management function needs to monitor the latency performance and report any anomaly, so that timely root cause analysis can be performed.

Note the following important distinction with respect to collecting performance statistics for network elements in MANETs versus wireline networks. While there exists a network element at the OSI physical layer (namely, the cable or wire) in wireline networks, in MANETs, there is no such analogue. Thus, whereas in wireline networks the performance management function can poll the transceiver at the end of a fiber to detect the presence or absence of a signal, no such network element with an associated signal exists at the physical layer in MANETs. Instead, the *effect* of the losses at the physical layer due to a broken link (caused by jamming, for example, or terrain obstructions such as a mountain) manifests itself as packet loss at the higher layer, which will then need to be taken into account by a root cause analysis engine to detect the source of the problem. Needless to say, pinpointing the particular faulty interconnection or link in MANETs becomes very challenging.

8.3.2   Collection of Performance Statistics for Network Services

In addition to monitoring and collecting performance statistics of individual network elements as discussed in Section 8.3.1, another important task of the performance management function is to monitor the performance (and as a consequence the "health") of a variety of MANET services. Note that the fault management function will monitor these services to ensure that the services are up and running. This can be done via regular polling of the processes that provide the services. Thus, while the fault management function monitors whether the service is "alive" or not, the performance management function monitors the "level of service" offered by a given server, thus ensuring the proper functioning of the given system. This cooperation between fault and performance management operation underscores, once again, the importance of an integrated network management system. Contrast this with the typical stovepiped implementation of NMSs for traditional wireline networks.

Below are some examples of (a) key network services that need to be in place in mobile ad hoc networking environments and (b) the role of performance management in monitoring these services.

- *Name–address Translation/Resolution Service*: This service is akin to the popularly used DNS service in wireline networks. The concept here is to associate a user-identifiable node name with a network-identifiable node name, which is used by the underlying MANET to route information packets through the network. A key performance metric associated with name–address resolution servers is the time taken for address resolution. The role of the performance management function is to monitor this service and raise an alarm if the time taken for address resolution exceeds a policy-defined threshold value.

· *Mobility Management Service*: Mobility is a fundamental characteristic of MANETs. Recall from Chapter 1 that MANETs do not have the concept of an infrastructure with some nodes being fixed and others being mobile. Instead, every node in a MANET can potentially be a mobile node. Thus, MANETs require mobility management techniques that help keep track of nodes' mobility so that any two MANET nodes can continue to remain in communication, despite random mobility. While the focus of this book is not on the details of mobility management, note that a variety of mobility management techniques are being researched for MANETs. Examples of these include SIP-based mobility management [Camarillo, 2002] and Mobile IP [Perkins, 2002]. The role of performance management here is to ensure that the mobility management service is functioning as expected. More specifically, based on the type of mobility management technique, the performance management function monitors specific performance metrics, and will generate alarms when policy-specified threshold values are exceeded. For example, if a SIP-based mobility management technique is employed, one performance metric will be the time taken to register the new IP address of a node with its SIP server, once node movement has taken place.

· *Quality of Service (QoS) Management Service*: Providing appropriate QoS to the various applications that use a MANET is an extremely important function. Recall that MANETs are expected to support a variety of applications with widely varying QoS requirements. For example, mission-critical applications (akin to platinum services in commercial networks) will require very stringent guarantees in terms of delay and loss; voice applications and other real-time applications will require tight delay guarantees but somewhat less stringent loss assurances; high-resolution image transfer applications that are non-real-time in nature can tolerate delay but are very sensitive to loss; and so on. The above requirements, coupled with the dynamic and unpredictable nature of MANETs (i.e., unpredictability both in terms of randomness in mobility and in terms of the bandwidth fluctuations due to extraneous factors such as the environment or even jamming), imply that fairly sophisticated QoS assurance mechanisms are needed that can cope with uncertainty while sustaining the required service guarantees. As will be explained in Section 8.4, this is achieved by designing an efficient, distributed QoS management solution that performs admission control. Now, since the QoS servers themselves are hosted on network elements, it is critical to ensure that both the services and the servers that host the QoS solutions are functioning as expected. The role of performance management is to monitor the QoS servers to ensure that certain performance thresholds are not crossed. Some examples of these thresholds are as follows:

- The response times of the QoS server remain below policy-defined threshold values, where the response time is the time taken to make an admission control decision (see Section 8.4).
- The number of flow preemptions remains below a policy-defined threshold.
- The number of flow request rejections remains below a policy-defined threshold.
- *Session Management Service*: Session management is another important MANET service. Session management provides the capability to locate users in a MANET, even when they move from node to node, for the purpose of establishing communications (such as voice calls or chat sessions). Users are associated with fixed user-friendly names, and the session management service is responsible for mapping these names to the IP addresses of the hosts where these users are located. When a user moves, the session management service registers the new location of the user, thus enabling other users to continue to locate this user for session establishment using the user's name. While the focus of this book is not on the details or design of session management techniques, note that it is important that the session management implementation provide the required level of service to the network users. The role of the performance management function is to monitor the associated servers by computing key performance metrics such as session setup time, time taken to register the location of a user, and so on. Policy-defined thresholds are defined for each of these metrics, and any threshold crossings are reported as alarms to the fault management function.

8.3.3 Policy-Controlled Monitoring

The previous sections described monitoring of network elements and services in MANETs. Recall that the distributed architecture of the network management system (NMS) for a MANET means that network monitoring is performed on every MANET node by the NMS instance on that node. Now, the scarcity of over-the-air bandwidth in MANETs requires that the amount of management information sent over the air needs to be kept at a minimum. Thus, although performance statistics are periodically collected on every node, they should not be sent to other nodes unless absolutely necessary. The determination of what type of information should be sent to other nodes on a periodic basis, as well as the frequency of information dissemination, is controlled via policies as follows. Policies are created that define how information should be aggregated and filtered prior to over the air dissemination. Furthermore, policies are also defined to specify how frequently this information is sent over the air. Information is therefore sent by a node to its parent node in the management hierarchy (refer to Chapter 5 for a discussion of the management hierarchy) based on these policies.

The value of using policies to specify the frequency of reporting management information is that other policies can be used to adjust this frequency *automatically*, based on the congestion status of the network. Section 8.4.4 of this chapter discusses a mechanism for QoS assurance for MANETs that derives estimates of network congestion based on throughput measurements. These estimates of network congestion can be used to throttle the frequency of reporting of management information over the air if the network is highly congested; and conversely, if the network is not congested, policies can be used to automatically increase the frequency of reporting. The relevant policies are illustrated by examples below.

- **Policy 1 (configuration policy):** Set a reporting frequency:
  - *Action*: Configure Performance Management function to report performance statistic X every 60 seconds.
- **Policy 2 (configuration policy):** Set a performance crossing threshold:
  - *Action:* Configure Performance Management function to generate a threshold crossing alert when the number of rejected flows exceeds n within time window m.
- **Policy 3 (ECA policy):** Specify an automated change in reporting frequency based on congestion status of the network. Note that the performance management function is responsible for generating a threshold crossing event based on the previous policy.
  - *Event*: Threshold crossing alert (number of rejected flows crosses policy-specified threshold).
  - *Condition*: None.
  - *Action*: Configure Performance Management function to report performance statistic X every 300 seconds.
- **Policy 4 (ECA policy):** Specify an automated change in reporting frequency back to the original value based on clearing of the previous threshold crossing alert.
  - *Event*: Clearing of the threshold crossing alert (alert for number of rejected flows crossing policy-specified threshold is cleared).
  - *Condition*: None.
  - *Action*: Configure Performance Management function to report performance statistic X every 60 seconds.

The first two policies above are configuration policies. The first one configures the performance management function to report information about a specific variable at a specified frequency. The second policy configures the value of the threshold at which the performance management function should generate a threshold crossing alert for a given congestion measure. The third policy is an

ECA policy that automatically modifies the reporting frequency based on the generation of a threshold crossing alert. Finally, the fourth policy is also an ECA policy that sets the reporting frequency back to the original value when the threshold crossing alert is cleared. Similar policies can be defined based on other performance measures and threshold crossings.

The above capability is a very powerful one, since it enables self-adjustment of the network monitoring system based on network status. The network management system can thereby adapt to changing network conditions in an automated fashion. This is a critical capability for MANETs, since network conditions are expected to vary dynamically in a MANET. The use of policies makes this automated adaptation possible.

8.4   AUTOMATED END-TO-END SERVICE QUALITY ASSURANCE IN MANETS

An important aspect of performance management in MANETs is the provision of service assurances to high-priority applications, sometimes at the expense of lower-priority applications. This section provides some background about this problem, including the particular challenges in providing end-to-end service quality in MANETs. An approach to providing quality of service using *dynamic throughput graphs* [Poylisher et al., 2006] is discussed in detail.

8.4.1   Background

Quality of Service is the ability to ensure that high-priority traffic has the highest probability of message completion to intended users, within dynamic limits of ad hoc network resources. Message completion must take into account the communication performance requirements of applications, such as low delay, low loss, or high throughput. Given the reality that the amount of traffic to be sent over a network may exceed its capacity, the need for techniques to provide end-to-end QoS assurances in MANETs is both obvious and critical. Note that wireline networks are typically overprovisioned, and they usually operate under benign operational environments as compared to MANETs; thus, these networks do not normally suffer from typical MANET problems such as unpredictable reduction in capacity due to fading, jamming, mobility, and so on. For these reasons, the issue of QoS assurance is far more simple in wireline networks than in MANETs. This critical QoS functionality is performed by the performance management process. Furthermore, due to the presence of applications that may need different levels of assurances (i.e., different types of guarantees on delay and/or loss), any efficient QoS mechanism should also be capable of providing different QoS assurances to different types of traffic. Finally, when bandwidth is scarce, applications should receive QoS based on their relative priorities.

As an example, MANETs usually support a set of applications called *mission-critical* applications, which require very stringent guarantees on delay and/or loss due to their critical nature. This is akin to a platinum service subscriber in commercial wireline networks, where the subscriber expects "top quality" service delivery (i.e., low delay and loss). On the other end of the spectrum, there exist routine messages that are sensitive to neither delay nor loss; that is, they are more tolerant of varying network delays and losses. Of course, there also exist applications, such as non-real-time image transfer, that are more sensitive to loss than delay. Each of these different types of services require different levels of service assurance.

The design of end-to-end service assurance mechanisms in MANETs poses many challenges, underlining the need for novel QoS assurance mechanisms. These challenges are highlighted next, followed by a description of a policy-based QoS solution for MANETs.

8.4.2 Challenges in Providing Quality of Service Assurances for MANETs

As was noted in Chapter 7, wireline networks largely avoid soft failures via overprovisioning of network capacity. For the same reason, network-level QoS management mechanisms are largely unnecessary, because there is sufficient bandwidth for all applications. Thus, the deployment of QoS mechanisms in commercial wireline networks has been mostly confined to some limited DiffServ marking, or has been omitted altogether. While wireline networks do offer a spectrum of services ranging from "platinum" to "bronze," such services typically encompass a range of services that include different levels of customer attention, guarantees about service availability (i.e., limited service outage guarantees, backed by money-back guarantees), and so on. In case additional capacity is required to support growth in demand in wireline networks, capacity is typically added via HITL (Human In The Loop) involvement either via

- A trigger "Reallocation of Inventory Assets" to Network Inventory Management or
- An "Additional Capacity Request" trigger to a Network Development and Planning function,

as illustrated in Figure 8.1. While additional capacity can obviously be added, the above process is very time-consuming. For example, the process of laying down new fiber-optic cables is a lengthy and expensive process, with substantial use of manual labor.

The situation is very different in MANETs due to the scarcity of over-the-air bandwidth and due to the fact "spare" capacity is not readily available. In fact, in MANETs, it is very often the case that the amount of network capacity is not sufficient for all the applications that need to use the network. This is because MANETs usually have to be deployed on demand, without a lengthy planning phase. Furthermore, the types and intensity of traffic and the network usage patterns are hard to predict. Therefore there is a critical need to provide QoS assurance mechanisms that ensure that the highest-priority traffic receives better treatment than lower-priority traffic. This is a direct consequence of the fact that adding network capacity to deal with additional application demands is not always an option in MANETs.

The design of QoS mechanisms for MANETs is fraught with a number of challenges. Listed below are some of the most significant of these challenges.

8.4.2.1 Dynamic Network Topology. While the design of end-to-end QoS assurance mechanisms for communications networks is itself a challenging problem, the challenges are further compounded in MANETs due to the absence of a stable network topology. Not only are the locations of network elements highly variable (due to mobility), but also the number of network elements in the network may vary dynamically due to nodes shutting down to conserve power, nodes moving out of range, nodes going down, and so on.

Furthermore, the network links themselves have highly variable capacity that may be affected by environmental conditions such as weather, terrain, and so on. These links are characterized by different speeds, with the differences sometimes being orders of magnitude. A direct consequence of the underlying link speeds is the amount of delay incurred by a packet in transit, which in turn directly impacts the end-to-end QoS that can be assured to the corresponding traffic flows. This poses a challenge for QoS management, as the QoS management mechanism needs to take into account all of this random variability when providing QoS assurances to applications.

8.4.2.2 Lack of Visibility into Network Topology. Not only is the network topology dynamic, it is also mostly unknown to the network management system. This is due to the encryption-related (namely, red and black) network separation issue discussed in earlier chapters. Since almost no management-related information about the encrypted (black) network segments is allowed to flow to the red side and since the network management system resides on the red side of the network, the black network topology is not visible on the red side in real time. This creates the challenge of having to deal with a network that cannot be directly observed. As shall be seen in the ensuing sections, this lack of observability is handled by collecting network measurements on the red side of the network to infer the state of the black network.

8.4.2.3 Wide Range of QoS Requirements. Even though the presence of applications whose QoS requirements range over a wide spectrum of delay and loss requirements is not unique to MANETs, it is still a challenge to be able to support such applications in a severely constrained resources environment. As an example, certain mission-critical applications in MANETs may require very stringent delay and loss guarantees. However, the limited and variable network capacity, coupled with variable topology and link bandwidth, makes managing QoS for services on an end-to-end basis rather challenging.

8.4.3 Related Prior Work on QoS Assurances in Communication Networks

It should be noted that there exists a body of work in the area of admission control for static wireline networks. Such functionality is implemented by a *Bandwidth Broker* [Nichols et al., 1999]. Bandwidth Brokers are used to control admission of flows of different service classes into a network, based on the network state. The network state is typically estimated based on knowledge of the network topology (which is largely static), network routes (which are again relatively static), and link capacities (which are well known). The fact that such information about the network is readily available makes the QoS assurance problem relatively trivial for these networks. For MANETs, the knowledge of network topology, routes, and link capacities is neither available in real time (due to red-black separation) nor very useful (since all of these quantities vary dynamically). Thus a radically new approach is needed for MANETs.

There exists some prior work in the area of QoS in communications networks that have characteristics similar to the MANET architecture presented in Chapter 5. Given the premise that black network information is largely unavailable, the approaches reviewed below are all based on the principle of *measurement-based admission control* (MBAC). In a nutshell, the MBAC approach uses red-side network measurements to attempt to infer the state of the black network (also called *opaque* networks in the literature) and, in particular, to determine whether additional traffic can be admitted into the network.

The work described in [Valaee and Li, 2002] uses "time-delay" measurements to assess the congestion status of opaque networks with the use of active probes. Broadly speaking, there are two types of measurements that can be collected in networks. They are "time-related" measures (also referred to as "latency measurements" in the literature) and "information-loss-related" measures (also referred to as "throughput measurements" in the literature). While the pros and cons of both of these approaches are described in more detail in Section 8.4.4.3.1, the following prior work has been done in this area. The authors in Valaee and Li [2002] employ latency measures by introducing time-stamps in packets and requiring timestamp information to be exchanged explicitly between various nodes. This is essentially an "active probe" approach (the various probing approaches along with their pros and cons are discussed in Section 8.4.4.3.2) whereby new packets are sent for the purpose of communicating timing information. The various network nodes use incoming time-stamp information to compute end-to-end delays, and they use this measure to infer the state of the underlying network. For example, if the computed delays are large, this is indicative of a "bad" network, whereas short delays indicate a "good" network.

While this is a valid approach for certain types of networks, such as wireline networks where bandwidth is plentiful, it suffers from the following severe drawbacks in the MANET environment:

- It is expensive in terms of the overheads introduced in order to derive latency estimates.
- It does not consider multiple service (traffic) classes.

Hence such an approach, although acceptable for the environments that it has been proposed for, cannot be used—nor extended without having to undergo major transformations—for MANETs.

In Grossglauser and Tse [1999] and Breslau et al. [2000], MBAC schemes are used to characterize the current network load. While their algorithms have been shown to perform well in terms of characterization of the current network load, they are not applicable to MANETs, since they assume complete knowledge of and control over the elements in the path of the data packets. Recall that the MANET architecture under consideration in this book contains opaque, or black, network segments that allow very limited management monitoring and control in real time. The terms "black network" and "opaque network" are used interchangeably in this chapter to refer to the encrypted network segment.

8.4.4 QoS Assurance for MANETs Using Dynamic Throughput Graphs

The unique characteristics of MANETs require novel QoS assurance techniques to support the variety of applications expected in such networks. This section describes an adaptive QoS assurance approach that can be used to provide end-to-end service assurances to the wide spectrum of applications that are expected to be transported by the MANETs under consideration.

8.4.4.1 System Model. Figure 8.3 provides a schematic of the network environment under consideration. It shows three unencrypted (red) routing domains labeled AD_x, AD_y and AD_z, where AD stands for Administrative Domain. It also shows one encrypted (black) routing domain (AD_A). As illustrated in this figure, the unencrypted domains are wired and the encrypted domain is wireless. The boundaries of the MANET nodes are shown in the figure; note that every MANET node fully contains one red administrative domain and is part of the black administrative domain. The wired networks are labeled "User Network: AD_x," "User Network: AD_y," and "User Network: AD_z" to emphasize that these are the networks from which application flows originate and terminate. The unencrypted (red) user networks communicate

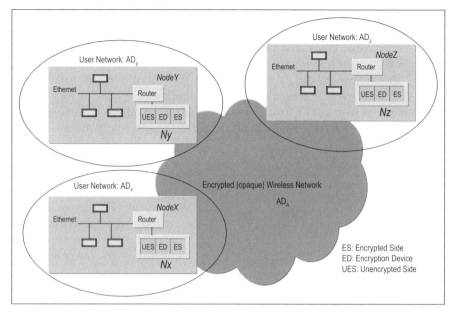

Figure 8.3. Sample red–black network.

with each other through a wireless network that is labeled "Encrypted (opaque) Wireless Network" in Figure 8.3. More specifically, the user networks are connected to the encrypted (opaque/black) wireless network via network elements labeled Nx, Ny, and Nz. As illustrated in Figure 8.3, each of these nodes has an "Unencrypted Side" (UES), an "Encryption Device" (ED), and an "Encrypted Side" (ES). The UES essentially performs the required packet processing at the originating (red) side. The ED, as its name implies, encrypts packets before they get transmitted to the encrypted side (ES) and enter the opaque network. The reverse process takes place at the destination node.

Observe that the properties of the end-user network segments and the intervening opaque network segments are significantly different. More specifically, user networks utilize wireline LAN technology such as gigabit Ethernet LANs, whereas the black network is wireless. The opaque nature of the black network means that limited network visibility or control is available to red-side network management applications. One aspect that is known about the opaque network is the type of radio and physical layer waveform that are being used, from which it is possible to infer the theoretical maximum capacity of the network—that is, the raw speed in bits per second. As an example, given an Ethernet LAN with speeds of 3, 5, or 11 Mb/s, it is possible to infer that the maximum capacity on the physical cannot exceed 3, 5, or 11 Mb/s, as the case may be. However, note that this maximum capacity is very different from "available capacity." This is because the available capacity on a given link is the capacity that is "seen" by the applications using that link and is a quantity that fluctuates randomly over time. For wireless links, the situation gets even more complicated, because here even the maximum capacity is not known, even though it is bounded by the theoretical maximum capacity. The maximum capacity for wireless links fluctuates over time because of a combination of several factors such as environmental characteristics (absorption, fading, jamming) and the amount of cross-traffic traversing the link, all of which are inherently stochastic processes themselves.

The only assumption made here is that the black network supports the Differentiated Services (DiffServ) model described earlier in Chapter 3. Thus the presence of differentiated support for different types of traffic classes, ranging from delay and loss-sensitive to best effort, is assumed. A DiffServ-based QoS approach is preferable to an RSVP-based approach, since the latter is not scalable to thousands of connections that are anticipated to exist in the MANET environment. Also, RSVP relies on explicit resource reservations for each traffic flow along the path of the flow; since paths change dynamically in MANETs, this approach would not provide much value. Note that DiffServ relies on IP packets being marked with a DSCP (DiffServ Code Point). When packets are encrypted (using HAIPE, IPSec, etc.), the DSCP from encapsulated inner packet is copied into the IP header of the encapsulating packet, thus allowing information about the QoS treatment to be accorded to each packet to leak into the black network.

A final observation is that the underlying opaque network is bandwidth-constrained, and therefore any management overhead arising from measurements that must be sent over the air must be kept as low as possible.

8.4.4.2 QoS Management Architecture. An important requirement of a QoS assurance solution is that it be able to adapt itself to the underlying network dynamics. This precludes any QoS mechanism based on static rules that dictate how much traffic of different types can be admitted into the network, since such static rules cannot take into account the fact that the network capacity changes dynamically in MANETs. The following three components are required to provide service assurances and soft failure management in MANETs:

- *Admission Control Function (ACF)*: The purpose of this function is to implement an efficient *admission control* algorithm that judiciously admits application traffic flows based on the QoS requirements and priorities of the applications and some knowledge of the network state.
- *Quality Adjustment Function (QAF)*: A *quality adjustment* solution is required to adapt to the dynamics of the underlying network by modifying the guarantees provided to current applications based on their priorities as needed.

- *Measurement Collection Function*: The purpose of this function is to collect measurements for the benefit of the previous two functions—that is, the ACF and the QAF.

Admission control functionality involves determination of whether a flow can be allowed into the network. This decision must take into consideration both (a) the QoS requirements of the flow to be admitted and (b) the state of the network. *Quality adjustment,* on the other hand, is responsible for ensuring that admitted flows continue to receive the required QoS guarantees based on their priorities, even under changing network conditions. This might in some cases involve reacting to a degradation in network state by preempting certain low-priority flows in order to sustain assurances to high-priority/mission-critical applications. When used in the above manner, the QAF is in essence a reactive control mechanism. However, note that the QAF can also be used proactively (e.g., if the QAF is triggered based on "expected" network behavior as derived from past history), and the proposed methodology does not preclude such a proactive use of QAF. The remainder of this chapter focuses solely on the reactive use of QAF.

In order to enable the ACF and QAF to make admission and preemption decisions that are based on the condition of the underlying network, information must be gathered about the network. This is the job of the Measurement Collection Function. The collected information can then be processed to construct some sort of representation of the underlying network that in turn can be used as the basis for the ACF and QAF decisions. Thus, central to the design of the ACF and QAF are the following:

- *Information Collection (Performed by the Measurement Collection Function)*: Information must be collected about the network in order to feed the QoS assurance decision-making process.
- *Information Processing and Use*: Once information has been collected, it must be processed and algorithms must be used to support ACF and QAF decisions.

These aspects are discussed in the next two sections. It should be noted that these two phases are concurrent and that information collection is performed continuously, even as the collected information is used.

8.4.4.3 *Measurement Collection Function.* The information collection performed by the Measurement Collection Function involves two key considerations:

- The type of information to collect—that is, what information is needed about the network in order to allow the QoS Admission Control Function and the QoS Adjustment Function to make accurate decisions.

- The information collection mechanism—that is, what mechanism should be used to collect the type of information identified in the preceding phase. Given the very small amount of bandwidth available in the opaque network for management traffic, it is critical for the information collection mechanism to minimize the management traffic overhead that it introduces into the network.

These considerations are discussed in turn in the next two sections.

8.4.4.3.1 Type of Information to Collect. The first consideration listed above with respect to the information collection phase is the type of information that needs to be collected in order to provide an effective QoS solution. Note that the network under consideration can be broadly categorized into two types of network segments: unencrypted network segments and encrypted network segments. The unencrypted or red network segments are typically wired gigabit Ethernet LANs and hence not anticipated to be bottlenecks. Thus QoS mechanisms are not needed to deal with traffic on these segments. Limited information about the encrypted, or black, network segments can be passed across the HAIPE devices to their interconnecting red enclaves due to security restrictions. However, the black network not only has restricted capacity, but is also the multiplexing point for many red enclaves and is therefore the anticipated bottleneck in the end-to-end path of an application flow across the network.

In view of the above, the information collected needs to reflect the state of the black network in such a manner as to facilitate the decisions related to QoS admission control and adjustment. The state of the MANET, however, is a complex function of many factors, including link dynamics, number of nodes in the network, mobility patterns, traffic patterns, and so on. It can quickly be seen that while a state descriptor that includes all of the above can provide a great amount of information, it could also result in an inefficient and even infeasible solution. Additionally, due to the stochastic nature of the network, a complex state descriptor that is a function of the above-mentioned items will be very difficult to derive unless otherwise oversimplifying and unrealistic assumptions are made. For example, factoring in mobility will require either *a priori* knowledge of or assumptions regarding the expected mobility pattern of all of the network nodes, or else will require expensive location tracking devices constantly transmitting location information (and consuming bandwidth in the form of overhead messages) at run time amongst all the communicating nodes. Similarly, usage of link dynamics assumes *a priori* knowledge of all possible terrain types in which the communications network will be deployed, which again is not realistic.

Thus there exists a canonical tradeoff with regard to computational complexity, feasibility, and accuracy. Furthermore, since the underlying network is not guaranteed to be in a "steady state," it may even be counterproductive to maintain very detailed state information. This is because the information may change by the time it is propagated throughout the network. In light of the above, the following key parameters have been identified to provide the information required to perform QoS actions (namely, judicious admission control and efficient quality adjustment):

- Bandwidth consumed (also referred to as throughput) between ingress–egress pairs.
- Latency between ingress–egress pairs.

These items are discussed in more detail below.

8.4.4.3.1.1 THROUGHPUT MEASUREMENTS. The measure under consideration here is throughput between ingress–egress pairs, where ingress and egress nodes refer to the MANET nodes on which flows originate and terminate, respectively. Depending on the direction of the flow, each of these nodes assumes the role of an ingress or egress node.

As an example, consider a flow F_{xz} from $Node_X$ to $Node_Z$ that originates in $Domain_x$ and terminates in $Domain_z$. In this case, $Node_X$ is the ingress node and $Node_Z$ is the egress node. Likewise, for a flow F_{yz} from $Node_Y$ to $Node_Z$ that originates from $Domain_y$ and terminates in $Domain_z$, node $Node_Y$ is the ingress node and node $Node_Z$ is the egress node. If on the other hand a flow F_{zx} from $Node_Z$ to $Node_X$ originates from $Domain_z$ and terminates in $Domain_x$, then node $Node_Z$ is the ingress node for flow F_{zx} and $Node_X$ is the egress node for flow F_{zx}.

Throughput measurements are computed as follows: For every ingress–egress node pair, keep count of:

- *Sent Packets*: The number of packets per DSCP that exit from the ingress node over a certain policy-defined interval.
- *Received Packets*: The number of packets that enter the egress node after traversing the opaque network segment, during the same interval.

A gross estimate of the throughput over that interval can then be computed using the ratio of the sent packet count to the received packet count. This ratio is used to estimate the characteristics of the path between the ingress–egress node pair over the opaque network segment over that period of time. Such throughput measurements are used as an estimate of the "state" of the opaque network. The term "state" is used in a qualitative manner here, reflecting the network "condition," where the condition of a network essentially refers to the amount of loss observed in the network. For example, if the ratio of received packets to sent packets is close to 1.0, then the path between the ingress–egress pair over that interval can be considered "good." On the other hand, if the above-mentioned ratio is much smaller than 1.0, this is indicative of some problem on that path over that time interval. Furthermore, if the problem persists over a certain number of intervals (again, this number can be a policy-driven, configurable interval), this is an indication of some persistent problem on the given network path. Recall that since the network management system resides on the red side of the network, and limited management information exchange is allowed between the red and black network segments, the best that can be done is to make inferences about the paths through the black network segment and subsequently make appropriate admission control and quality adjustment decisions on the red side.

At this point the observant reader may ask the following question: Given the packet-switched nature (versus circuit-switched) of the network, coupled with the mobile and wireless aspects of the opaque segment, the actual paths taken by packet flows between an ingress–egress pair may not be the same over a period of time; hence how can one make judgments on the "path" between the ingress–egress node pair? This is a very pertinent question; indeed, the paths between a given ingress–egress node pair may change, and there is no guarantee whatsoever that the path between an ingress–egress node pair will remain the same over a certain period of time. In fact, the actual path taken by packets between an ingress–egress node pair will vary over time, and it is a complex function of the network load, routing type, and environmental conditions such as fading and propagation that in turn influence the available links. However, this impediment is common to any measure, be it a throughput-based measure or a latency-based measure (as will be explained next), and is due to the inherent mobile wireless nature of the opaque network segment and the fact that the opaque network does not allow the management system to "look" inside it in real time in order to take management decisions. In fact, it is precisely due to this lack of information that techniques (be they throughput-based measurements or latency based measurements) have to be devised that can be used to (a) learn about the state of the opaque segment and (b) take suitable actions in the event of problems.

Another point that may be picked up by an observant reader is that although the knowledge that an ingress–egress path is lossy is very useful, this knowledge does not provide a complete picture about why the path is lossy. The losses could stem from either (a) congestive loss due to queue/buffer overflows or (b) link losses caused by environmental effects. While the former is something that can be addressed by a network engineer or management system, the latter is not something that can be controlled. However, the ability to discern between these two cases is helpful in making management decisions. For example, in the former case, the network management system can try to preempt lower-priority flows to make room for higher priority/mission critical applications.

Finally, before looking at the next measure (namely, the latency-based measure), consider the overheads incurred via this approach. The throughput-measure-based approach will require the exchange of the measurements over the air (i.e., from the egress node to the ingress node), so that the ingress node can compute the ratio and then take suitable actions, based on the value. For example, if the value of the ratio is much smaller than 1.0, certain policy actions such as preemption of lower priority flows to make room for higher-priority flows may need to be taken. These measurements can be sent periodically (again based on a policy-defined interval) from the egress node to the ingress node. This in turn introduces overheads, since the measurement information has to be sent back to the ingress node periodically. In order to keep this overhead low, measurements should be aggregated among the various service classes/DSCPs and should be sent back to the ingress node at configurable intervals. This is discussed further in Section 8.4.4.3.2.

8.4.4.3.1.2 LATENCY MEASUREMENTS. Another measure that can be used to learn about the condition/state of the opaque network segment is a latency-based measure. As its name indicates, this measure captures the delays incurred along a certain ingress–egress path. One way to measure latency is to times-tamp the packets when they enter the opaque network via the ingress node, and again when they exit the opaque network segment at the egress node. This method therefore requires a special set of operations on every packet; that is, each packet has to be opened and a timestamp added to it by the ingress node at which the packet enters the opaque network segment. A timestamp field needs to be introduced for this purpose. At the egress node, the time of receipt must be noted in order to be able to compute the time difference between the time of receipt and the time of sending; this difference is the latency incurred through the opaque segment. The egress node computes this latency on a per DSCP basis, and it periodically sends latency information back to the ingress node. This provides the ingress node with information about the latency incurred by different packets with different DSCP markings, when traveling from the ingress to the egress node.

Just like its counterpart, the throughput measure, such a latency measure can be used as a basis for making admission control and quality adjustment decisions. However, latency measurements tend to impose a higher overhead (as compared to throughput measurements) on the network, both in terms of increasing the size of every packet (by introducing a timestamp) and in terms of the latency measurements that need to be sent over-the-air from egress to ingress node. Of course, just as with throughput measurements, the frequency of messages sent over the air from egress to ingress nodes can be varied, to control the overhead. However, unlike throughput measurements, for latency measurements to be effective, timeliness of measurements is a critical factor. This means that there is a need for more frequent over-the-air exchanges from egress to ingress node to communicate path latencies.

8.4.4.3.1.3 COMPARISON: THROUGHPUT VERSUS LATENCY MEASUREMENTS. The estimation of the available capacity across an opaque network results in a very expensive design in terms of overheads. Furthermore, due to the stochastic nature of the network, such a measure is very difficult, and even impractical to obtain. The topic of available network capacity in MANETs is one that is still in an active area of research. While the seminal work of Gupta and Kumar

[2000] tries to estimate capacity in a wireless network, it is restricted to homogeneous networks with homogeneous traffic demands. The links in realistic MANETs such as the ones under consideration are far from homogeneous in terms of the link capacities. Furthermore, the traffic demands exhibit a wide spectrum of characteristics in terms of bandwidth, delay, and loss requirements, and again are far from being homogeneous. The work of Gupta and Kumar has spurred a variety of research efforts, but none of them address realistic MANETs of the types described in this book. An understanding of network capacity and their scaling laws is far from complete and is an active area of ongoing research.

Given this, some other means must be used for estimating the state of the opaque network. Both throughput as well as latency measures require periodic over-the-air (OTA) exchanges between ingress–egress nodes, to communicate throughput or latency measurements. More specifically, the egress node must send back an aggregate (aggregated over various service types/DSCP values) measure of the byte-count or latency value back to the ingress node, which will then compute the required loss/delay measures. In both cases, the frequency of the OTA message exchanges will determine the overheads incurred. This frequency can be dynamically controlled using policies. Canonical trades exist with regard to the frequency of the messages being sent (and hence the overheads incurred) and the quality of the resulting estimates of network state. The number of messages with either of the methods can be tuned and present a similar burden on the network from the perspective of bandwidth overheads.

However, in addition to OTA message exchange, the latency measure involves opening every packet to insert and extract timestamps. Such an operation can soon become very expensive and even counterproductive for latency-sensitive applications such as mission-critical applications. While optimizations such as random packet timestamping or timestamping every kth packet, where k is a parameter that can be introduced to reduce the latency overhead, the latency measure nevertheless introduces extra complexity and overhead as compared to computation of throughput measures.

In an attempt to provide the most effective yet least cost metric, keeping both bandwidth and complexity to a minimum, the following considerations were used. Strictly speaking, the throughput between ingress–egress pairs and the latency between the corresponding ingress–egress pairs are not completely independent of each other. In fact, there exists a very predictable relationship between throughput and latency. For example, in the region for which the throughput (bandwidth usage) between the ingress–egress nodes has not reached its saturation point, the latency between the ingress–egress pair will be low. However, when the throughput (bandwidth usage) between the ingress–egress nodes reaches a certain maximum operating point and subsequently begins to drop (i.e., has reached the "knee" of the throughput curve as illustrated in Figure 8.4), the latency will begin to rise. Furthermore, as the throughput continues to decline beyond this knee, the latency will continue to increase. Thus, using throughput measurements alone, latencies can be inferred without actually having to explicitly measure them. Of course, with such an inference, an absolute measure of the latency will not be available; however, note that these absolute values are not really needed.

Additionally, note that while throughput and latency are related, the relationship is not necessarily linear. In particular, latency buildup can occur much before significant loss is observed, thus making the throughput measure a somewhat delayed indicator of latency buildup. However, given the additional complexity incurred in obtaining latency measurements, and bearing in mind the need for low complexity/low overhead methods, an approach that uses throughput measures to infer the state of the opaque network will be described in the following sections. Sections 8.4.4.4.3 and 8.4.4.4.4 describe the efficacy of a throughput-based measure in assisting the performance management system to provide QoS assurances in heterogeneous red–black MANETs.

Although the latency-based measurement technique is not explored in this book, it should be noted that such measurements do hold the potential to add insights on providing QoS assurances in heterogeneous MANETs. Thus, the use of throughput measurements to help the performance management function provide QoS assurances in MANETs should be viewed as one possible approach, with a latency-based approach being another possibility.

As a final remark before discussing information collection mechanisms, note that yet another possible approach to making measurements is a hybrid approach that combines throughput-based measurements with a limited amount of latency-based measurements to infer the state of the opaque network. More specifically, latency measurements could be performed on a small subset of applications, namely, real-time services such as voice streams and real-time video streams, thus restricting the added latency-related time-stamping and complexity to a limited subset of applications, while employing throughput measurements for all of the remaining applications. As always, a hybrid solution that is at neither extreme end of a spectrum holds potential, but will have to be tuned to obtain good performance; that is, the subset of applications for which latency measurements are used should be selected carefully, and it will depend on the traffic mix or information exchange requirements of the MANETs under consideration.

8.4.4.3.2 Measurement Collection Mechanisms. A straightforward method for collecting measurements from any communications network is to use probe packets. The basic principle behind the probe-based approach for the type of networks under consideration is as follows: Red packets are exchanged between ingress–egress nodes. Recall that "red" packets are unencrypted packets that originate on the red side of the network; these packets are encrypted at the source, or ingress node, and sent over the air as "black" packets. They are then decrypted upon arrival at their destination, or egress node, to recover the original red packets. Based on how the probes are exchanged in the forward direction (between ingress and egress nodes), there exist the following three categories of probes:

- Active probes
- Passive probes
- Hybrid probes

Each of these is discussed in more detail below.

8.4.4.3.2.1 ACTIVE PROBES. In this approach, small fixed-size red control packets are sent periodically between the various ingress–egress node pairs to obtain the desired information (throughput or latency) along the path between the given ingress–egress node pair.

The advantages of active probing are:

- No extra latencies are introduced for application packets, since timestamp information does not have to be inserted into or extracted from application packets.
- The network state estimates are generally up-to-date, since information is available when an application flow starts at an ingress router. This is because active probing is performed proactively so that measurements are constantly collected and maintained, irrespective of whether there is any application traffic flowing through the network.

A major disadvantage of active probing is that it adds to the network bandwidth overhead. In addition, since probe traffic is sent regardless of whether traffic is flowing in the network, it may be the case that measurements are being collected that are not even needed! Furthermore, wireless bandwidth fluctuations may result in the wireless bandwidth becoming extremely small. This can result in the probe traffic itself introducing latency into the network, causing performance degradation.

8.4.4.3.2.2 PASSIVE PROBES. In the passive probe approach, no explicit packets are introduced into the network in the forward direction. Reports from the egress nodes to the ingress nodes, however, may be sent in one of two forms: either (a) as explicit packets or (b) by piggybacking on existing application traffic by simply adding the information in the options field in the packet header.

The advantages of passive probing are:

- It does not add to the network bandwidth overhead when piggybacking is used.

- In the case where explicit reports are sent back to the ingress router, the overheads are nonzero, but are about 50% smaller than the overheads introduced by active probing.

One disadvantage of passive probing is that the use of piggybacking adds to the latencies experienced by application traffic. Note that this disadvantage can be overcome if piggybacking is not used; in such a case, there will be no additional latencies introduced. However, if piggybacking is not used, then some overheads will need to be introduced in order to transmit measurement information.

Finally, note that while passive probing lends itself very well to throughput measures, its efficacy is questionable for latency measurements. While in theory passive probing with piggybacking can be used for latency measurements, in practice, this approach introduces additional latencies that could potentially become an impediment in terms of timeliness of the decisions resulting from QoS control mechanisms. This is because passive probing with piggybacking involves opening every packet, inserting a timestamp at the sending end, and performing a reverse set of operations (opening every packet and extracting timestamp information) at the receiving end. Ironically, such latencies will introduce additional latencies, thus almost defeating the purpose of QoS management, since critical latency-sensitive real-time applications are the ones that benefit the most from latency-based measurements! This is the reason why in practice, active probing is preferred for use with latency measurements.

In the case of throughput measurements, passive probing works as follows. Each ingress node keeps count of the byte count over a given interval for each of the DSCPs (service classes) that originate from it, for every egress node. The corresponding egress nodes send their respective measurements of the received byte counts over the same interval back to the ingress node—either piggybacked on application messages if such messages exist, or via explicit measurement packets. Of course, to reduce overheads, the frequency of the measurement packets from the egress to ingress nodes can be controlled via policies, which in turn implies that a measurement packet could carry information for the past n measurement intervals, where n is a parameter.

8.4.4.3.2.3 HYBRID PROBES. This approach attempts to maximize the merits of the passive and the active approaches while minimizing their demerits. Two possible approaches can be used for hybrid probing:

- **Approach 1:** Use explicit (active) probe packets in the forward direction for delay-sensitive traffic classes. For loss-sensitive classes, use application packets to piggyback information such as timestamps for latency measurements. In this approach, explicit report packets are sent from egress to ingress nodes for all classes.

- **Approach 2:** Use application packets to drive measurements when available; otherwise, use explicit probe packets for each traffic class. As with Approach 1, use explicit report packets from egress to ingress nodes for all classes.

Approach 1 provides guaranteed accuracy for the high-priority and delay-sensitive traffic. At the same time, this approach results in potentially excessive management traffic. However, this overhead will not be as high as that incurred by active probing, because of the reduced number of explicit probe packets from ingress to egress nodes that are sent with this approach.

Approach 2 provides guaranteed accuracy for all traffic classes. This is at the cost of extra processing required to keep track of sent probes, extra latencies incurred in order to process application packets, and potentially excessive management traffic overhead.

8.4.4.3.2.4 SUMMARY: WHICH APPROACH WORKS BEST? Based on the qualitative pros and cons of each of the active probing, passive probing, and hybrid probing approaches described above, the use of throughput measures using the passive probing approach with explicit measurement reports from the egress to ingress node results in the lowest overheads and implementation complexity. The active probing approach introduces excessive overheads and is therefore not an acceptable approach. Observe that the hybrid approach introduces the additional complexity of maintaining different kinds of measurement information for different types of traffic flows, unlike the passive probing approach, while the passive and hybrid approaches both generate a similar amount of overhead traffic. Thus the passive probing approach holds better potential than the other two approaches and is therefore the method that is recommended for collecting measurements, as will be described in Sections 8.4.4.4.3 and 8.4.4.4.4.

8.4.4.4   Information Processing and Usage. Having collected data that will provide information about the condition or state of the underlying MANET, the next phase is to construct techniques to process and use this information efficiently. This section deals with information processing and usage. In other words, how does the information collected during the information collection phase get utilized by the ACF and QAF in their decisions? This section describes an approach based on the collection of throughput measurements that uses distributed algorithms for admission control and quality adjustment. These algorithms enable the ACF and QAF to make service manipulation decisions such as flow preemption, downgrading, throttling, and so on, that have the best possible outcome for the network applications, given the current network dynamics. The algorithms work by constructing *Dynamic Throughput Graphs (DTGs)* from the ingress–egress measurements that are used to assist the ACF and QAF in making decisions.

8.4.4.4.1 Dynamic Throughput Graphs Overview. Dynamic Throughput Graphs (DTGs) can be thought of as adaptive representations of the state of the opaque network between ingress–egress pairs, where the adaptation is based on the dynamics of the underlying network. In their most general form, DTGs may be defined based on both bandwidth and latency measurements. More specifically, in the ideal case, the bandwidth-based DTG is a plot of the throughput of the traffic flows between a given ingress–egress pair as a function of the load on the network. Here the network load is defined as the number of bits per second sent by the ingress node, and the throughput is defined as the number of bits per second received at the egress node. Similarly, in the ideal case, the latency-based DTG is a plot of the latency experienced by a flow between a given ingress–egress pair as a function of the load on the network. Latency is the time (in seconds) that elapses between the sending and the receiving of a packet; latencies are collected and averaged over a time interval for all packets transmitted and received in that time interval. Figure 8.4 illustrates a graph that provides both throughput and latency information for a single traffic class for a wireline network carrying TCP traffic. Note that although this graph is for TCP traffic over a wireline network, the general observations about throughput, latency, and network load remain the same for wireless MANETs. A few salient properties of the graph are discussed next.

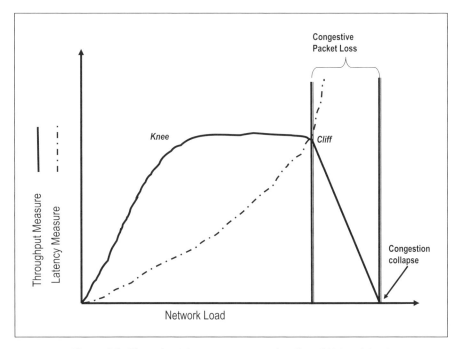

Figure 8.4. Throughput–latency curve as a function of Network load.

First, observe that as the load increases, so does the throughput (represented by the solid curve in Figure 8.4), as expected, up until a certain point called the knee. After the knee, the throughput increases far more slowly as compared to the load, indicating packet losses. If the load is further increased, then at a point called the *cliff*, the throughput collapses, resulting in the loss of close to all transmitted packets. A similar behavior is seen with regard to the latency graph (represented by the dashed curve in the figure). In this case, the delay increases slowly until the knee, then a little more rapidly until the cliff; and when the load is increased beyond the cliff, the delay becomes unbounded. As can easily be seen, there is a correlation between the bandwidth throughput graph and the latency graph.

This correlation is leveraged in the following approach. Rather than using both latency and throughput measurements, only throughput measurements are used. These measurements are captured via throughput graphs, and do not explicitly measure latency, so as to significantly reduce overheads. Note that there are multiple such graphs, as throughput measurements are collected per DSCP, per ingress–egress node pair. Due to the stochastic nature of the MANETs, the ingress–egress throughput measurements will vary as a function of time. Thus the throughput graphs must be updated dynamically, leading to the term *Dynamic Throughput Graphs (DTGs)*. In essence, the basic properties of the dynamic throughput graphs are as follows:

- Delay and packet loss start increasing gradually at the knee.
- At the cliff of the graph, increasing packet loss and delays result in congestion collapse.

The DTGs therefore capture the "state" of the MANET. Recall that the opaque network segment is the bottleneck segment. The DTGs therefore shed light on the condition or state of the path between the ingress–egress node pair for the currently active traffic classes. This information can then be used by the ACF when deciding whether or not to admit traffic into the network in a certain service class. For example, qualitatively speaking, if a certain ingress–egress path exhibits very poor throughput, then the ACF either will have to deny new application requests or will need to ask the QAF to determine whether any lower priority flows exist that can be preempted to make room for an incoming higher priority request.

8.4.4.4.2 Constructing Dynamic Throughput Graphs. This section describes how DTGs are constructed in MANET environments. Each time a flow is initiated between an ingress–egress node pair, the following actions occur. Based on a policy rule, the Measurement Collection Functions (MCFs) on ingress and egress nodes decide upon a byte collection interval. This interval could be a fixed interval, defined *a priori* and used as a default value by all nodes, or could assume different values via the use of policies that set this interval based on the dynamics of the MANET. As can be easily seen, an *a priori* fixed interval may not be desirable for MANETs due to their inherently stochastic nature and the widely varying characteristics of applications that need to use the network. For example, a sustained dip in network bandwidth may require reducing the volume of network management traffic, resulting in larger measurement intervals. However, there may be certain critical latency-sensitive applications that need a more up-to-date view of the network than other noncritical, non-latency-sensitive applications. Since DTG measurements are collected based on DSCP classification, policies to set smaller collection intervals for certain DSCP classes may need to be in place. The flip side here is that although the use of different collection intervals across different traffic classes (DSCPs) will help in reducing overheads, it will also increase the complexity of implementation. Another parameter that could be varied based on policies is the DTG window size—that is, the number of measurements maintained at any point in time in a DTG. If a MANET node has limited memory, then a smaller measurement window may be preferable, although this would adversely affect the quality of the DTGs themselves. All of the above are merely illustrative examples underscoring the need for policy-driven QoS management in MANETs.

The MCF at the egress point sends a report to the ingress node at the end of the reporting interval, containing the count of bytes received for this flow from the ingress node in the reporting interval. The ingress MCF stores this information, and the process repeats itself. If a flow belonging to another class starts in the meantime, then the same process is applied to that flow as well; that is, the MCFs at the ingress and egress nodes for this new flow will start exchanging byte count information. As a result, the MCF at the ingress node will be able to compute the ratio of bytes received to bytes sent within that reporting interval. Note that there exists one such reporting interval for every service class[1] of flow that originates at an ingress node. Also, recall that the MCF also has information about the number of bytes sent into the network during every reporting interval—that is, the ingress byte count.

When the MCF has an adequate number of data points (i.e., ingress–egress byte counts), the MCF computes polynomials that describe the network state based on the input and output data gathered. Techniques such as multivariate regression can be used to estimate the parameters of these polynomials. For example, a multiclass DTG for an ingress–egress pair is defined as a function of n variables, whose value is a vector of size n as shown below:

$$[i_1, i_2, \ldots, i_n] = DTG_f(o_1, o_2, \ldots, o_n), \tag{1}$$

where o_k is the output (egress data rate) and i_k is the input (ingress data rate) in class k. DTG_f is a function that can be estimated with an n-dimensional nonlinear (e.g., polynomial) least-squares fit to the measurements, based on

---

[1] *Note*: The terms "service class" and "DiffServ class" are used synonymously in this book.

the Levenberg–Marquardt method [More, 1977] implemented in a number of numerical computing libraries (e.g., the GNU Scientific Library (GSL)). If a polynomial fit is used, the estimation results in n polynomials of n variables of a given degree d. Once the polynomials have been determined, the ingress node can construct the DTG. This DTG can then be used for both ACF and QAF decisions. It should be noted that a certain minimum number of data points are necessary to produce good DTG estimates. Using k to denote the minimum number of data points needed, it is most effective to let k be a policy-driven parameter. Finally, while it is important to have a certain minimum number of data points to produce good DTGs, it is also equally important to have wide enough range of data points to yield good curve-fitting results. In other words, if most of the data points are close together, then approximation errors introduced by the curve-fitting techniques can get larger.

The resulting DTG is used to compute the projected allowable input values for a class, given the measured and/or projected values of the output for all classes. These projected values can then be used when making ACF decisions, as explained in Section 8.4.4.4.3. Note that the polynomial estimations mentioned above do not pose significant processing overheads, and therefore the resulting computational overheads are acceptable on most hardware. Therefore, such a technique is applicable on even resource-sensitive nodes such as handheld computers. Recall that a MANET is composed of heterogeneous network nodes, and therefore it is important that the above approach remain within the computational limits of all nodes.

The minimum number of data points needed in a computation window as well as the size of reporting intervals are all policy-defined parameters. The choices of "good" window sizes, measurement intervals, and polynomial fits to apply are parameters that are configured via policies based on a larger optimization objective. The larger optimization objective can assume many forms, with an example of one optimization function being the minimization of OTA QoS management traffic.

Given that the MANET state varies dynamically, the corresponding DTGs vary over time. Thus, the DTGs need to be periodically updated to capture the most recent network state. The updates to DTGs occur in a manner similar to their construction. Each ingress node pair maintains a sliding window of size WS of the reports received by the ingress node for every egress node; here WS is a parameter that is computed based on the stochastics of the network, the reporting intervals, and the desired responsiveness of the QAF to network fluctuations. The DTG update mechanism typically waits for a period of time to collect a certain number of data points that correspond to a given WS. Once a set of data points corresponding to a WS is obtained, the data points within the window are used to update the DTG polynomials. More specifically, a new set of polynomials is computed based on this new set of data points and a new DTG is constructed which in turn may be used to trigger potential QoS reconfigurations (QAF actions) as discussed later in this chapter. Every time a new measurement is received from an egress node for a given DSCP, the corresponding DTG is updated by discarding the oldest measurement and replacing it with the latest measurement received.

Once the DTG has been updated to reflect the new measurements (i.e., new polynomials have been computed thus resulting in a new instance of the DTG), a check is performed to measure the difference between the old and new DTGs. If the difference in the least-squares sense exceeds a certain threshold, this implies that the DTGs have changed significantly, which in turn implies that there has been a change in the underlying environment. Examples of significant changes in the underlying environment include a sudden change in the attenuation properties caused either by hostile actions (e.g., jamming) or environmental problems (e.g., cloud cover, rain, change in vegetation, etc.). In some cases, this can also indicate that a significant change has occurred in the network (e.g., network split/merge). Such deviations are provided to the fault management system to aid the latter in any ongoing root cause analysis.

A more detailed explanation of how to compute DTGs using an interpolation method[2] is provided below. In a nutshell, the DTG for a given *service class C* between a given *ingress node A* and a given *egress node B* is a graph that allows computation of the expected throughput for the given class of traffic between these ingress and egress nodes for a specified number of input bytes. The DTG is obtained as follows:

- Node A counts the total number of bytes of class C sent from node A to node B over preconfigured periodic time intervals. Similarly, node B counts the total number of bytes of class C received from node A over the same preconfigured periodic time interval and sends this information to node A. For example, if the time interval is 1 minute, then every minute, node A computes the number of bytes of class C sent in the last minute to node B, and every minute, node B computes the number of bytes of class C received in the last minute from node A. Node B sends this number back to node A every minute.

- At the above time interval (e.g., every minute), the ingress node A calculates a *current QoS operating point* from the its latest set of measurements of

  - the number of bytes of class C sent over the last time interval to node B (let this number be x) and

  - the number of bytes of class C received by node B from node A in that time interval (let this number be y).

  The QoS Operating Point is defined as the ratio y/x. Thus, if the number of bytes sent and received is roughly equal (i.e., there is little or no loss in the network), then this ratio will be close to 1; and if there is a great

---

[2] Note that the interpolation method is yet another way (in addition to polynomial fit) to compute DTGs. It has been explained in detail here due to ease of implementation.

deal of loss in the network, then the ratio will be lower than 1. In other words, the smaller this number, the greater the amount of loss in the network.

- The ingress node A maintains a record of the last n QoS Operating Points, and it uses these points to create a *graph* (this is the Dynamic Throughput Graph, or DTG) by interpolation between the set of n QoS operating points. The *x-axis* of this graph represents the number of bytes sent, and the *y-axis* of this graph represents the number of bytes received. At the periodic interval used for computing QoS operating points, the DTG is updated by including the latest QoS operating point. Once a certain number of points have been collected, the oldest QoS operating point is dropped whenever a new QoS operating point is added. This provides a sliding window for computing the DTG and ensures that the most recent measurements are used to compute a DTG.

Once the latest DTG has been computed, it is used to determine whether there is sufficient available capacity to accept new QoS flow requests, and to determine whether existing flows are consuming more bandwidth than can be supported by the network. These functions are performed by the ACF and QAF respectively, as discussed next.

8.4.4.4.3 Using Dynamic Throughput Graphs for Admission Control. Recall that Dynamic Throughput Graphs (DTGs) capture the state of network paths through the opaque network between a given ingress–egress node pair for a given service class. More specifically, the DTGs capture the throughput averaged over a certain time interval between an ingress–egress node pair, with the understanding that the path between the ingress–egress node pair may actually (and very likely will) vary over time. However, such a measure of throughput provides the red-side network management function with an estimate of the state of paths through the opaque network. By building the input–output byte-count relationship, the quality of opaque network paths are estimated with respect to the amount of loss on these paths. Additionally, by understanding the relationship between throughput and latency captured via the DTG, the network management system can make a judgment not only about the loss characteristics of the path, but, to a certain extent, about the latency aspects of the path too. As noted earlier, although the relationship between throughput and latency is not very tight—that is, latency buildup can occur prior to loss buildup—the DTGs nevertheless do provide insight into the latency or delay characteristics of paths between the ingress–egress node pairs. The network management system can then use this information to assist in judiciously admitting new applications that request entry into the network as described below.

DTGs are used by the Admission Control Function to determine whether there is sufficient available capacity to admit new QoS flow requests. The algorithm used by the ACF is described below.

1. Given a new flow request, the ACF first determines whether the sum of the requested bandwidths of all the flows for the requested class of service (including the incoming flow) is within the "quota" for that class of service (assuming that every service class is allocated a certain quota of the overall projected bandwidth). Such a quota is defined for every service class in order to prevent starvation of lower-priority classes, and it applies to the entire network. However, ensuring that all the flows of a class satisfy the quota over the entire network will lead to unnecessary overhead. Hence the quota is applied to every node. Note that this is an optimistic approach; the conservative approach would be to use a small fraction of the quota as a limit on the flows of a class at a node. If the quota is exceeded, the flow request is treated as described in Step 5 below.

2. If the flow request is within the allocated quota, then the ACF looks up the DTG for the requested class of service between the ingress node and the specified egress node. This DTG is used to extrapolate the projected traffic that is expected to be *received* at the egress node based on the projected outgoing traffic that is *sent* from the ingress node, by using a DTG lookup. The projected outgoing traffic is the sum of the bandwidth that is being requested by the new flow request, and the total bandwidth requested by all the currently admitted flows in this class; let this number be X. Then the projected amount of traffic that is expected to be *received* at the egress node is obtained by looking up the value on the *y-axis* of the DTG corresponding to the value X on the *x-axis*. Note that this lookup may involve extrapolation of the graph if the value X is not within the range of the past QoS operating points' values for *x*.

3. If the projected amount of received traffic is Y, then the ratio Y/X (i.e., the ratio of the projected incoming traffic to the projected outgoing traffic based on the DTG extrapolation) is referred to as the *projected QoS Operating Point*.

4. This projected QoS operating point is used by the ACF to make its admission decision. The ACF accepts the flow request if the projected QoS Operating Point is greater than or equal to a predefined DTG *threshold*, which is a policy-defined parameter that lies between 0 and 1 for each traffic class. This essentially implies that the network is expected to be in a stable state *after* accepting this request, based on its projected input/output values. The DTG threshold captures the amount of loss that is acceptable for that traffic class; for example, if a given class consists of only VoIP traffic, then a 5% packet loss is tolerable. The corresponding DTG threshold is 0.95; and the projected QoS operating point must be greater or equal to 0.95 for a flow to be admitted in this case.

5. If the ACF determines that the projected QoS Operating Point is less than the corresponding DTG threshold, or if the flow would exceed the

quota for the class (as described in Step 1), then the ACF consults pre-defined configuration policies to determine whether it should:

- Reject the flow request;
- Accept the flow request at a lower priority, should "room" be available for the flow in a lower priority class (this is again determined by using the DTGs as explained earlier); or
- Admit the flow as best effort traffic.

The observant reader will notice that there is a bootstrap issue here. What happens when the system starts and no DTG has been constructed because no data points have been received? In such a scenario, the ACF simply admits flow requests, since the bandwidth pipes are just beginning to get filled. Equivalently, this scenario can be viewed as being similar to operating on the linear portion of a default DTG—that is, below the DTG threshold—where all incoming requests are accepted. At the same time, the ingress–egress nodes begin collecting measurements for admitted flows, so that they can start to gather data points and thereby construct DTGs.

As an example, consider a flow request starting from an ingress node A to an egress node B. Assume that there is no other flow between A and B at this time in the class into which this new flow would be placed, if admitted. Also assume that there have been no flows in this class in the recent past. This means that the ACF at the ingress point does not have any history on which to base its decision. In this case, the ACF will decide to admit the flow into the network. The MCF at the egress point will send a report at the end of the reporting interval about the number of bytes received in the reporting interval. The ingress MCF will store this information. The process continues, until sufficient data points, in the form of byte counts, are gathered. The MCF uses these measurements to compute the QoS operating points (as described earlier) that capture the network state based on the gathered input and output data, thus constructing its DTG for use in subsequent admission request decisions.

Once a flow has been admitted, the MCF collects measurements about the flow to keep the DTG up to date. These measurements are then used for quality adjustment, in case network conditions change. Such adjustments are performed via the quality adjustment function (QAF), which is described next.

8.4.4.4.4 *Using Dynamic Throughput Graphs for Quality Adjustment.* The QAF depends on periodic throughput measurements being sent by every egress node to the ingress node. As described earlier, these periodic measurements are used to update the DTGs, in order to capture the most recent network state. The QAF periodically checks whether the network is in the stable region (see step 1 later in this section), using the latest DTGs. Based on these DTGs, if the network condition is determined to be stable, the QAF does not need to take any action.

On the other hand, if the network is not in the stable region, the QAF needs to identify flows that are candidates for downgrading, so that the network can be brought back to the stable region. The meaning of "downgrade" is defined based on policy and can range from preemption of the flow (i.e., blocking packets from the flow from entering the opaque network) to assigning a lower priority DSCP to the flow. In addition, several policies about which flows to downgrade can be specified. As an example, the search for flows to be downgraded can be restricted to flows within the service class for which instability is detected. As another example, an ordering of service class priorities could be specified, and the search starts with the lowest-priority service class and works upwards until it reaches the class in which instability is detected. A more detailed discussion of flow preemption is provided in the next section. In each service class, the QAF identifies the flows to downgrade. It stops downgrading flows when a sufficient number of flows have been identified for downgrading. The resultant network state can then be expected to be in the stable regime.

The QAF works as follows:

1. The QAF periodically checks the current QoS Operating Point against the predefined DTG threshold.
2. If the current QoS Operating Point is above the DTG threshold, the Quality Adjustment Function does nothing until the next periodic interval.
3. If the current QoS Operating Point is below the DTG threshold, the Quality Adjustment Function needs to perform preemption to avoid deterioration of the overall network throughput.
4. First, the Quality Adjustment Function needs to determine how much bandwidth to preempt by locating the QoS Operating Point on the DTG (interpolated based on past QoS Operating Points) that satisfies the DTG threshold. Since the QoS Operating Point corresponding to the DTG threshold gives the maximum amount of outgoing traffic that is allowed and the current QoS Operating Point gives the actual amount of outgoing traffic that is in the network, the amount of bandwidth that must be preempted is the difference between the two outgoing traffic amounts.
5. Next, the Quality Adjustment Function must preempt a sufficient number of flows to meet the required bandwidth reduction based on the existing preemption policy.

As an example, assume that the DTG threshold is defined as 0.9, and the current operating point is computed at 0.8 based on incoming traffic of 4400 Kb/s and outgoing traffic of 5500 Kb/s. This will trigger the Quality Adjustment Function to perform preemption. It first looks up a QoS Operating Point on the DTG that is greater than or equal to 0.9. Assuming that in this case the incoming traffic and outgoing traffic corresponding to the selected QoS

Operating Point are 4500 Kb/s and 5000 Kb/s, respectively, then the amount of bandwidth to be preempted is computed to be 500 Kb/s (i.e., 5500 Kb/s – 5000 Kb/s).

Recall that network stability as measured via the DTGs uses measurements of packet losses only. More sophisticated definitions of network stability can be constructed in case explicit latency measures are added via dynamic latency graphs (DLGs).

8.4.4.4.5 A Note on Flow Preemption. Flow preemption was discussed in the preceding sections at a high level. In this book, the *preemption* of a flow means that a flow that was earlier marked with a DSCP for which some level of QoS was being provided (e.g., EF or AF) gets "downgraded" in some fashion, by changing its DSCP marking either to Best Effort or to some other DSCP that is considered "lower" in importance than the original DSCP. In other words, the packets of the preempted flow are marked differently than before. Section 8.4.4.4.4 discussed flow preemption and how different strategies might be used for determining (i) what flows to preempt and (ii) how to treat the preempted flows. This section takes a look at these two issues in a little more detail.

In a DiffServ-based network, a number of traffic classes will be defined. Although the DiffServ standards allow for 64 traffic classes, in practice it is expected that the number of traffic classes will be far smaller. This is because the overhead in configuring and maintaining separate queues for a large number of traffic classes results in diminishing returns at some point. The DiffServ standards define some standard traffic classes that provide different per-hop behaviors in an IP network. The first of these is Expedited Forwarding (EF) [Davie et al., 2002], which is designed for traffic with stringent requirements on delay and is suited for real-time applications such as voice and video, which are delay-intolerant. For such applications, packets that arrive later than a certain bound are of no use; also, such applications can tolerate some amount of loss. The second set of traffic classes is the Assured Forwarding (AF) group [Heinanen et al., 1999], which is designed for traffic that cannot tolerate loss but can tolerate some amount of delay. Examples of such applications are file transfer, e-mail, and so on, which do not have stringent delay guarantees, but require that all sent packets are received at the destination. Packets that do not require QoS are not marked and are said to belong to the Best Effort class.

In a MANET, given that there is almost always insufficient bandwidth for all the applications that users would like to run, there is a need to prioritize traffic at a more granular level than can be done by the standard EF and AF classes. For example, there may be several applications that need to send video streams (which are typically in the EF class) over the air; however, some applications may be more important than others. Thus, it may be desirable to have multiple EF classes, possibly with different DSCPs, for video of different priorities. The differentiation of flows based on priority enables the QAF to preempt lower-priority video flows at the expense of higher-priority video flows. This capability is known as MLPP (Multi-Level Precedence and Preemption).

In order to implement MLPP, the QAF needs a specification of which flows can preempt other flows. For example, it may not be desirable to allow a video flow to preempt an FTP flow. On the other hand, it is certainly desirable to allow a higher-priority video flow to preempt a lower-priority video flow. Thus, configuration policies need to be in place that specify what flows (characterized by DSCP and priority) can preempt other flows.

An example is shown in Figure 8.5. The table on the left in Figure 8.5 shows the assignment of traffic types and priorities to DSCPs, while the table on the right specifies which flows can be preempted by which other flows. Note that the table uses the standard military MLPP-related priority values [National Communications System Technology & Standards Division, 1996] of "*flash-override*," "*flash*," "*immediate*," "*priority*," and "*routine*," in descending order of priority, for illustration purposes. The table on the right in Figure 8.5 is used by the QAF when determining which flows to consider for preemption when quality adjustment is required. For example, based on the first table, if a video flow with a priority of "*flash-override*" requests admission, it should be assigned DSCP 22. Now, if the Admission Control Function determines that this flow cannot be admitted without preempting other existing flows, then the flows that will be considered for preemption will be flows with DSCPs 20 and 21, in that order. If a data flow with priority "*immediate*" requests admission, it should be assigned DSCP 17. If the ACF needs to preempt other flows to admit this flow, the only flows that can be preempted are other flows with the same DSCP, but with lower priority, namely "*routine*" and "*priority*."

Flow Type	Flow Priority	DSCP
Network Control	Flash-override	23
Video	Flash-override	22
Video	Flash	21
Video	Immediate, Priority, Routine	20
Data	Flash-override	19
Data	Flash	18
Data	Immediate, Priority, Routine	17
Voice	Flash-override	16
Voice	Flash	15
Voice	Immediate, Priority, Routine	14

DSCP	Can preempt
23	17, 18, 19, 20, 21, 22, 14, 15, 16
22	20, 21
21	20
19	17, 18
18	17
16	14, 15
15	14

Figure 8.5. Examples: DSCP mapping and preemption policies.

TABLE 8.1   Preempted Flow Treatment Configuration Policies

Original DSCP	Downgrade DSCP
23	0
22	21, 20, 0
21	20, 0
20	0
19	18, 17, 0
18	17, 0
17	0
16	15, 14, 0
15	14, 0
14	0

The second issue is the treatment of the preempted flow. Again, configuration policies must be specified that tell the QAF how to remark the packets of preempted flows. As discussed earlier, packets of preempted flows could be marked as Best Effort or could be assigned a new DSCP, that is somehow accorded less preferential treatment in the network. Sample configuration policies specifying how to treat preempted flows are shown in Table 8.1. This table is used by the QAF to determine how to remark a flow upon preemption—for example, a flow whose DSCP was 21 can be preempted and downgraded down to DSCP 20, if there is room in the corresponding traffic class; or else it must be downgraded and remarked as Best Effort (DSCP = 0) upon preemption.

8.4.4.4.6   Putting It All Together: Implementation of QoS Assurance. The preceding sections described the details of admission control and quality adjustment algorithms that use red-side measurements to make admission and preemption decisions for application traffic flows requiring QoS. This section describes the implementation of these algorithms, by describing the steps that need to be performed for QoS assurance. The following steps are discussed:

- Application flow QoS request
- Application flow preemption

8.4.4.4.6.1   APPLICATION FLOW QOS REQUEST. At a high level, the steps performed for a QoS request are the following:

- An application that needs QoS for a traffic flow sends a request to the ACF within the NMS instance on the local node. The request contains relevant information about the traffic flow, including the flow priority, source and destination IP addresses and ports, application type, required bandwidth (including burst sizes), and flow duration (i.e., duration for which QoS is being requested).
- The ACF looks up the DSCP that should be assigned to this flow, based on the information passed in the QoS request. A mapping of flow type/

priority to DSCP (as shown in Figure 8.5) should be configured ahead of time within every NMS instance on every node.

- Having determined the DSCP that should be assigned to the flow, the ACF uses the DTG-based admission control algorithm described earlier to determine whether or not to admit the flow.
- If the Admission Control Function determines that there is sufficient bandwidth in the network to admit the flow:
  - The ACF returns a status message to the requesting application, indicating that the flow has been successfully admitted.
  - The ACF configures the source host for the flow to mark packets belonging to this flow with the appropriate DSCP. It also configures the source host to police the flow, using the bandwidth and burst sizes specified in the flow request.
  - The ACF updates its database with information about the admitted flow.
- If the ACF determines that there is *not* sufficient bandwidth in the network to admit the flow, there are two possibilities:
  - If there exist flows in the network that can be preempted (based on the specification of which flows can be preempted, as discussed in Section 8.4.4.4.5 and illustrated in Figure 8.5), then one or more flows (based on the bandwidth required) are preempted. Preemption takes place based on the preemption flow treatment policies (see Table 8.1). If a flow is to be preempted by assigning it a new DSCP that is not the Best Effort DSCP, then the ACF must first check whether or not the new DSCP can accommodate this flow. If not, the next DSCP specified in the preemption flow treatment policies is tried, and so on, until a DSCP can be found where the flow can be admitted. The worst case here is that the flow will be downgraded to Best Effort. The ACF informs the preempted flows about the preemption, and it reconfigures the source hosts for the preempted flows so that flow packets are marked with the new DSCP. Any policers configured for those flows are also updated, or removed in the case where the flow is downgraded to Best Effort.
  - If there do not exist any flows in the network that can be preempted in favor of the new flow, then a response is returned to the requesting flow, denying admission to the flow. The result of the denial is that the flow may be sent as Best Effort.

8.4.4.4.6.2 APPLICATION FLOW PREEMPTION. Application flow preemption may occur under the following circumstances:

- When the ACF has no room to admit an incoming flow request and therefore must preempt one or more existing flows (this case was discussed in the previous section as part of the QoS request discussion).

- When the congestion state of the network crosses a policy-specified threshold: This case is known as *Quality Adjustment.*
- When the duration of a flow that was granted QoS expires.

The second and third cases are discussed below.

8.4.4.4.6.2.1 Quality Adjustment. Given the fact that the capacity of a MANET can vary dynamically, it is necessary for any viable QoS solution for MANETs to periodically assess the congestion state of the network and to preempt flows as needed if the congestion status of the network exceeds a policy-defined threshold. As discussed in Section 8.4.4.4.4, this function is performed by the Quality Adjustment Function (QAF). The following steps are performed for quality adjustment:

- The QAF periodically checks whether the network is in the stable region—that is, whether the current QoS operating points are within the policy-defined DTG thresholds, for all the DTGs.
- If not, the QAF identifies which flows should be preempted (based on the preemption policies), and it preempts them as described in Section 8.4.4.4.5.

8.4.4.4.6.2.2 Flow Duration Expiration. When a flow requests QoS, it must specify a duration for which QoS is needed. When this duration expires, the flow is preempted, as described earlier. The process followed is:

- When a flow is admitted, the QAF is informed about the flow and its duration.
- When the flow duration expires, the QAF preempts the flow as explained earlier.

8.4.4.5 Interdomain QoS Management. The subject of interdomain policy management was discussed in Chapter 6. The question of how to provide end-to-end QoS assurances for traffic originating in one administrative domain and terminating in a different administrative domain was deferred to this chapter, since an understanding of admission control was required for this discussion. Consider the situation where two different administrative domains D_1 and D_2 need to support end-to-end QoS assurances for application traffic flows. Assume that each domain supports DiffServ-based QoS and that domain D_1 implements the QoS solution described in this chapter. Since domain D_1 does not have visibility into the congestion status of paths in domain D_2, an SLA must be defined between the two domains to regulate the amount of traffic per traffic class that is allowed to go from D_1 to D_2 and vice versa. For simplicity, assume that there is only one border node in D_1 that connects D_1 and D_2. This border node must enforce the traffic limits prescribed by the SLA.

One possible solution to the QoS control problem is the following. Assume that the NMS on the border node can obtain statistics about the amount of traffic flowing from domain D_1 to D_2 in each traffic class. Note that this requires passing a small amount of black-side monitoring information from the black side to the red side on the border node. The ACF on the border node periodically broadcasts a short message to all the ACFs on all the nodes in domain D_1, informing them about the amount of capacity remaining for traffic flows from domain D_1 to D_2. This number is computed by the ACF on the border node by subtracting the actual amount of traffic flowing in each traffic class into domain D_2 (obtained from black-side monitoring) from the upper bound specified in the SLA for this traffic class. Each ACF within domain D_1 uses this number to determine whether or not to admit a flow request that will terminate in domain D_2.

Note that this approach suffers from the drawback that all nodes in domain D_1 may all admit a large amount of traffic destined for domain D_2 at around the same time, resulting in excess traffic being sent to the border node. However, the same risk exists for the DTG-based admission control approach, because it may happen that multiple nodes in the network simultaneously start sending traffic that transits through the same bottleneck node, creating sudden congestion in the network. Here the border node becomes responsible for policing traffic traversing into domain D_2. The nodes in domain D_1 will eventually receive messages from the border node about the lack of available capacity. In the case where the amount of traffic being sent by nodes in domain D_1 to domain D_2 exceeds the SLA, the periodic broadcast message from the border node will contain a negative number for the available capacity, thus providing an indication to nodes in domain D_1 that they should reduce the amount of traffic being sent to domain D_2.

The situation becomes more complicated if there is more than one border node between the two domains. Routing policies can be configured so that predefined percentages of the traffic from domain D_1 to D_2 are routed via different border nodes. Although this is a challenging task in and of itself [Caesar and Rexford, 2005], automated tools are available to assist in evaluating the effect of changing link weights dynamically to achieve this effect [Feamster et al., 2004]. In this case, the SLA-defined bandwidth is split among the border nodes based on these percentages. Broadcast messages are sent by each of the border nodes as described before, indicating the remaining amount of capacity at each node (computed as before); and nodes within domain D_1 admit traffic as long as there is capacity remaining within any border node.

8.5 SUMMARY

This chapter presented issues, models, and operations related to MANET performance management. As in the case of fault management, the performance management model needs to be integrated with the remainder of the

FCAPS functionalities and with a policy infrastructure in order to be able to adapt to the dynamics of MANETs. In particular, the capability to control performance monitoring using policies is a very useful feature, since it enables self-adjustment of the network monitoring system based on network status. The network management system can thereby adapt to changing network conditions in an automated fashion.

The key contribution of this chapter is the description of a QoS management mechanism that provides admission control and quality adjustment for MANET environments. While service assurance in wireless networks is in itself a challenge, the random mobility, fluctuating links, and lack of visibility into the network topology in MANETs complicate the matter further. A distributed QoS management solution was presented that makes use of dynamic throughput measurements to estimate and infer the state of encrypted black network segments. These QoS mechanisms can deal with both (a) admitting new application traffic flows into the network and (b) reallocating network resources among the admitted applications so that the high-priority applications continue to receive the negotiated QoS levels when the underlying MANET resources dwindle or fluctuate.

9

SECURITY MANAGEMENT

This chapter presents security-related topics that are relevant for managing the security of MANETs. A brief overview of MANET security management is provided in Section 9.1. This is followed by a description of the key security management functions and related operations process models for MANETs in Section 9.2. The remainder of this chapter then expands upon each of the security management functions identified in Section 9.2. Section 9.3 discusses the use of security policies for role-based access control and provides an overview of a standards-based language used for expressing role-based access control policies. Section 9.4 discusses key management, including an overview of encryption, public, and private keys; the use of a public key infrastructure for distributing public keys; and other key distribution and revocation mechanisms. Section 9.5 provides an overview of communications security using the IETF-defined IPSec standards for securing Internet communications. This section also discusses the use of a HAIPE (High Assurance Internet Protocol Encryptor) device for encryption, the concepts of red–black separation in MANETs, and the implementation of multiple levels of security via cross-domain guards. Finally, Section 9.6 introduces the topic of intrusion detection, and it also discusses an intrusion detection architecture for MANETs. A brief summary is provided in Section 9.7.

9.1 OVERVIEW

The security management component of a network management system is responsible for managing the overall security of the network. Just like the other FCAPS management components (i.e., configuration, fault and performance management), security management for MANETs is more complicated and challenging than for wireline networks, for reasons cited in the previous

Policy-Driven Mobile Ad hoc Network Management, by Ritu Chadha and Latha Kant
Copyright © 2008 John Wiley & Sons, Inc.

three chapters. In particular, the wireless nature of MANETs enables easier eavesdropping and renders the MANET as a whole more vulnerable to intrusion-related attacks. Furthermore, the presence of multiple levels of security (MLS) with restrictions on how information is allowed to be exchanged across security domains introduces additional challenges. These challenges, combined with typical MANET characteristics such as bandwidth scarcity and random mobility, gives rise to the need for an *adaptive* security management architecture for MANETs. Additionally, the security management operations in MANETs have an intricate relationship with the other FCAPS tasks due to need for (a) on-the-fly reconfigurations in response to security attacks and (b) integrated root cause analysis in response to network problems that cause disruption of services. In light of all of the above, policy-driven models for MANET security management operations that have the potential to adapt to the dynamics of the underlying MANET are presented in this chapter.

9.2 SECURITY MANAGEMENT FUNCTIONS AND OPERATIONS PROCESS MODELS

This section describes the operations process models for the security management task in MANETs. To better understand the operations processes related to security management, it is useful to first take a look at the key security functions and services, which are listed below.

- *Controlling Access to Critical System Resources*: This involves monitoring and restricting access to critical system resources. Examples of critical system resources include servers (e.g., name resolution server, QoS server, and mobility server, to name a few) and managed network elements (e.g., routers, switches). As can be easily seen, tampering with any of the above-mentioned resources can result in undesirable consequences. As an example, if the name resolution server can be accessed by a malicious user, such a user could cripple the entire network by corrupting the name–address resolution service in such a way that none of the applications can find and communicate with each other. The security management component therefore has the important task of controlling access to such critical system resources.

- *Securing Communications via Encryption*: This involves providing appropriate encryption algorithms so as to secure communications that occur over wireless links. Recall that wireless links are very amenable to "eavesdropping." Thus, efficient encryption techniques are critical to securing such links.

- *Providing for Authentication and Nonrepudiation*: Authentication involves checking that users are indeed who they say they are. More specifically, proper authentication guarantees that an unauthorized user will not be able to masquerade as an authorized user, thus protecting the system from Byzantine attackers. Also, authentication allows for the use of digitally signed files to ensure that data files have not been tampered with. Non-repudiation, on the other hand, is the ability to guarantee that a user cannot claim to not have done something that the user did actually do.

- *Protecting against Network Attacks*: This involves monitoring the network for any security-related attacks and providing appropriate solutions should the network be compromised. An example of a security-related attack that was also mentioned in Chapter 7 while discussing MANET fault scenarios is a denial of service (DOS) attack. Typically, in a DOS attack, a malicious user starts to flood the network with unwanted (junk) traffic, thereby denying legitimate network users the use of the underlying network and/or its services. For example, a DOS attack may result in denying new applications use of the network or in denying existing users access to services (e.g., name resolution) provided by a network server (achieved by keeping the network server busy serving "idle" queries).

- *Implementing Multiple Levels of Security (MLS)*: This involves providing different levels of security, based on the level of protection required by different types of data and applications. Multiple levels of security are typically required in military settings where data of different classifications (such as top secret, secret, unclassified) may coexist on a single platform. Such data must be kept separate—on different LANs—which are either physically disconnected or where data are allowed to cross from one LAN to another only via a *cross-domain guard*, which is programmed with rules about what data are allowed to cross from one LAN to another. Data of one classification may also be allowed to transit over a network of a different classification, provided that it is encrypted prior to entering the other network.

Having provided a brief overview of the key security management functions that need to be performed for MANETs, the next topic addressed is how the above-mentioned security management tasks fit in with the overall network management architecture for MANETs. Figure 9.1 shows a schematic of the security management tasks and their relationship to other management components.

The left-hand side in Figure 9.1 shows a set of example triggers that are input to the security management task. Note that these input triggers stem both from other management components (e.g., the trigger from the configuration management task in Figure 9.1) and from security-related components (e.g., from the component labeled "Security Anomaly Detection"). For example, the trigger on the top left-hand corner in Figure 9.1 is from the configuration management task, whereby the latter requests the security management task for permission to access a protected resource before it actually performs the (re)configuration operation. The configuration request from the

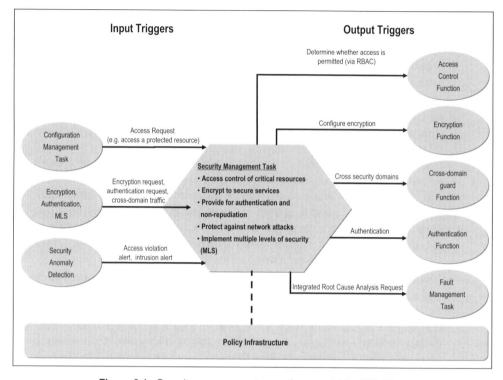

Figure 9.1. Security management operations model for MANETs.

configuration management task may in turn have originated from either (i) the configuration management task itself or (ii) the configuration management task in response to another management task. As an example of (i), the configuration management task could be invoked to perform system initializations during mission deployment. As an example of (ii), one sample outcome of the root cause analysis function of the fault management task might be to reconfigure the DNS server, in which case the trigger from the configuration management task shown in Figure 9.1 may be in response to a hard or soft failure event.

Another example of an input trigger is the case when there is a request to configure encryption, or a request to authenticate an incoming user. This is shown as a trigger in to the security management task from the bubble labeled "Encryption, Authentication, MLS" on the left-hand side in Figure 9.1. Another such trigger is the sending of traffic from one security domain to another; such traffic must be examined to verify whether its content satisfies predefined rules about what traffic is allowed to cross security domains. A final example input trigger is the case when there is a security-related anomaly detected (e.g., from a device that houses the network intrusion detection system (NIDS), as explained in Section 9.6.1.1). This is shown as a trigger from the bubble labeled "Security Anomaly Detection" in Figure 9.1.

On its output side, the security management task essentially performs the functions listed within the central hexagonal box. Thus, for example, the output trigger to the "Access Control Function" shown on the upper right-hand corner of Figure 9.1 is essentially a request to perform the required access control checks in order to accomplish a task that requires access to a protected resource. An example of such a task is a configuration request, which should only be performed by authorized entities. Similarly, the output triggers to the functions "Encryption Function" and "Authentication Function" are requests to configure a certain requested service encryption and to verify user authentication, respectively. In order to accomplish the above, the security management task configures the appropriate encryption algorithm or invokes the authentication handshake, respectively.

The output trigger labeled "Cross Security Domains" into the cross-domain guard function is a trigger to perform the required checks prior to permitting cross-domain traffic. The cross-domain guard function must verify that any predefined rules about what traffic is or is not allowed to cross from one security domain to another are satisfied prior to allowing traffic across domains.

The output trigger labeled "Integrated Root Cause Analysis Request" into the fault management task is a trigger that is generated by the security management task and sent to the fault management task when a security-related anomaly is observed (e.g., an alert generated by the NIDS). Observe that one of the key functionalities of the security management task is to protect against network attacks. To perform this function, the security management task first processes the security-related results (e.g., duplicate suppression and filtering) and then forwards security alarms to the fault management task for further analysis of the problem, via an integrated root cause analysis, as indicated in Figure 9.1.

9.3 SECURITY POLICIES

Chapter 2 introduced policy terminology and discussed *access control policies*, which are used to control access to protected resources. This section begins with an overview of role-based access control, and describes XACML (eXtensible Access Control Markup Language), an XML-based language standardized by OASIS (Organization for the Advancement of Structured Information Standards) for expressing access control policies.

9.3.1 Role-Based Access Control

One of the tasks that a Security Management component needs to perform is controlling access to certain resources, such as managed network elements, servers, bandwidth, and so on. There are several approaches to access control, the most popular of which today is Role-Based Access Control (RBAC). RBAC allows an administrator to define a series of *roles* and to assign them to *subjects*, which are the entities that may need to access protected resources. The protected resources themselves are categorized as being of different *types*. Finally, the administrator defines which roles have access to which object types. This saves the administrator from the tedious job of defining permissions per user, since permissions are associated with roles rather than with individual users. The use of the role paradigm allows each user to have multiple roles; also, each role can have multiple users. When a user is assigned a role, the user automatically obtains all the access permissions associated with that role. Note that different entities can administer the mappings of roles to permissions and users to roles.

Other access control paradigms used in the past include DAC (Discretionary Access Control) and MAC (Mandatory Access Control). With DAC, the subject has complete control over the objects that it owns and the operations that it performs on those objects. With MAC, the administrator manages access controls and defines access policies, which users cannot modify. The advantage of RBAC over MAC and DAC is that it saves the administrator from the tedious job of defining permissions per user.

Within the RBAC framework:

- A *user* is a person or entity that needs to access protected resources, or *objects.*
- A *role* is a collection of job functions.
- An *operation* represents a particular mode of access to a set of one or more protected RBAC *objects.*

Roles can be organized into hierarchies so that permissions can be inherited based on the role hierarchy. A sample role hierarchy is shown in Figure 9.2. Here, a hierarchy of roles is shown for doctors. Every specialist is also a doctor; and every cardiologist and rheumatologist is a specialist. Thus any access that is granted to a user with role = doctor will automatically also be granted to any user with role = specialist, cardiologist, and rheumatologist.

The use of RBAC simplifies access control specifications and reduces the number of permissions that need to be kept track of. However, this simplification does come at a price, which is an increased complexity of access control checks. An engine that checks whether an entity is allowed to access a resource has to determine which role(s) a user belongs to, look up the role hierarchy, and finally look up the access control rules to arrive at a decision.

RBAC has fostered a change from the traditional MAC/DAC dichotomy to policy-driven access control models. Access control policies are used to define which roles have access to which resources. At a high level, an access control policy defines the following:

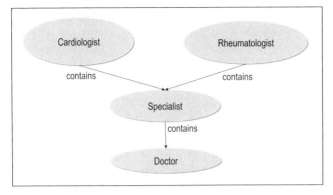

Figure 9.2. Illustration of role hierarchy.

- **Targets:** These include
  - *Resources*: These are the protected resources.
  - *Subjects*: Identity of the entity attempting to access the resources.
  - *Actions*: Operations that the subject wants to perform on the protected resources.
- **Effects:** This defines whether or not access is allowed, and takes the values "Permit" or "Deny."
- **Conditions:** Additional conditions can be defined on the access, if it is conditional. As an example, access to a specific resource by a set of users could be permitted only at specific times of day.

9.3.1.1 XACML (eXtensible Access Control Markup Language).

XACML [Moses, 2005] is an OASIS standard that describes both a policy language for access control and an access control decision request/response language, both written in XML. The policy language is used to describe general access control requirements, and has standard extension points for defining new functions, data types, combining logic, and so on. The request/response language lets users formulate queries to ask whether or not a given action should be allowed, and interpret the result. The response always includes an answer about whether the request should be allowed using one of four values: Permit, Deny, Indeterminate (an error occurred or some required value was missing, so a decision cannot be made), or Not Applicable (the request cannot be answered by this service).

XACML defines rules as

- Targets
- Effects
- Conditions

A target includes

- Resources
- Subjects
- Actions

An Effect is either "Allow" or "Deny." Conditions are predicates and attributes defined in the XACML specification.

XACML defines the concept of policies and policy sets. A *PolicySet* is a container that can hold other Policies or PolicySets, as well as references to policies found in remote locations. According to the terminology introduced in Chapter 2, XACML policies fall into the category of access control policy. A XACML policy represents a single access control policy, expressed through a set of rules. These policies are described in XML. Each XACML policy document contains exactly one Policy or PolicySet root XML tag.

If multiple policies exist that apply to a given target, there is a potential for modality conflicts (see Chapter 4 for a description of modality conflicts). A Policy or PolicySet may contain multiple policies or rules, each of which may evaluate to different access control decisions, leading to a need for mechanisms to resolve such conflicts. XACML incorporates mechanisms for resolving modality conflicts. Conflicts are resolved by allowing specification of conflict resolution mechanisms called *Combining Algorithms*. Each such algorithm represents a different way of resolving conflicts and combining multiple decisions into a single decision. Algorithms can be specified for combining the results obtained from policies (called Policy Combining Algorithms) or for combining the results obtained from rules (called Rule Combining Algorithms). An example of a combining algorithm is the *Deny Overrides* Algorithm, which says that no matter what, if any evaluation returns a "deny" decision, or if none of the evaluations return a "permit" decision, then the final result is also "deny." These combining algorithms are used to build up increasingly complex policies. The XACML standard defines seven standard combining algorithms. Additional combining algorithms can be written by users and included in policy specifications for conflict resolution.

The basic policy enforcement mechanism used in XACML is illustrated in Figure 9.3. A requestor, which is an entity that needs to access a protected resource, submits a request for access to the *Policy Enforcement Point* (PEP). The PEP is a piece of software that is responsible for receiving access requests; in effect, it is the "gatekeeper" to the protected resource. The PEP sends the request to a *Policy Decision Point* (PDP). The PDP then looks up the relevant policies in a policy repository, and it evaluates whether the request should be permitted or denied, based on the applicable policies. The decision (permit or deny) is returned to the PEP, which then allows or denies access accordingly.

In order to illustrate the use of XACML, the Figures 9.4 to 9.6 show a policy that contains two rules. The policy applies to requests to access the resource

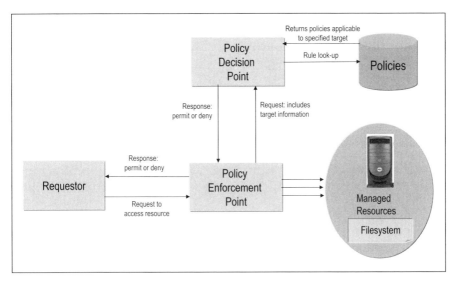

Figure 9.3. XACML policy enforcement.

```
<Policy PolicyId="SamplePolicy" RuleCombiningAlgId="urn:oasis:names:tc:xacml:1.0:rule-
    combining-algorithm:permit-overrides">

<Target>
<Subjects> <AnySubject/> </Subjects>
<Resources>
    <ResourceMatch MatchId="urn:oasis:names:tc:xacml:1.0:function:string-equal">
        <AttributeValue
        DataType="http://www.w3.org/2001/XMLSchema#string">Server1</AttributeValue>
        <ResourceAttributeDesignator DataType="http://www.w3.org/2001/XMLSchema#string"
        AttributeId="urn:oasis:names:tc:xacml:1.0:resource:resource-id"/>
    </ResourceMatch>
</Resources>
<Actions> <AnyAction/> </Actions>
</Target>
```

Figure 9.4. XACML policy: Target.

called "Server1." Recall that a XACML rule contains targets, effects, and
conditions. Figure 9.4 shows the specification of the target for this policy. A
target includes resources, subjects, and actions. Here both the subjects and the
actions are specified as "any"; that is, the policy applies to any subject and
allows any action by that subject. The protected resource here is Server1, as
specified by the "Resource" tags. Another important point to note is that the
"Policy" tag specifies a conflict resolution mechanism, which is the "permit-
overrides" rule. This means that if any of the rules evaluates to "permit," then
access is permitted.

In Figure 9.5, the first rule of the policy is shown. This rule has both subjects
and resources specified as "any"; that is, the rule applies to any subject and
any resource. The action here is a "login"; and a condition is specified that

```
<Rule RuleId="LoginRule" Effect="Permit">
<Subjects> <AnySubject/> </Subjects>
<Resources> <AnyResource/> </Resources>
<Actions>
    <ActionMatch MatchId="urn:oasis:names:tc:xacml:1.0:function:string-equal">
    <AttributeValue DataType="http://www.w3.org/2001/XMLSchema#string">login
    </AttributeValue>
    <ActionAttributeDesignator DataType="http://www.w3.org/2001/XMLSchema#string"
    AttributeId="ServerAction"/>
    </ActionMatch>
</Actions>
</Target>
<Condition FunctionId="urn:oasis:names:tc:xacml:1.0:function:and">
    <Apply FunctionId="urn:oasis:names:tc:xacml:1.0:function:time-greater-than-or-
    equal"
    <Apply FunctionId="urn:oasis:names:tc:xacml:1.0:function:time-one-and-only">
    <EnvironmentAttributeSelector DataType="http://www.w3.org/2001/XMLSchema#time"
    AttributeId="urn:oasis:names:tc:xacml:1.0:environment:current-time"/>
    </Apply>
    <AttributeValue DataType="http://www.w3.org/2001/XMLSchema#time">09:00:00
    </AttributeValue>
    </Apply>
    <Apply FunctionId="urn:oasis:names:tc:xacml:1.0:function:time-less-than-or-equal"
    <Apply FunctionId="urn:oasis:names:tc:xacml:1.0:function:time-one-and-only">
    <EnvironmentAttributeSelector DataType="http://www.w3.org/2001/XMLSchema#time"
    AttributeId="urn:oasis:names:tc:xacml:1.0:environment:current-time"/>
    </Apply>
    <AttributeValue DataType="http://www.w3.org/2001/XMLSchema#time">17:00:00
    </AttributeValue>
        </Apply>
</Condition>
</Rule>
```

Figure 9.5. XACML policy: Login rule.

```
<Rule RuleId="FinalRule" Effect="Deny"/>
</Policy>
```

Figure 9.6. XACML policy: Catch-all rule.

applies only if the Subject is trying to log in between 9 am and 5 pm. The purpose of this rule is to determine whether a given subject should be allowed to log in or not. This rule only applies when the requested action is a login action.

Figure 9.6 shows the second rule of the policy. This rule is a "catch-all" rule that is used if the first one doesn't apply. This rule always returns a "deny." It should be noted that rules are evaluated in order.

The reader should note that the rule-combining algorithm that was specified (permit overrides) will result in permitting access if either of the two rules permits access. Since Rule 2 will never permit access, access will be permitted if and only if the first rule explicitly permits access.

In summary, the chief advantage of XACML is that it was designed to be generic and is not designed for any specific environment or kind of resource.

Policies that can be used by many different kinds of applications can be written, making policy management easier. It allows distributed policy definition; policies can be written that refer to other policies kept in arbitrary locations. The result is that rather than having to manage a single monolithic policy, different people or groups can manage portions of policies as appropriate, and XACML can combine the results from these different policies into one decision. The base language can be extended and there are standards groups working on extensions and profiles that will hook XACML into other standards like SAML (Security Assertion Markup Language) and LDAP, which will increase the number of ways that XACML can be used. Finally, XACML is a standard that will aid interoperability once it is widely adopted.

9.3.2 Firewall Access Control Lists

An important aspect of securing any network is to configure firewalls that control and filter what traffic is permitted to enter or exit the network. Firewall functionality is either (a) provided on stand-alone hardware devices or (b) hosted on network equipment such as a router or even a general-purpose host computer. The only requirement for firewall placement is that all traffic that needs to be filtered must pass through the firewall, so that the firewall can examine the traffic and drop any traffic that must be blocked, based on the specified firewall rules.

This section provides an overview of firewalls and firewall rules, and discusses the use of firewalls in MANETs. In particular, the aspects of firewall configuration in MANETs that differ from firewall configuration in traditional wireline networks are pointed out.

9.3.2.1 Types of Firewalls. The most basic functionality of a firewall is to (a) filter packets by examining each packet transiting the firewall and (b) either drop it or transmit it, based on a set of firewall rules. The firewall rules specify packet header matching rules that the firewall software applies to each IP packet to determine whether or not to drop the packet. These rules typically involve examination of the source and destination IP addresses, source and destination ports, and protocol of the incoming IP packet. If these header values match any of the firewall rules, and the rule states that the traffic should be denied, then the firewall blocks the packet (i.e., does not transmit it).

Firewalls have evolved greatly since their inception in the 1980s, when firewalls provided the simple packet filtering functionality described above. The next generation of firewalls provided *stateful inspection* of packets, where the *state* refers to the state of a connection. Thus when a packet is received at the firewall, the firewall uses its knowledge about the state of existing connections to determine whether this packet is part of an existing connection or is the initiation of a new connection. This information is helpful in detecting and blocking certain types of attacks that use existing connections.

The next level of sophistication provided by firewalls is *deep packet inspection*, which is a filtering technology that is aware of several standard protocols (such as TCP, DNS, FTP, HTTP, etc.), and can detect known attacks carried out using these protocols. The difference between deep packet inspection and simple packet filtering is that in the former, the payload of an IP packet is examined, as opposed to just examining its header. Firewalls that provide deep packet inspection are basically providing intrusion detection capabilities (see Section 9.6 for a discussion of intrusion detection) as well as packet filtering capabilities. Deep packet inspection uses both signature-based analysis and statistical techniques for detecting anomalies. Further discussion of intrusion detection is deferred to Section 9.6.

9.3.2.2 *Firewall Rules Specification.*
Firewall packet filtering rules are often referred to as firewall *policies*, since they capture the network administrator's policy about what traffic should and should not be allowed in the network. According to the definition of policies in Chapter 2, firewall rules are configuration policies. This may be confusing to some readers, because firewall rules are often also called *access control lists*, which may be confused with the *access control policies* defined in Chapter 2. However, the two concepts are quite different and should not be confused. An access control list for a firewall typically contains the following information:

- Source IP address
- Destination IP address
- Source port
- Destination port
- Protocol
- Permit or Deny

Any number of the above fields can be specified as wild cards, indicating that the rule should match all possible values for that field. Typically, multiple access control lists are defined in a firewall and are evaluated in the order in which they are defined. For example, consider the following two firewall rules:

- Rule 1:
  - Source IP address: 192.168.0.0/16
  - Destination IP address: 10.1.1.0/24
  - Source port: *
  - Destination port: *
  - Protocol: *
  - Permit or Deny: Permit

- Rule 2:
  - Source IP address: *
  - Destination IP address: *
  - Source port: *
  - Destination port: *
  - Protocol: *
  - Permit or Deny: Deny

Rule 1 above states that any IP packet with a source address belonging to the subnet 192.168.0.0/16 and a destination address belonging to the subnet 10.1.1.0/24, with any source and destination ports and protocol, should be permitted. The second rule states that all IP packets should be denied, regardless of their header contents (since all the firewall rule fields other than the "deny" are wild cards). These rules are evaluated in the order of their specification. Thus, the net result of defining these two rules is that the only traffic that will be permitted through the firewall is traffic with a source address belonging to the subnet 192.168.0.0/16 and a destination address belonging to the subnet 10.1.1.0/24. This traffic is permitted based on the first rule. For all other traffic, since the first rule does not apply, the second rule will be applied and will result in all traffic being denied.

Note that the order of the firewall rules is critical here; if the order of the two rules above is reversed, then all traffic will be blocked at the firewall, since the first rule (Rule 2 above) would apply to all traffic. The second rule (Rule 1) would never be evaluated, because the first rule would be applied to all traffic.

9.3.2.3 Firewalls for MANETs. The issue of defining firewall rules for MANETs is examined next. Referring back to the MANET architecture from Chapter 5, recall that all application traffic originates and terminates on the red side of the network, with the possible exception of a limited amount of traffic that crosses the red–black boundary for management purposes, as will be discussed in Section 9.5.5. Thus firewalls must be in place on every MANET node, on the red side of the network, to filter application traffic that enters and exits the node, as shown in Figure 9.7.

The cross-domain guard shown in the figure filters traffic that crosses the red–black boundary, and the firewall filters traffic that enters and exits the red side of the MANET node, prior to encryption.

The one point that needs to be kept in mind when configuring firewalls for a MANET, which is different from standard wireline networks, is the fact that firewall rules may need to be dynamically reconfigured due to mobility. An example of this is the following: Suppose that a MANET is being used for a military mission, where certain information is only allowed to flow between certain groups of MANET nodes, based on their location or other criterion. For example, information from a group of sensors may only be allowed to be

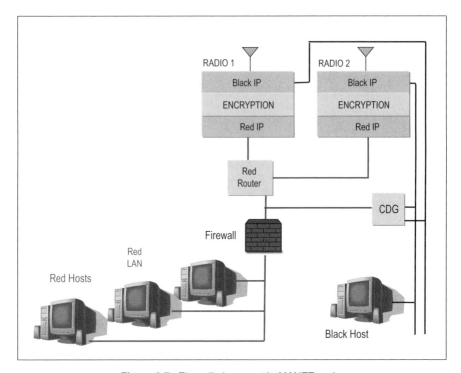

Figure 9.7. Firewall placement in MANET node.

sent to a MANET node that is responsible for controlling these sensors and managing them. Firewall rules would be put in place to ensure that the sensor communications can only flow between the managing MANET node and the group of sensors, and nowhere else. Now, if the MANET node managing the sensors moves away and is replaced by another MANET node, this new MANET node must be able to communicate with the sensors; and the former node (that moved away) should no longer be allowed to communicate with the sensors. This requires a dynamic modification of the firewall rules on multiple MANET nodes. Thus mechanisms must be in place to perform this dynamic reconfiguration. This can be accomplished via appropriate policies that trigger the reconfiguration based on mobility events.

Firewalls may also need to be dynamically reconfigured to cut off network intrusions. If a network intrusion is detected (see Section 9.6 for a discussion of intrusion detection), there may be a need to isolate a compromised or malicious MANET node from the rest of the network. This is easily achieved by configuring new firewall rules for this purpose. Again, this can be accomplished by the use of a policy that triggers the reconfiguration based on a root cause event generated by the Fault Management root cause analysis component, as discussed in Chapter 7. A scenario illustrating this was provided in the fault scenarios section of Chapter 7.

9.4   KEY MANAGEMENT

One of the prime security considerations in a MANET is the need to encrypt over-the-air traffic. The wireless medium is intrinsically prone to snooping and attack, given its susceptibility to eavesdropping. In order to secure wireless traffic from eavesdropping, encryption is required. The field of cryptography is a vast and rich field of research, and an in-depth treatment of cryptography is beyond the scope of this book; however, in the remainder of this section, a high-level overview of the field is provided in order to give the reader an understanding of (a) the functions that need to be provided to secure a MANET and (b) the key management mechanisms that need to be put in place to support these functions.

There are several aspects of security that needs to be considered in any type of network, but are essential for wireless networks. In this section, mechanisms for securing electronic communications are discussed. Again, the purpose of this section is not to provide a comprehensive treatment of cryptography and related algorithms; rather, the section provides a flavor of the security-related capabilities that need to be put in place and managed for MANETs. In other words, the objective is to describe the "*what*" rather than the "*how*" of communications security mechanisms. The reader who is interested in the technical details of the implementations of cryptographic algorithms is referred to the large body of literature on the subject [Stallings, 2003].

In the remainder of this section, mechanisms for achieving confidentiality, message integrity, and user authentication will be described. First, a quick introduction is provided to some encryption and keying concepts that will be needed for explaining how to ensure security of over-the-air communications.

9.4.1   A Quick Overview of Cryptography

Encryption is the process of encoding data in a way that makes the original content of the data undecipherable. There exists a large body of algorithms that can be used to encrypt—and, conversely, to decrypt—data. Since the algorithms themselves are usually published and therefore well known, it is imperative that they be parameterized in some way that allows the result of encryption to be undecipherable. Therefore these algorithms rely on the existence of one or more *keys*, which are secret parameters that are used in conjunction with the algorithms to encrypt and decrypt data.

When data are sent from one MANET node to another, it must be encrypted at the source and decrypted at the destination. Thus, the sender and receiver must coordinate ahead of time with respect to the encryption keys that will be used for encryption/decryption. In general, the sender and receiver can use either the same key (known as *symmetric keys*) or different keys (known as *asymmetric keys*).

The use of symmetric keys requires both the sender and receiver to be in possession of the same key, which in turn requires the existence of a mechanism to agree on a key. One way of sharing keys is to manually place a physical copy of a key on every pair of nodes that needs to communicate. This is naturally a laborious, manual, and error-prone process. Another way is to use some electronic mechanism for sending key information to other nodes. Since this sending of keys is done outside of the normal encrypted channel, some other means must be used for securing this data exchange, which leads to further complications. Another problem with using symmetric keys is that a given user must use a different key for each party that it communicates with and must therefore keep track of multiple keys.

Public key cryptography, which makes use of asymmetric keys, was developed in order to address the above difficulties with sharing symmetric keys in a safe and reliable fashion. Using public key cryptography, users can communicate securely over an insecure channel without having to agree upon a shared (symmetric) key beforehand. This approach makes use of *asymmetric* keys. Asymmetric keying still requires coordination between the sender and receiver, as will be explained below; however, the sender and receiver do not need to exchange actual key information over the air, since they do not need to be in possession of the same key. An analogy that can be used to understand the use of symmetric and asymmetric keys is the following. Suppose that a sender S needs to communicate with a receiver R. If S and R are using symmetric keys, then S must encrypt the message with its key K and send the message to R. R must then use the same key K to decrypt the message, and can also use the key K to encrypt its reply and send it to R. If S and R are using asymmetric keys, however, S and R have different keys K_s and K_r. S and R do not need to know each others' keys; however, using the analogy of physical locks and keys, they do need to be in possession of each others' "locks." Thus S and R must have exchanged their "locks" L_s and L_r ahead of time. When S sends a message to R, S uses R's lock L_r to "lock" (or encrypt) the message. When the message reaches R, R can use its own key K_r to "unlock" the lock L_r. Similarly, when R responds to S, it uses S's lock L_s to encrypt the message. When the message reaches S, S uses its key K_s to "unlock" the lock L_s.

In summary, when using symmetric keys, the sender S and receiver R must be in possession of the same key K; whereas when using asymmetric keys, the sender S and receiver R must be in possession of their own keys K_s and K_r, respectively (commonly known as their *private* keys, since they are never shared), and they must also be in possession of each others' "locks" L_r and L_s (commonly known as their *public keys*, since they are sent to communications partners). The important difference to note here is that when using asymmetric keys, the sender and receiver never need to send their private keys to each other, which reduces the chance of the keys themselves being compromised when they are sent over the air. Although public keys are called keys, an easier way to think of them is to think of them as *locks*, as explained above. Public

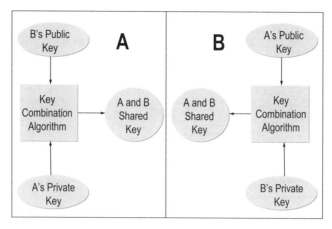

Figure 9.8. Creation of shared session key.

keys do need to be exchanged between communicating parties, and extensive work has been done to develop mechanisms for doing so, as will be discussed later in this section.

One important point to note here is that algorithms that make use of symmetric keys are much faster than those that use asymmetric keys. Thus a combination of symmetric and asymmetric keys is sometimes used as follows: A shared "session key" is generated by one of the communicating parties. This key is then encrypted using the public keys of all communicating parties. Each party then uses its own private key to decrypt the session key. This key is then used in conjunction with a fast symmetric encryption algorithm for communication. Another mechanism for generating a shared session key is illustrated in Figure 9.8 with an example of two communicating parties. Here a shared session key is obtained by combining the public and private keys of A and B as shown in Figure 9.8.

9.4.2 Confidentiality

Confidentiality refers to protection of a message from snooping by others. Confidentiality of a message is assured if no one other than the intended receiver(s) can obtain the contents of the message. A pair of public and private keys can be used for this purpose, as illustrated in Figure 9.9. In this figure, A is the sender and B is the receiver. A wishes to send a secret message to B. Assume that A is already in possession of B's public key; A could have obtained this public key via a PKI (Public Key Infrastructure, discussed further in Section 9.4.4) or other means. A uses an encryption algorithm along with B's public key to encrypt the secret message, and it sends the message over the air to B. When B receives this message, she decrypts it using her private key to retrieve the original message sent by A.

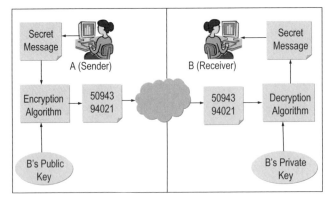

Figure 9.9. Message confidentiality.

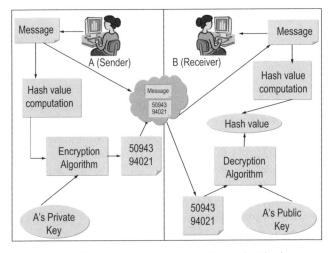

Figure 9.10. Message integrity and user authentication.

9.4.3 Message Integrity and User Authentication

Message integrity refers to the protection of a message from tampering by others. Message integrity is assured by ensuring that the intended receiver(s) of a message can verify whether the message was tampered with or not when they receive it. User authentication refers to a mechanism for ensuring that the receiver of a message can verify that the message was sent by a particular sender.

Figure 9.10 illustrates the mechanism used for assuring both message integrity as well as user authentication. Here A computes a hash value of the message (this hash value is also called *message digest*) and encrypts the hash with her own private key before sending it to the receiver B. B uses A's public key to decrypt this hash value, and in parallel she computes the hash value of the message herself. If the hash value sent by A and the hash value computed by B are the same, this implies that A's private key was used to encrypt it; this in turn guarantees that A is truly the sender of the message, thus providing user authentication. Furthermore, the fact that the hash value computed by B is the same as the hash value computed by A implies that the message has not been tampered with during transmission, thus providing message integrity. Note that this mechanism relies on the assumption that no one other than A is in possession of A's private key.

The encrypted hash value of the message (produced by A) is also known as A's digital signature for this message. Note that a digital signature is always relative to a message; in other words, the digital signature of an entity differs from message to message. This is slightly counterintuitive, since a person's signature is always the same, regardless of which message the person is signing.

Note the difference in the encryption process in this example, as compared to the previous example depicted in Figure 9.9. In Figure 9.9, the encryption performed by A used *B's public key*, which is generally available to anyone who wishes to communicate with B. The corresponding decryption could only be performed using *B's private key*, which is held only by B. In the example here (Figure 9.10), though, the encryption performed by A uses *A's private key*, which only A possesses. The corresponding decryption can be performed using *A's public key*, which is generally available to anyone who wishes to communicate with A.

Note that in the above scenario, the actual message was transmitted in clear text. If confidentiality of the message is required in addition to authentication and message integrity, then A would encrypt the message using B's public key after having performed the above steps, before sending the encrypted message and the encrypted hash value over the air to B. B would perform the additional step of first decrypting the message using B's private key, before decrypting the hash value using A's public key and computing the hash value of the message for comparison with the transmitted hash value.

There are several reasons why a hash value of a message would be signed (i.e. encrypted) rather than the entire document:

- *Efficiency*: The signature of the hash value of the message is obviously shorter than the message itself and therefore encryption of the hash value will take less time than encryption of the entire message.
- *Message Integrity*: If a large message is to be encrypted, it must be split into several smaller parts, each smaller than the length of the private key. Each part is then signed and sent separately to the receiver. This creates the following problem: The receiver cannot tell whether any of the message parts have been lost during transmission. If the hash of the message is used for encryption, its value can be used to verify that the message was received in its entirety.

9.4.4 Public Key Infrastructure

The difference between public keys and private keys was explained in the previous section. Since public keys must be shared with communications partners, there is a need for an infrastructure for sharing these keys. The *Public Key Infrastructure (PKI)* is a means for allowing sharing of public keys via a trusted third party. Even though public keys are not secret, it is important to protect against impostors; for example, if a malicious party M wants to disrupt communications between two entities A and B, it could provide each of these two entities with its own public key in place of their public keys without their knowing it. If A sends a secret message to B using M's public key instead of B's public key, then M could eavesdrop, intercept the message, and decrypt it using its own private key (since the message was encrypted using M's public key). Thus it is critical to ensure that public keys are distributed in a way that ensures that the party providing the keys is trusted. PKI uses a trusted third party for distributing keys and ensuring that the keys that are distributed do indeed belong to the right entities. PKI makes use of *certificates* to distribute keys. The trusted party providing the PKI service is responsible for verifying the identity of users and issuing a digital certificate; the latter is a signed message certifying that a particular key belongs to the specified user. Digital certificates are discussed in detail in the next section.

The next section describes ITU-T and IETF standards for PKI, along with some of the issues that arise in implementing these standards for MANETs; Section 9.4.4.2 presents alternate key distribution and revocation approaches; and Section 9.4.4.3 describes the "web of trust" paradigm.

9.4.4.1 X.509 and PKIX. X.509 is an ITU-T PKI standard initially issued in 1988 that specifies standard formats for public key certificates and a certification path validation algorithm. X.509 was tightly coupled with the ITU-T X.500 directory standard, which defines the protocols and information model for a directory service that is independent of computing application and network platform. First released in 1988 and updated in 1993 and 1997, the X.500 standard defines a specification for a rich, distributed directory based on hierarchically named information objects (directory entries) that users can browse and search (see Chapter 3 for more details about X.500). X.509 specifies a hierarchical organization of certificate authorities (CAs) that issue certificates. These certificates bind an entity's public key to a Distinguished Name (using X.500 notation) or to an e-mail address or other identifier. Later versions of the X.509 standard allowed the use of nonhierarchical certificate authorities as well. The meaning of *hierarchical organization* of certificate authorities is explained below.

Since the X.500 directory standard was not widely implemented and deployed, due partly to the complexity of the OSI standards, the IETF launched an effort to standardize a public key infrastructure. The IETF PKIX (Public Key Infrastructure X.509) working group was created in 1995 to develop standards aimed at supporting an X.509-based PKI. The chief contribution of this working group has been to define X.509 certificates and certificate revocation list standards [Housley et al., 2002], which are described next. The PKIX working group is also responsible for updating these standards as related standards efforts such as LDAP evolve.

9.4.4.1.1 X.509 Certificates. The following process is followed to use a digital certificate. Let's say that a party A needs to communicate securely with party B and therefore needs B's public key. It requests it from a certificate authority (CA), which is a trusted entity that issues digital certificates. In the example shown in Figure 9.11, the CA is a fictitious company called *Certificas*, which is the issuer of the certificate. The subject (i.e., the party whose public key is being provided by the CA) here is another fictitious company called *Securitas*, with common name (CN) www.securitas.com. The public key for Securitas is listed following the label "Subject Public Key." The "Certificate Signature" is the CA's way of digitally signing this certificate, to assure the integrity of the certificate. It is obtained by computing an MD5 hash of the first part of the certificate and encrypting it with Certificas' private key. Figure 9.11 shows all of the contents of a sample digital certificate, which includes, in addition to the fields described above, version and serial numbers, the algorithms used for the certificate, and its validity period.

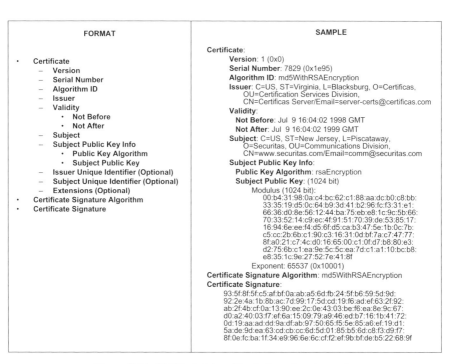

Figure 9.11. X.509 digital certificate format and sample.

An important feature of PKI is that digital certificates such as those shown in Figure 9.11 can be transmitted electronically, over insecure channels. This means that these certificates are subject to eavesdropping and tampering. As seen before, eavesdropping is not an issue for public keys, since the purpose of public keys is to reveal them to others for use in communications. However, tampering is an issue because, as explained before, if an intruder can somehow intercept party B's public key and substitute its own public key in its place, then party A will use the intruder's public key to encrypt its communications with B. Not only will B be unable to decrypt the messages sent by A, but also—and more importantly—the intruder will be able to decrypt any message sent by A, provided that he is able to eavesdrop on the traffic sent from A to B. Thus PKI provides a mechanism to verify that certificates are not tampered with during transmission. The CA that issues a certificate must *digitally sign* the certificate to assure message integrity (refer to Section 9.4.3 for an overview of message integrity) by computing a hash of the first part of the certificate and encrypting it with its own private key. When this certificate is received by the requestor, it must validate the certificate as follows:

1. First, it must obtain the public key for the CA (in Figure 9.11, this is the public key for Securitas).
2. Next, it uses Securitas' public key to decode the certificate signature on the first certificate. The decoded signature is a clear text form of the MD5 hash of the first part of the certificate.
3. It computes the MD5 hash for the first part of the received certificate.
4. It compares the results from (2) and (3). If they are identical, the certificate is valid.

The observant reader will note that Step 1 introduces a problem. The public key for Securitas must presumably be obtained via another digital certificate, which itself requires validation that in turn may require yet another digital certificate, and so on. This is where the concept of *hierarchy* of CAs comes into play. The public key for a given certificate authority may be obtained from another certificate authority that is its parent in this hierarchy of CAs, and so on, resulting in a *certificate chain.*

So what happens when the root of the tree is reached? In this case, the issuer and the subject of the digital certificate will be the same, resulting in a self-signed certificate. There is no way to verify the validity of this certificate. For this reason, such a certificate must be *manually configured.* In other words, this certificate cannot be transmitted electronically. Note that root CA certificates may be recognized by browser vendors such as Microsoft and Netscape, and they may be supplied along with web browsers and e-mail clients. This means that these root CA certificates are typically preconfigured into the web browsers and other client software that make use of certificates, making it possible to start using certificates issued by other CAs. Since there are relatively few root CAs, the expectation is that manual configuration of certificates of root CAs will be an operation that is performed very infrequently and therefore does not impose an undue burden on the system operators.

9.4.4.1.2   X.509 Certificate Revocation. Certificate revocation is a process of dealing with compromised private keys by informing all the concerned parties that they should stop using the corresponding public keys. Given that a public key may be very widely distributed, this can be a nontrivial process. The X.509 standard defines a process for revoking certificates based on a *certificate revocation list* (CRL). Each Certificate Authority must periodically issue a CRL that contains a list of all revoked certificates, identified by their serial numbers. This CRL is signed by the CA or by the issuer of the CRL. Now, when an entity wishes to use a certificate, it must acquire a recent CRL and check whether the certificate that it wants to use is on the revocation list contained in the CRL. CRLs are issued periodically and certificate users are expected to request the latest CRL from a CA prior to using any certificate, using the same distribution mechanisms as for certificates. This puts the responsibility for checking whether a certificate is valid or not on the user of the certificate. A drawback of this method is the potential latency in determining that a certificate is no longer valid. Since CRLs are issued periodically, there may be a time lag between the time that a certificate becomes invalid and the time when a new CRL is made available with this information, which may be several hours or even days later.

9.4.4.2   *Distribution of Certificate and Revocation Information.* The previous section described the PKIX key distribution and revocation standards. The distribution and revocation of certificates is an extremely important aspect of using a certificate-based key management approach in MANETs. As mentioned earlier, the high degree of eavesdropping that can potentially occur in MANETs, coupled with the low bandwidth availability in noisy wireless links, renders the task of certificate distribution and revocation rather challenging in MANETs. In this section, different methods of distributing these certificates are discussed that are not based on PKIX.

Without loss of generality, the distribution of certificate and revocation information can be classified into the following methods: (a) sender-based, and (b) receiver-based. The following subsections discuss these approaches and their general merits and demerits with respect to delays incurred and bandwidth usage.

9.4.4.2.1   Sender-Based Approach. In a sender-based approach, the senders have the onus of proving themselves as legitimate. Using a mechanism similar to that described earlier for PKIX, each sending node must get a trusted CA to issue a certificate on its behalf. The node then sends this certificate (referred to below as Cert-ID) to nodes that it communicates with. The

Cert-ID confirms that the public key that the sender is distributing indeed belongs to it. The sender must also prove that its key has not been revoked by the Revocation Authority (RA) by supplying a certificate of nonrevocation (Cert-NonRev). The certificate of nonrevocation contains the identifier of a node and a timestamp, indicating that the certificate issued by that node is valid as of the time of issue of the Cert-NonRev. The validity of the Cert-NonRev is determined by the frequency with which these certificates are provided by the sender. Thus the difference between this approach and the PKIX approach is that in the latter, nodes must communicate with the CA to retrieve public keys of entities that they wish to communicate with; and also, they must communicate with the RA to periodically retrieve the CRL so that they know which certificates have been revoked. With the sender-based approach, however, nodes receive key and revocation information directly from the peer node that they wish to communicate with. This can be an advantage in MANETs because it guarantees that keying and revocation will be available if the two nodes that need to communicate have network connectivity with each other. This is because they are not dependent on a third node, the CA/RA (which may be temporarily unreachable due to mobility or other factors), to communicate with each other.

The time taken to establish a session includes the time taken to send the certificates Cert-ID and Cert-NonRev to the receiver, which is typically of the same order as the length of these certificates. It is assumed that the sender gets these certificates before it begins session establishment, so the time taken for the sender to access its certificates does not contribute to the session establishment time.

In terms of the bandwidth usage to distribute the certificates, one must account for both the sender periodically getting valid nonrevocation certificates from the RA, as well as the sending of the certificates from the sender to the receiver. In general, the sender-based approach bandwidth usage is proportional to many variables: (i) the length of the certificates, (ii) the number of hops to reach the RA and CA (a variable that depends on the topology and number of RAs/CAs and their placement and network loading), (iii) the number of sessions and nodes participating in communications (a variable depending on the traffic pattern in steady state), and (iv) the frequency at which revocation information is refreshed. In its simplest form, the size of the Cert-ID is proportional to the size of the public keys (usually in the order of a small number of kilobits).

With regard to the frequency at which the revocation information is refreshed, note that there exist canonical arguments with regard to the tradeoffs involved. As expected, frequent refreshing involves higher bandwidth penalties, but provides more accurate information in terms of certificate validity; while infrequent refreshing provides a lower bandwidth penalty but less accurate information about validity. Based on the MANET deployment objectives and bandwidth available, this will typically be a policy-controlled parameter.

9.4.4.2.2 Receiver-Based Approach. As its name implies, in a receiver-based approach, the onus of confirming the "legitimacy" of received keys rests with the receiving node. This mechanism is in some way simply the inverse of the sender-based approach. Thus in this case, the receiving node will contact the CA/RA to verify that the sender of the accompanying information is valid. However, unlike the sender, it is assumed the receiver cannot pre-fetch the information.

In terms of the delay, the receiver-based approach adds the delay of a round trip from the receiver to the RA and/or CA to the session initiation delay. Here the assumption is being made that the receiver obtains the required key management related information (certificate verification and revocation information) within one round trip; that is, no extra handshaking trips are required. Note that whereas current standards such IKEv2 [Kaufman, 2005] typically require two round trips, the assumption of one round trip in MANETs is a reasonable assumption, because customized key management techniques will be available due to the need for optimization. Assuming that the key management between the sender and receiver is just one round trip, this additional round trip can approximately double the delay. The actual delay due to the round trip between the receiver and the RA will be proportional to (i) the number of hops to reach the RA and back (a variable that depends on the topology and number of RAs and their placement and network loading), (ii) the link speed, and (iii) the processing time within the RA (a variable that depends on the loading of and number of entries within the RA, look-up mechanisms, and RA processor speed).

The bandwidth usage in the receiver-based approach is proportional to (i) the certificate size, (ii) the number of hops to reach the RA (a variable that depends on the topology and number of RAs and their placement and network loading), and (iii) the number of sessions and nodes participating in communications (a variable that depends on the traffic pattern in steady state).

With regard to key revocation, note that, in general, key revocation can get expensive in terms of both computational requirements and delays. Bearing in mind some of the important issues with regard to MANET environments—namely, low bandwidth availability, limited computational resources, and potentially high communications latencies—a combinatorial set-selection approach that involves sharing certificates can be used to improve performance. The combinatorial set-selection approach works by grouping users into sets of "good" users and performing revocation based on these sets of users. As an example, a set of "good" users would be "all users within domain X." With a certificate-based approach, this would involve assigning a public key for sets of users. For the purpose of key revocation, one Cert-NonRev would carry revocation information for a collection of users that belong to the same set of "good" users. Thus a receiver would only need to retrieve one such Cert-NonRev for the entire set of "good" users. Such a mechanism has the potential to result in significant savings in bandwidth and delay, all of which are critical in MANET environments. However, one drawback of this mechanism is that it assumes that all the members of a given set are equally trustworthy, which in turn requires a judicious selection of "good user set."

Additionally, if a distributed trust model (e.g., a mechanism analogous to a web of trust, as discussed in Section 9.4.4.3) is available whereby nodes ask other nearby nodes for trust information and use a majority quorum to for trust decisions, this may also be used with the receiver-based approach for revocations. Stated very briefly, in a distributed trust model, rather than going to an RA, the receiving node simply queries a set of "trusted" nodes/neighbors about the validity of the sender's keys/certificates. The receiver then uses a quorum-based approach (i.e., establishes a majority quorum) to certify good versus bad senders. Such a distributed trust model is also very useful and will provide robustness to the system if the RA were to be compromised, since otherwise the consequences of a "bad" RA can be very detrimental to the system/mission objectives.

9.4.4.2.3 Distribution of Revocation Information via Broadcast. As its name implies, in a broadcast approach, information about certificates that have been revoked is broadcast by the RA to the entire MANET. Thus, with the broadcast mechanism, every node will maintain a list of the currently revoked nodes in the entire system. Whenever a message is received, information about the validity of the sender of the message is obtained by a local look-up at each node.

A broadcast can be periodic, event-driven (e.g., when the RA changes the revocation list), or a combination of the two. The information distributed can be a "delta" (i.e., including only new information), or it could include a complete list of revoked nodes in the system.

In terms of delay, the broadcast approach (like the sender-based approach) adds virtually nothing to the session initiation delay, since every node is assumed to have all the certificates/revocations before a session is initiated. The only additional delay is the time to perform a local look-up within the destination node's own database.

The bandwidth requirements to distribute the certificates are fairly complex and depend on many factors. The bandwidth requirement is potentially a function of (i) the rate at which nodes are revoked, (ii) the time since keys were last changed, (iii) the total number of nodes and subnets/domains in the domain, and (iv) the degree of node mobility. Notice that for revocations, the number of nodes revoked will be a monotonically increasing function as the lifetime of the keys increases, until keys are too old to be valid. It is assumed that the lifetime of a key is a parameter in the original certificate. Clearly, there is a tradeoff in setting this time, since a small value can result in a lot of key generation while large values can mean broadcasting (and storing) a lot of revocation information.

The storage costs for certificates are proportional to the total number of nodes in a network. Even if only revocations are broadcast, the size of the required storage will increase until keys expire (see discussion in previous paragraph). Thus the revocation information that needs to be stored by every node within the system is a monotonically increasing list (where no dynamic re-keying is permitted).

Another important contributor to the large bandwidth overhead associated with the broadcast-based approach is the fact that the broadcast message has to potentially reach every node within every subnet/domain in the entire network. Nodes entering a particular subnetwork must get this revocation information as quickly as possible in order to avoid any compromises. Furthermore, if the networks group and re-group themselves very randomly and frequently—which is typical in a MANET environment—the broadcast approach may not scale gracefully to large network sizes.

9.4.4.3 Web of Trust Paradigm. The past few sections have described the use of certificates and how PKI provides a hierarchy of certificate authorities that can establish that a given public key belongs to a given entity. An alternative to this hierarchical system is the *web of trust* paradigm that is used in the popular PGP (Pretty Good Privacy) standard. The IETF OpenPGP working group is developing standards for algorithms and formats of PGP processed objects, as well as the MIME framework for exchanging them via e-mail or other transport protocols. The web of trust concept is based on the concept that any user can "vouch for" the validity of the binding between another entity and that entity's public key. Anyone wishing to communicate with this entity can then decide whether or not to trust the provided public key, based on the number of people who have voucher for the validity of the key and whether these people are trusted or not. This means that, rather than having a few certificate authorities responsible for signing certificates, individual users can vouch for whether a certificate is valid or not. This added flexibility comes at the cost of increased risk, since not all individuals may be trustworthy.

9.5 COMMUNICATIONS SECURITY

The previous section discussed how communications can be secured using cryptography at the application layer, and it also discussed the required key management mechanisms to support these capabilities. The discussion was rather general, and the mechanisms discussed can be applied at different layers of the protocol stack. In this section, protocols and standards for securing communications at the IP layer are discussed. In order to standardize IP security, the IETF developed a suite of protocols for securing IP communications by using encryption and authentication mechanisms. The idea was to standardize these protocols so that implementations from different vendors could interoperate and would provide a standard set of services. The resulting set of standards was called IPSec (IP Security). In a nutshell, IPSec provides the following capabilities:

- *Securing Packet Flows*: IPSec provides encryption for either the payload of an IP packet or the entire packet.
- *Establishing Keys*: IPSec defines protocols for key establishment mechanisms.

Each of these are discussed in turn.

9.5.1 Securing Packet Flows

IPSec provides two modes of operation, which are explained here: *transport mode* and *tunnel mode*. When used in transport mode, IPSec provides for encryption of the payload of every IP packet. The original IP packet header is retained, but the payload is replaced with an encrypted version of the original payload at the IPSec gateway. This means that the original IP header is visible to eavesdroppers. Tunnel mode provides for encryption of the entire IP packet, including the IP header. Since the packet cannot be transmitted in this form to its destination (because intermediate routers need to be able to see the IP header to be able to route it), a new IP header must be created for the encrypted packet, via a process called *encapsulation*. This new IP header typically contains the IP address of the originating IPSec gateway as its source address, along with the IP address of the destination IPSec gateway as its destination address. Other IP header fields are created appropriately or are copied over from the original IP header. The DSCP is one of the notable fields in the original IP header that is copied over to the new IP header, to enable DiffServ treatment of the encrypted packet in the network core.

The intent of transport mode was to provide end-to-end security between two communicating hosts. The hosts would be responsible for encrypting and decrypting the IP payload. Tunnel mode, on the other hand, was intended for securing networks by tunneling traffic between two gateways that would provide the encryption as well as encapsulation functionality for the secured networks. Traffic would flow unencrypted from hosts in one network to their gateway; the gateway would encrypt and encapsulate packets and send them to its counterpart gateway on the destination network; and the destination gateway would decapsulate and decrypt traffic and forward it to the destination host.

IPSec provides authentication as well as confidentiality mechanisms. It defines Authentication Header (AH) and Encapsulating Security Payload (ESP) mechanisms. AH provides authentication for the entire IP packet, including payload and header. AH protects the IP payload and all header fields of an IP datagram except for mutable fields—that is, those that might be altered in transit. Mutable, therefore unauthenticated, IP header fields include TOS, Flags, Fragment Offset, TTL, and Header Checksum. The Encapsulating Security Payload (ESP) extension header provides origin authenticity, integrity, and confidentiality protection of a packet. ESP also supports encryption-only and authentication-only configurations, but using encryption without authentication is strongly discouraged. Unlike the AH header, the IP packet header is not accounted for.

9.5.2 Key Establishment

IPSec defines the IKE (Internet Key Exchange) protocol. IKE [Piper, 1998; Maughan et al., 1998; Harkins and Carrel, 1998; Kaufman, 2005] uses a *Diffie-Hellman* [Diffie and Hellman, 1976] key exchange to create a shared session key. The Diffie–Hellman algorithm allows two users to agree on a shared secret key using insecure communications, without having shared any prior keying information. This shared session key can then be used for communications using a symmetric encryption algorithm.

ISAKMP (Internet Security Association and Key Management Protocol) defines procedures and packet formats to negotiate *Security Associations*. A Security Association (SA) contains all the information related to the securing of communications between two entities, such as the IP addresses of the two communications endpoints, the encryption algorithms used, the security keys, and so on. The concept of a Security Association (SA) is somewhat abstract, but fundamental to IPSec. A Security Association is similar to a TCP or UDP socket. According to RFC 2401 [Kent and Atkinson, 1998]:

- "A Security Association (SA) is a simplex 'connection' that affords security services to the traffic carried by it." At least two SAs are required to secure bidirectional communications. It is, however, perfectly valid to secure communications in only one direction using one SA.
- "A security association is uniquely identified by a triple consisting of a Security Parameter Index (SPI), an IP Destination Address, and a security protocol (AH or ESP) identifier." The SPI is used to distinguish between different SAs terminating at the same destination and using the same IPSec protocol. In each SA, the IPSec sender will need to include the SPI chosen by the IPSec receiver in every packet sent using the SA; otherwise, incoming packets will be dropped.

Security associations may operate in either "Transport" or "Tunnel" mode. In transport mode, modifications are made to the original IP packets. In tunnel mode, source packets are taken whole and encapsulated inside a new IP packet. Traffic to be passed over a given SA may be selected based on source IP address, destination IP address, name (user ID or system name—e.g., using DNS or X.500), data sensitivity level (IPSO/CIPSO labels), transport layer protocol (e.g., TCP, UDP, or "OPAQUE" for ESP), and source/destination ports. SAs also have a defined lifetime, which may be based on time or data volume.

Figure 9.12 illustrates different IPSec modes. In this figure, two security associations, SA_1 and SA_2, are shown between Host A and Host B. These security associations operate in transport mode and originate and terminate

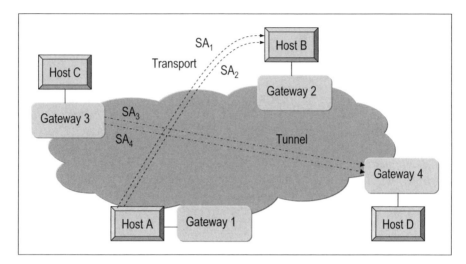

Figure 9.12. IPSec modes and security associations.

at the hosts where the traffic originates and terminates, respectively. Two other security associations, SA_3 and SA_4, are shown between Gateway 3 and Gateway 4. Here, SA_3 and SA_4 operate in tunnel mode, where the tunnel originates at Gateway 3 and terminates at Gateway 4.

ISAKMP defines payloads for exchanging key generation and authentication data. These formats provide a consistent framework for transferring key and authentication data. It is important to note that ISAKMP is distinct from IKE and is independent of the key generation technique, encryption algorithm, and authentication mechanism.

9.5.3 HAIPE

A HAIPE (High Assurance Internet Protocol Encryptor) is a Type 1 encryption device that complies with the United States Department of Defense's HAIPE Interoperability Specification. A Type 1 encryption device is one which has been certified by the United States National Security Agency (NSA) to be suitable for encrypting classified U.S. Government data. The HAIPE specification is based on IPSec, but includes certain enhancements, including the ability to encrypt multicast data using a preplaced key.

9.5.4 Securing a MANET

In order to secure MANET traffic, all traffic going out over a wireless medium must be encrypted. This requires the use of IPSec or similar encryption (e.g., HAIPE) to encrypt traffic before it exits a MANET node. Referring back to the MANET Reference Architecture in Chapter 5, recall that manned and unmanned platform nodes have wired networks on board, including multiple hosts on one or more LANs. The traffic on these LANs can be transmitted in the clear (i.e., unencrypted) while it is on the platform, but must be encrypted before exiting the platform. Borrowing from military terminology, unencrypted IP traffic on the platform LAN is called *red* traffic, whereas encrypted traffic that is sent over the air is called *black* traffic (see Poylisher et al. [2006]). The platform LANs are often referred to as red networks, and the wireless network is referred to as the black network. Typically, IPSec tunnels must be built between red IP ports in different platforms. IPSec encrypts red IP packets, encapsulates them with appropriate black IP addresses and sends them to the black IP side that routes them to the destination via the black network. Similarly, when a platform receives a black IP packet destined to a local red IP address, it removes the black IP header, decrypts the red IP packet, and delivers it to the appropriate platform router that forwards the packet to the red IP host. These concepts are illustrated in Figure 9.13.

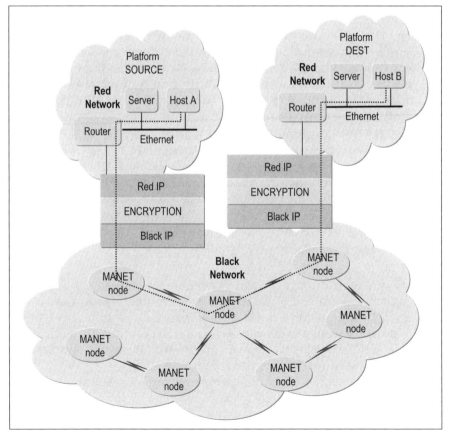

Figure 9.13. MANET red–black architecture.

In order for the above architecture to allow seamless communications in the face of mobility and network dynamicity, additional functionality needs to be provided to support mobility. As was discussed in Chapter 6, IP addresses for nodes on the red networks shown above do not need to change, even in the face of mobility. This is because only the black IP addresses (i.e., those that are configured on the wireless interfaces) need to change when nodes move from one routing domain to another. However, when black IP addresses change, there is a need to update red-to-black address mappings throughout the network, so that red-side flows can continue to be routed to their destinations over the black network.

In order to understand how this works, a detailed example is provided below, using a HAIPE or similar device. When application traffic flows from one MANET node to another (see the application flow from Host A on platform SOURCE to Host B on platform DEST in Figure 9.13 above), the following steps occur:

1. Traffic packets leaving the source host (Host A on Platform SOURCE in Figure 9.13) arrive at the HAIPE device on its platform.
2. The HAIPE device looks up the destination IP address in the packet header. This destination IP address is the address of a host on the red side of some other platform (Host B on Platform DEST in Figure 9.13). It looks up the corresponding black IP address of the HAIPE device fronting that platform (for now, assume that every HAIPE device has access to such a mapping).
3. The HAIPE device on platform SOURCE encrypts the sent packet (based on a security association with platform DEST) and creates a new header for the outgoing packet, with its own black IP address as the source IP address and the black IP address of the HAIPE device on platform DEST as the destination IP address.
4. The packet arrives at the HAIPE device on platform DEST and is decapsulated and decrypted. The resulting IP packet is sent to its destination host (Host B on platform DEST in Figure 9.13).

Going back to Step 2, note that every HAIPE device must have access to the mapping between red and black IP addresses. In other words, given a red destination IP address for a traffic flow, it must be able to find out which HAIPE device to send this traffic to, so that it arrives at the right destination. So, how is such a mapping created and maintained? Below, the mechanisms provided by HAIPE for this purpose are discussed.

Even if the initial red–black IP address mapping for a MANET could be created manually and configured ahead of time into all the HAIPE devices, there is a need to update this mapping every time a MANET node moves into a new routing domain and changes its black IP address. Furthermore, if new devices join the red network, the mapping of their red IP addresses to the black IP address of the HAIPE on their platform needs to be provided to all HAIPE devices in the network, so that they can route traffic to such devices. In order to deal with this need, HAIPE version 3 provides (see Nakamoto et al. [2005]) the following:

- The use of RIP to discover the red network behind the HAIPE.
- A mechanism for HAIPE devices to communicate with each other to discover red-to-black address mappings.

Nakamoto et al. [2005] discusses a scalable architecture for discovering red-to-black address mappings using HAIPE. The paper describes a DNS-like referral model, where servers scattered throughout the network maintain mapping information. When a HAIPE device comes up, it uses RIP to discover the red networks for which it encrypts and tunnels traffic to other nodes, and then it summarizes and registers this information with a *PED* (Peer Enclave Discovery) node. When it needs to look up a red-to-black address mapping, it queries its nearest PED for this information; if the PED does not have the requested information, it can refer the requesting HAIPE device to another PED that may have the information. The referral chains resemble those implemented by DNS. This approach provides scalability in the same way that DNS does, as new PEDs can be added as the network grows.

9.5.5 Cross-Domain Information Flows

The previous section provided a detailed description of a typical security architecture for MANETs. However, as discussed in the security considerations section of Chapter 6, in order to be able to manage the black network using an NMS on the red network, there is a need to pass information from the black network to the red network, and vice versa. Network management for the MANET will be performed on the red networks, and network elements and services on the black network will have to be configured and monitored by network management components on the red networks. This means that there needs to be a way to send information across the red–black boundary. However, any information that is thus sent needs to be carefully controlled, in order to ensure that there is no undesirable leakage of information into the wireless network that could open up the network to vulnerabilities and attacks.

In order to restrict the information that is allowed to cross the red–black boundary, *cross-domain guards (CDGs)* are used to control the information flow across this boundary. The function of a CDG is to (a) inspect every message that traverses the red–black boundary and (b) confirm that the message conforms to a predefined set of rules about what information is allowed to cross the boundary. This predefined set of rules is typically rather restrictive, and it constrains the set of messages allowed to cross the guard based on size, length, content, frequency, and so on. It is desirable to be able to define the data that are allowed to cross the CDG as precisely as possible. This means that limited messages, fixed data fields and sizes, as well as predetermined field values, are preferred.

Although CDGs and firewalls are similar in concept, the reader is cautioned not to confuse the two. Whereas a firewall is transparent to the two endpoints of a traffic flow, a CDG is not; traffic that must traverse a CDG is actually sent explicitly to the CDG, where it is inspected and then sent to its final destination if its contents conform to stored rules. Furthermore, a firewall merely inspects certain fields in a packet header or in part of the payload of the packet, whereas a CDG functions at the application layer and inspects application traffic.

The flow of data across the CDG is depicted in Figure 9.14. Monitoring information from the black side radios is collected by a monitoring component on a black host connected to the radio. Such monitoring data are aggregated and filtered before sending a subset of data across the CDG to the red-side NMS. The reason for the aggregation and filtering is that the amount of data sent across the CDG must be as small as possible and must conform to the rules defined about what information is allowed to cross the CDG boundary. The red-side NMS receives the monitoring data, and it uses these data to drive its management decision-making process. The reverse process is followed to

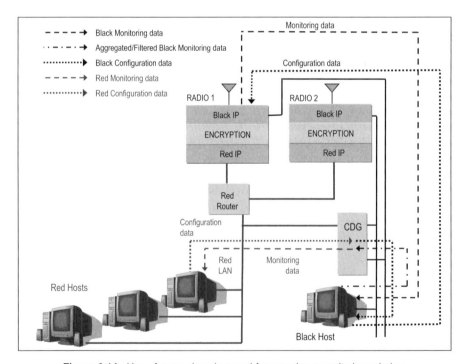

Figure 9.14. Use of cross-domain guard for crossing security boundaries.

send data to the radios to configure them. Configuration data originate from the red-side NMS, and they are sent across the CDG to the black host, and from there they are sent to the radio.

An important point to note is that no equipment on the black network should have knowledge about red-side IP addresses. This complicates communications from the black to the red side (e.g., to send back monitoring information) because, in effect, it requires sending information to an unknown address! The way this is dealt with is by having the sender on the black side (in Figure 9.14, this is the host labeled "Black Host") address the information to the CDG, along with an agreed-upon identifier (e.g., "NMS"); the CDG is responsible for mapping this identifier to a red-side IP address. In this example, the CDG would know which red-side device hosts the NMS and would send the appropriate messages to the NMS, after having verified that their contents conform to the specified rules.

In addition to being able to control information exchange across red–black boundaries, the need for exchanging information between networks belonging to different security domains may also arise. As an example, networks at different security classification levels may need to exchange data. As mentioned in Section 9.2, multiple levels of security are typically required in military settings where data of different classifications (such as top secret, secret, unclassified) may coexist on a single platform. Such data must be kept on different LANs that are physically disconnected or, alternatively, where data are allowed to cross from one LAN to another only via a CDG, which is programmed with rules about what data are allowed to cross from one LAN to another. This is accomplished in the same way as described above for red–black information passing, the only difference being that multiple CDGs are required, one for each pair of LANs that need to exchange information (see Hubbard [2002] for a discussion of multilevel security and cross-domain solutions). Data of one classification may also be allowed to transit over a network of a different classification, provided that it is encrypted prior to entering the other network.

9.6   INTRUSION DETECTION

Intrusion detection is the process of identifying malicious behavior that is intended to cause harm to the network and/or the applications running on the network. An intrusion could be caused by an insider (i.e., someone who legitimately has access to a given network) who misuses his or her access privileges to disrupt the network; or by an outsider who is not a legitimate user of the network and who disrupts the functioning of the network by various means. Intrusions have always been a concern in IP networking, even for wireline networks; however, the challenges of intrusion detection in wireless networks are even greater, because there is no such thing as physical security for such networks. In wireline networks, physical access to network resources, such as cables, routers, end hosts, and so on, can be secured to prevent any kind of intrusion. For example, an enterprise could set up a private wired network within an office building; and as long as all of the network components remain physically secure, it is not possible for an external intruder to gain access to such a network. The situation is quite different for wireless networks, for the simple reason that there are no "wires" to secure! Thus physical security alone cannot guarantee the integrity of a wireless network. The situation becomes even more complex for an ad hoc network, which has no fixed infrastructure.

In this section, an overview of intrusion detection for MANETs is provided. The next section provides a brief overview of existing intrusion detection techniques for the Internet. This is followed by a discussion that is more specific to MANETs and provides a description of an intrusion detection architecture for MANETs.

9.6.1   Intrusion Detection Techniques

Intrusion detection is a rich field of research. There exists a large body of literature that discusses different approaches and tools for detecting intrusions in IP networks. This section provides a quick overview of the different existing approaches to intrusion detection. The purpose of this section is not to provide an in-depth treatment of intrusion detection, since this topic could easily fill up a book in its own right; however, a high-level overview of the different approaches is provided to give the reader a flavor of the techniques that are currently being used for intrusion detection. The reader is referred to Anjum and Mouchtaris [2007] for a discussion of security mechanisms for ad hoc networks.

There are two basic approaches to intrusion detection: network-based intrusion detection and host-based intrusion detection. Each of these are discussed separately below. Both techniques have certain strengths and weaknesses; in practice, it is usually necessary to use a combination of both techniques to adequately protect a system from attack.

9.6.1.1 Network-Based Intrusion Detection. Network-based intrusion detection techniques operate by capturing traffic on a network segment and analyzing the *content* of the traffic as well as *traffic patterns* to determine whether the traffic exhibits any anomalies. In order to capture traffic on a network segment, a network-based intrusion detection system (NIDS) must be hosted on a network element that can examine all the traffic transiting the network segment. Thus a NIDS is typically attached to an entry/exit point of a network segment, such as a router or a firewall.

Once traffic packets have been captured, they need to be analyzed for anomalies. This is performed in one of several ways. First, the contents of the packets themselves need to be examined and compared with known attack signatures. There are several types of signatures that can be used:

- *String Signatures*: Certain strings in network traffic could indicate potential attacks. String signature matching searches for known strings in network traffic that are indicative of attacks, based on existing knowledge of possible attacks. As an example, an NIDS could be configured to search for the string "/etc/passwd" in incoming traffic and to generate an alert when such a string is detected. The rationale behind doing this is that the system can detect intruders who are trying to gain access to the system password file for malicious reasons.

- *Port Signatures*: By keeping a watch on which ports are being accessed, an NIDS can detect suspicious activity and flag it. For example, certain ports are known to frequently be the subject of attacks (e.g., TCP port 23). In order to guard against such attacks, access to these ports is often not allowed; thus, any connection attempt to such ports could be an indicator of malicious activity.

- *Header Condition Signatures*: There are certain patterns that are illegal and/or dangerous in IP packet headers. A well-known example of this is a TCP packet where the SYN flag and the FIN flag have both been set. This is an illegal combination, since it indicates that a connection is being opened and closed simultaneously. This is probably the most well-known illegal combination of TCP flags. The SYN flag is used to start a connection, and the FIN flag is used to end an existing connection. Obviously, it does not make sense to set both flags at the same time, because this would mean that a connection is being opened and closed at the same time.

As stated above, in addition to examining the contents of traffic packets, NIDS tools are also used for analyzing traffic patterns. Here, the nature, frequency, source, destination, and so on, of traffic are analyzed, rather than the contents of the traffic. Such analysis is used to detect potentially malicious traffic patterns that could indicate a network attack.

It should be noted that the above techniques depend on the existence of a knowledge base that captures information about known attack signatures and vulnerabilities. The effectiveness of NIDS is therefore closely tied to the existence of accurate and up-to-date information about such signatures.

9.6.1.2 *Host-Based Intrusion Detection.* Host-based intrusion detection, as the name indicates, is targeted at the individual hosts in the network. Host intrusion detection systems (HIDS) reside on every host and are used to analyze log files, system calls, files, data generated by system audits, and so on, to look for anomalies. HIDS tools also examine traffic that enters and leaves a host. HIDS tools can be classified into two categories:

- *Host Wrappers or Personal Firewalls*: These tools are used to examine all connections to the host and all network traffic entering or leaving the host.

• *Agent-Based Software*: Such tools use agents that are installed on every host to monitor activities on the host that could lead to vulnerabilities, such as attempts to modify specific system files, root logins, disk usage, and so on.

As in the case of NIDS, the effectiveness of HIDS is also closely tied to the existence of accurate and up-to-date information about the types of patterns and behaviors that signal intrusions.

9.6.2 Intrusion Detection for MANETs

The previous section briefly described existing approaches to intrusion detection. This section takes a look at intrusion detection in the specific context of MANETs.

9.6.2.1 *Why Is Intrusion Detection Different for MANETs?* As was briefly touched upon at the beginning of this section, intrusion detection for wireless networks is more challenging than for wireline networks, owing to the nature of the wireless medium, which cannot be physically secured from access. Ad hoc networks further complicate the picture, due to the absence of any kind of infrastructure in these networks. This means that there is no one point at which network intrusion detection can be performed, since there is no single entry/exit point for such networks. Furthermore, given the unpredictability of the network topology and the network links, it is difficult to perform any statistical analysis of the traffic in a MANET to detect anomalies.

In addition to the above, the observations made in Chapter 5 about network management for MANETs apply equally to intrusion detection. Recall that the need for a highly distributed network management solution for MANETs was described in Chapter 5; such distribution is required due to (a) the need to locate decision-making and monitoring close to the nodes being managed and (b) the scarcity of bandwidth in a MANET. For the same reasons, an intrusion detection system for MANETs must be distributed. Many intrusion detection techniques—as described above—rely on analysis of logs, audit data, and other relevant information in order to detect anomalies that signal intrusions. Sending all of these data to one central location for analysis would pose an inordinate overhead for a MANET, where bandwidth is already scarce. Thus analysis must be performed locally as far as possible; and once local analysis has been performed, any relevant diagnoses can be shared with the relevant nodes. A suitable architecture for intrusion detection is discussed in the next section.

9.6.2.2 *Intrusion Detection Architecture for MANETs.* The architecture for intrusion detection can mirror that of the network management system. In Chapter 5, an architecture for a network management system for MANETs was described that is fully distributed, where network management instances

are organized hierarchically for the purpose of information dissemination. An intrusion detection component within such a network management system provides the required local intrusion detection capabilities, as well as the infrastructure for disseminating intrusion-related information to other nodes. Information about intrusions may need to be shared with other nodes in order to perform correlation of intrusion alerts across multiple nodes. For example, an intruder may use one MANET node to attack another; and it may not be possible to completely diagnose the attack using purely local information, either at the intruder node or at the attacked node. In such cases, it is important to have a strategy for sharing information across nodes, so that a higher-level view of the network intrusions can be used to diagnose problems that involve more than one node.

Another important facet of the intrusion detection architecture is its relationship with fault management. Recall that the fault management function must correlate both hard and soft faults and diagnose the root cause of any network alarms. In addition to such faults, any intrusion-related alarms must also be correlated with fault and performance alarms, to obtain a more complete diagnosis. As an example, the fault management component may receive alarms relating to poor performance in a certain area of the MANET. At the same time, there may be an intrusion alert emanating from one of the nodes in the same neighborhood. These two events need to be correlated, because it may be the intrusion that is causing the poor performance. Thus in order to accurately diagnose the root cause of any set of alarms, it is critical that security-related alarms be correlated with fault-related and performance-related alarms. Figure 9.15 shows the architectural approach.

Figure 9.15 illustrates the flow of information and the processing that is performed by each of the FCAPS and the Policy Management components of the NMS. Security alerts are generated by the intrusion detection function within the security management component, based on traffic data collected from the network, as well as other data such as log and audit data, and so on. Similarly, performance alerts are generated by the performance management component based on performance statistics collected from the network. The performance management component generates these performance alerts when the performance statistics collected from the network cross predefined thresholds or exhibit other anomalies. These alerts are sent to the fault management component, where they are correlated with fault alerts generated by network elements. Thus, the fault management component correlates alerts from Security Management, Performance Management, and Fault Management, and it generates one or more root cause events. The root cause events are sent to Policy Management, where stored policies are triggered and executed by the Policy Decision Point. Policies are created ahead of time and are used to define responses (via policy actions) to root cause events (which act as policy events). These responses typically result in reconfiguring the network as needed to address the root cause of the problem, if indeed the problem can be fixed by reconfiguration.

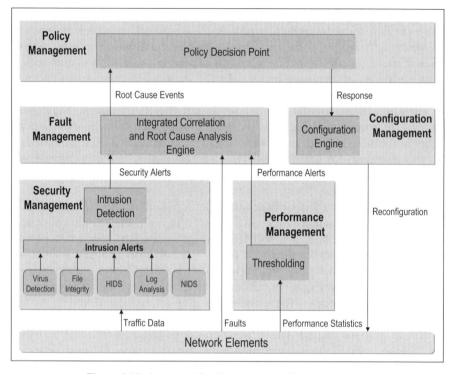

Figure 9.15. Integrated fault/intrusion detection architecture.

It may be that the root cause of the problem is a hard equipment failure, which can only be fixed by physically replacing the faulty equipment. In such cases, reconfiguration may not be necessary. However, in the case of soft failures, or in the case of certain types of network attacks, the problems can be addressed via suitable reconfiguration. An example of this was given in Chapter 7, where a DOS attack resulted in performance problems that were correlated with the attack diagnosis, and the problem was fixed by cutting off the source of the DOS attack from the rest of the network.

9.7 SUMMARY

Due to their wireless nature, MANETs are, in general, more susceptible to security violations than wireline networks, thus requiring stringent security measures to be put in place to secure wireless communications. The purpose of this chapter was to describe relevant security services and management functions for MANETs. It is important to keep in mind that many of the security concepts in place in today's wireline networks carry over to the MANET environment, including role-based access control (RBAC), cryptography, confidentiality, message integrity, and authentication. However, related services such as key management need to take into consideration the limited bandwidth available in MANET environments for key distribution and revocation.

In order to secure communications over a MANET, mechanisms such as IPSec or HAIPE need to be implemented. Furthermore, the exchange of data between different security domains (e.g., networks at different classification levels) must be carefully controlled in order to prevent undesired leakage of information from one domain to another. This is achieved by the use of cross-domain guards that are used to control the information flow between different security domains. A cross-domain guard inspects every message that traverses the boundary between different security domains in order to verify that the message conforms to a predefined set of rules about what information is allowed to cross this boundary.

Intrusion detection for MANETs is more challenging than for wireline networks, owing to the nature of the wireless medium, which cannot be physically secured from access, and due to the absence of any kind of infrastructure. This means that there is no one point at which network intrusion detection can be performed, since there is no single entry/exit point for such networks. Furthermore, given the unpredictability of the network topology and the network links, it is difficult to perform any statistical analysis of the traffic in a MANET to detect anomalies. An intrusion detection architecture for MANETs was presented that uses a distributed, hierarchical approach to intrusion detection that is integrated with fault correlation and root cause analysis.

10

CONCLUDING REMARKS

Characterized by their flexibility to be deployed "on-the-fly" and their capability to transport a wide spectrum of applications, mobile ad hoc networks (MANETs) are becoming the foundation for a variety of on-demand networks, in both the commercial and military arenas. Examples of on-demand networks in the commercial arena include networks that need to be set up in a traveling-exhibition/trade-show environment or even disaster areas, where communications need to established rapidly. Examples of on-demand networks in the military segment include the future network centric warfare (NCW) thrust, where once again there exists a critical need to deploy networks rapidly in either familiar or unfamiliar terrains. The power of MANETs stems from their flexibility—since they can be deployed without any fixed infrastructure—and from their dynamism and resilience to node failures and terrain-related obstacles.

However, the very set of features—namely flexibility, resilience, and dynamism—that make MANETs powerful also present considerable challenges for network design and management systems. While there has been a considerable amount of work and attention devoted to the development of mobile ad hoc networking technologies such as MANET routing, mobility, scheduling, and transmission techniques in the past decade, not much work currently exists in the area of MANET *management*. This can be attributed to the fact that one needs to first have a network before attempting to manage it! However, given that mobile ad hoc networking technology has gained a reasonable amount of attention and maturity, it is time to turn to investigating efficient MANET management techniques, so that the network once created can be *maintained*. This subject was the focus of this book.

The first question that one will then ask is, Why not apply the rich body of wireline network management techniques to managing MANETs? As mentioned throughout this book, mobile ad hoc networks differ significantly from

Policy-Driven Mobile Ad hoc Network Management, by Ritu Chadha and Latha Kant
Copyright © 2008 John Wiley & Sons, Inc.

their wireline counterparts. While these differences are not reiterated here, suffice it to say that these differences are fundamental in nature. For example, the functioning of the networking technologies that are used to construct the networks, the environment in which the networks have to operate, and the capabilities of and expectations from the two types of networks, namely, MANETs and wireline networks, differ fundamentally. Therefore, traditional network management techniques cannot be applied without significant enhancements to MANET management. More specifically, there is a requirement for *adaptive* network management that can take automated and efficient management decisions in response to the uncertainty and dynamics of the underlying network.

The next section provides a brief summary of the book, followed by a listing of the salient research issues that remain to be addressed in the area of MANET management.

10.1 SUMMARY

This book describes a policy-driven management paradigm that automates and adapts its management actions in response to the network dynamics. The approach deviates from the traditionally stovepiped network management operations by integrating the key MANET management functionalities. Having omitted a discussion of Accounting Management, due to its diminished role in MANETs, the book explains why the remainder of the FCAPS functions—namely, Fault, Configuration, Performance, and Security management functions—need to be integrated to provide a coherent set of network management actions and responses. The policy-based network management paradigm is shown to provide a powerful framework for integrating the FCAPS functionalities and for providing the critically needed automation to increase the capability for lights-out operations in MANETs.

Although policy-based management has been discussed in a number of texts (e.g., Verma 2000; Strassner 2003), none of these books addressed the subject of managing MANETs. As explained above, this subject is sufficiently different from the subject of network management for static wireline networks to spawn a new field of study. In addition, what has been lacking so far in the policy area is a detailed discussion of the intricacies of using policies in realistic mobile ad hoc network environments. Research papers on this subject typically discuss point solutions for management problems such as QoS management, without looking holistically at the problem of network management. This is one of the key contributions of this book.

Another important feature of this book is that it pulls together various types of information that are important for understanding how to manage MANETs. Wherever necessary, relevant related work is presented and explained in detail so that the reader can grasp the concepts necessary to be able to develop appropriate network management tools. In particular, the concept of red–black separation for the wireless network permeates much of the MANET management environment and is a topic whose impact on network management has not been presented in any existing textbook, although a number of papers have been written on specialized topics such as QoS management for red–black networks.

10.2 PROMISING RESEARCH DIRECTIONS

As mentioned in this book, the area of network management for MANETs is relatively new, and not much prior work exists in this field. The complexity of managing MANETs unveils extremely interesting (and challenging!) research topics. Therefore, in the course of writing this book, several research areas related to MANET management that merit attention were identified, some of which are new and some of which are already actively being researched. Although these areas were discussed in this book, the presentation was by no means intended to provide a complete solution, but rather was intended to spur interest in conducting further research in these areas. Some of the open research challenges in the field of mobile ad hoc network management are listed below. Note that the pointers to new research areas provided below are intended to serve as illustrative examples to stimulate research by interested readers.

- *Control and Optimization Function (COF)*: The need for a control and optimization function (COF) to assist with "smart" configuration operations was introduced in Chapter 6. As mentioned in that chapter, there exists a plethora of configuration-related "knobs" at the various OSI layers, with each knob presenting a wide variety of choices in terms of its possible values. In order to realize an efficient network, the various knobs have to be set in the "best" possible manner. However, the complexity of MANETs and the huge number of choices of values for configuration parameters underscores the need for a systematic set of techniques based on formal mechanisms to accomplish this task. A set of techniques based on Bayesian models and the use of an objective function to represent network performance were described. Based on formal optimization and learning techniques, the described COF can go a long way in realizing a truly adaptive system that has the ability to optimize system performance. However, while a framework was provided for such a COF, further research is needed in the area of the efficient optimization and learning algorithms that can be implemented via such a COF.
- *MANET Root Cause Analysis and Policy-Driven Self-Healing*: An extremely important and challenging aspect of fault management in MANETs is fault diagnosis and automation of self-healing actions. The need for fast and efficient fault diagnosis and root cause analysis (RCA)

algorithms that can pinpoint the root cause of an observed problem is critical. The algorithms must be fast in order to cope with the dynamic nature of MANETs, and they must be efficient so that they do not introduce unnecessary overheads in the already bandwidth-constrained MANET environment. While these are significant challenges in themselves, the challenges are compounded because many of the fault symptoms can get lost in a MANET, due to the high-loss environment resulting from wireless links. Thus the algorithms must not only be tractable, but must also be able to function in the face of missing or imperfect input information. To this end, Chapter 7 provided pointers to a large area of active research that uses probabilistic/stochastic approaches (e.g., Bayesian-based inferencing techniques) to assist with root cause analysis. While such stochastic approaches can accommodate and therefore function in noisy environments such as MANETs, they are notoriously intractable. For this reason, approximations and heuristics are required to make these techniques tractable. Needless to say, canonical trades exist with regard to the accuracy of the resulting solution and its tractability, or speed. This is an active area of ongoing research and is an exciting area requiring further work. Additionally, the automation of self-healing actions via policies to provide seamless service restoration is another critical aspect of fault management in MANETs. While this book provided a few illustrative examples to highlight both the importance of and difficulty associated with this critical task, the area of policy-driven self-healing in MANETs is another area for potential future research.

· *Quality of Service*: There is a critical need to provide QoS assurances to the applications in a MANET. Since MANETs operate under extremely tight bandwidth constraints, scarcity of bandwidth is a known problem that has to be addressed by prioritizing the traffic flows that are granted quality of service. However, as discussed in Chapter 8, providing QoS assurances in the dynamic MANET environment is extremely challenging, with the challenges only exacerbated by the red–black separation, since this separation results in insufficient information being available about the state of the black network. This lack of information makes it difficult to obtain the information required to support decision-making to be able to provide the required level of QoS guarantees. Thus, novel QoS algorithms and techniques are required that can provide the critically needed service assurances in MANETs. In this book, one possible approach based on dynamic throughput measurements was described. While this approach has been shown to hold potential and result in reasonable service guarantees, this is only the tip of the iceberg. Several challenges still persist; for example, how well would such an approach work in conjunction with an approach that uses dynamic latency measurements, which are, in general, more expensive to obtain in bandwidth-constrained MANETs? Also, how do such QoS assurance mechanisms fare in the presence of a mix of applications using connection-oriented (TCP) as well as packet-oriented (UDP) transport mechanisms? These are just sample questions that were identified during the course of this book and that still require further research.

- *Security Management*: Finally, as noted in Chapter 9, the wireless and ad hoc aspects of MANETs render them more vulnerable to intrusions than their wireline and fixed wireless counterparts, thereby unveiling a plethora of challenging research topics. Efficient key management is one such topic. Recall that key management (i.e., certificate management and revocation) is vital for the smooth functioning of a MANET, but can also become very expensive in terms of bandwidth penalties. A promising approach is the use of adaptive key management based on the current congestion status of the network. As an example, if the network is highly congested, the frequency at which certification revocation information needs to be refreshed can be reduced, in turn reducing the amount of network overhead imposed by sending certificate revocation information. Network congestion information can be inferred from the Dynamic Throughput Graphs described in Chapter 8. Similarly, due to the increasing sophistication of network hackers, there is an urgent need for accurate intrusion detection techniques that can function in the vulnerable wireless ad hoc networking environment. Since intrusion detection algorithms can be very bandwidth-intensive, adaptive methods can be used to trade off the accuracy of analysis with the amount of computing power available in different MANET nodes. Further research is required in these areas to provide solutions suitable for MANETs.

REFERENCES

Aggelou G. (2004). *Mobile Ad Hoc Networks: From Wireless LANs to 4G Networks*. McGraw-Hill Professional, New York.

Anjum F. and Mouchtaris P. (2007). *Security for Wireless Ad Hoc Networks*. Wiley-Interscience, New York.

Argyriou A. and Madisetti V. (2003). Performance evaluation and optimization of SCTP in wireless ad-hoc networks. In *Proceedings of the 28th IEEE International Conference on Local Computer Networks*.

Bacchus F. and Grove A. (1995). Graphical models for preference and utility. In Proceedings of the Eleventh Conference on Uncertainty in Artificial Intelligence, pp. 3–10.

Bacchus F. and Grove A. (1996). Utility independence in qualitative decision theory. *In Principles of Knowledge Representation and Reasoning 1996*, pp. 542–552.

Bandara A. K., Lupu E. C., and Russo A. (2003). Using event calculus to formalise policy specification and analysis. *In 4th IEEE International Workshop on Policies for Distributed Systems and Networks*, Como, Italy.

Bernet Y. and Pabbati R. (2000). Application and Sub Application Identity Policy Element for Use with RSVP. IETF RFC 2872, June.

Blake S., et al. (1998). An Architecture for Differentiated Services. RFC 2475, December.

Boyle J., Cohen R., Durham D., Herzog S., Rajan R., and Sastry A. (2000). COPS usage for RSVP. IETF RFC 2749, January.

Braden R., Clark D., and Shenker S. (1994). Integrated Services in the Internet Architecture: An Overview. IETF RFC 1633, June.

Braden R., et al. (1997). Resource ReSerVation Protocol (RSVP)—Version 1: Functional Specification. IETF RFC 2205, September.

Breslau L., Jamin S., and Shenker S. (2000). Comments on the performance of measurement-based admission control algorithms. In *Proceedings of Infocom 2000*.

Caesar M. and Rexford J. (2005). BGP Routing Policies in ISP Networks. UC Berkeley Technical Report, UCB/CSD-05-1377, March.

Policy-Driven Mobile Ad hoc Network Management, by Ritu Chadha and Latha Kant
Copyright © 2008 John Wiley & Sons, Inc.

Camarillo G. (2002). *SIP De-mystified*. McGraw-Hill Publications, New York.

Case J., et al. (1990). A Simple Network Management Protocol (SNMP). IETF RFC 1157, May.

Case J., et al. (1999). Introduction to Version 3 of the Internet-Standard Network Management Framework. IETF RFC 2570, April.

CCITT (1993). The directory. *CCITT Recommendations X.500–X.521*, ISO/IEC standard 9594.

Chadha R., et al. (2003). PECAN: Policy-enabled configuration across networks. In *Proceedings of the 4th IEEE International Workshop on Policies for Distributed Systems and Networks*, Como, Italy, June.

Chadha R., et al. (2004). Policy-based mobile ad hoc network management. In *Proceedings of the IEEE 5th International Workshop on Policies for Distributed Systems and Networks*, Yorktown Heights, NY, June.

Chadha R., et al. (2005). Scalable policy management for ad hoc networks. Invited paper, 24th IEEE Military Communications Conference (MILCOM).

Chadha R. (2006a). A cautionary note about policy conflict resolution. Invited paper, 25th IEEE Military Communications Conference (MILCOM).

Chadha R. (2006b). Beyond the Hype: Policies for Military Network Operations. In *International Conference on Systems and Networks Communications (ICSNC)*, October 29–November 3, 2006, Tahiti, French Polynesia.

Chan K., et al. (2001). COPS Usage for Policy Provisioning (COPS-PR). IETF RFC 3084, March.

Chan K., et al. (2003). Differentiated Services Quality of Service Policy Information Base. IETF RFC 3317, March.

Charalambides M., et al. (2006). Dynamic policy analysis and conflict resolution for DiffServ quality of service management. In *10th IEEE/IFIP NOMS*.

Chen S., Bambos N., and Pottie G. (1994). On distributed power control for radio networks. In *Proceedings, IEEE International Conference on Communications*, New Orleans, LA, May, pp. 1281–1285.

Chen W. and Jain N. (1999). ANMP: Ad hoc network management protocol. *IEEE Journal on Selected Areas in Communications*, **17**(8):1506–1531, August.

Chiang C.-Y., Levin G., Gottlieb Y. M., Chadha R., Li S., Poylisher A., Newman S., and Lo R. (2007). On automated policy generation for mobile ad hoc networks. In *Proceedings of the IEEE 8th International Workshop on Policies for Distributed Systems and Networks*, Bologna, Italy, June.

Chlamtac I., Farago A., and Zhang H. (1997). Time-spread multiple-access (TSMA) protocols for multihop mobile radio networks. *IEEE/ACM Transactions on Networking*, **6**(5):804–812, December.

Chomicki J., Lobo J., and Naqvi S. (2003). Conflict resolution using logic programming. *IEEE Transactions on Knowledge and Data Engineering*, **15**(1):244–249.

Clausen T. and Jacquet P. (2003). Optimized Link State Routing Protocol. IETF RFC 3626, October.

Cox P. T. and Pietrzykowski T. (1992). Causes for events: Their computation and applications. In *Proceedings of the 8th International Conference on Automated Deduction (CADE)*, pp. 608–621.

Damianou N., Dulay N., Lupu E., and Sloman M. (2001). The ponder specification language. IEEE Policy 2001.

Davie B., et al. (2002). An Expedited Forwarding PHB (Per-Hop Behavior). IETF RFC 3246, March.

Diffie W. and Hellman M. E. (1976). New directions in cryptography. *IEEE Transactions on Information Theory*, **22**:644–654.

DMTF (Distributed Management Task Force), http://www.dmtf.org.

DMTF CIM Standards, http://www.dmtf.org/spec/cims.html.

Droms R. (1997). Dynamic Host Configuration Protocol. IETF RFC 2131, March.

Droms R., et al. (2003). DHCP Failover Protocol. IETF Internet Draft, draft-ietf-dhc-failover-12.txt, March.

Durham D., et al. (2000). The COPS (Common Open Policy Service) Protocol. IETF RFC 2748, January.

Enns R., ed. (2006). NETCONF Configuration Protocol. IETF RFC 4741, December.

Eriksson J., Faloutsos M., and Krishnamurthy S. (2004). Scalable Ad Hoc Routing: The Case for Dynamic Addressing. INFOCOM.

Feamster N. and Balakrishnan H. (2005). Detecting BGP configuration faults with static analysis. In *Proceedings of Network Systems Design and Implementation*, May.

Feamster N., Winick J., and Rexford J. (2004). A model of BGP routing for network engineering. In *Proceedings of ACM SIGMETRICS*, June.

Fullmer C. and Garcia-Luna-Aceves J. J. (1997). Solutions to hidden-terminal problems in wireless networks. In *Proceedings, ACM SIGCOMM '97*, Cannes, France, 14–18 September.

Gamez J. A., Moral S., and Salmeron A. (2005). *Advances in Bayesian Networks*. Springer-Verlag, Berlin.

Garey M. R. and Johnson D. S. (1979). *Computers and Intractability: A Guide to the Theory of NP-Completeness*. W. H. Freeman and Company, New York.

Ghahramani Z. (1998). Learning dynamic bayesian networks. In C. L. Giles and M. Gori, eds., *Adaptive Processing of Sequences and Data Structures*. Lecture Notes in Artificial Intelligence. Springer-Verlag, Berlin, pp. 168–197.

Glesner S. and Koller D. (1995). Constructing flexible dynamic belief networks from first-order probabilistic knowledge bases. In *Proceedings of the European Conference on Symbolic and Quantitative Approaches to Reasoning under Uncertainty (ECSQARU '95)*, Lecture Notes in Artificial Intelligence. Springer-Verlag, Berlin.

Gopal R. (2000). Layered model for supporting fault isolation and recovery. In *Proceedings of Network Operations and Management Symposium (NOMS)*, Honolulu, HI.

Grossglauser M. and Tse D. (1999). Framework for robust measurement-based admission control. *IEEE/ACM Transactions on Networking*, **7**(3):293–309.

Gupta P. and Kumar P. R. (2000). The capacity of wireless networks. *IEEE Transactions on Information Theory*, **46**:388–404.

Harkins D. and Carrel D. (1998). The Internet Key Exchange (IKE). IETF RFC 2409, November.

Hamer L.-N., et al. (2003). Session Authorization Policy Element. IETF RFC 3520, April.

Hasan M., Sugla B., and Viswanathan R. (1999). A conceptual framework for network management event correlation and filtering systems. In Sloman M., Mazumdar S., and Lupu E., eds. *Integrated Network Management VI*, IEEE Publishing, New York, pp. 233–246.

Heckerman D. and Wellman M. P. (1995). Bayesian networks. *Communications of the ACM*, **38**(3):27–30.

Heinanen J., et al. (1999). Assured Forwarding PHB Group. IETF RFC 2597, June.

Herzog S. (2000). RSVP Extensions for Policy Control. IETF RFC 2750, January.

Herzog S. (2001). Signaled Preemption Priority Policy Element. IETF RFC 3181, October.

Holland G. and Vaidya N. H. (1999). Analysis of TCP performance over mobile ad hoc networks. In *5th Annual International Conference on Mobile Computing and Networking*, pp. 219–230, August.

Housley R., Polk W., Ford W., and Solo D. (2002). Internet X.509 Public Key Infrastructure Certificate and Certificate Revocation List (CRL) Profile. IETF RFC 3280, April.

Howes T. and Smith M. (1995). The LDAP Application Program Interface. IETF RFC 1823, August.

Ilyas M. (2002). *The Handbook of Ad Hoc Wireless Networks*. CRC Press, Boca Raton, FL.

Ingber L. (1989). Very fast simulated re-annealing. *Journal of Mathematical and Computational Modeling*, **12**:967–973.

Isoyama K., et al. (2000). Policy Framework MPLS Information Model for QoS and TE. IETF Internet Draft, December, http://quimby.gnus.org/internet-drafts/draft-chadha-policy-mpls-te-01.txt.

International Telecommunication Union (1996). Principles for a Telecommunications Management Network. ITU-T Recommendation M.3010, 1996.

Jensen F. V. and Nielsen T. (2001). *Bayesian Networks and Decision Graphs*. Springer-Verlag, Berlin.

Jha S. and Hassan M. (2002). *Engineering Internet QoS*, 1st ed. Artech House, Norwood, MA.

Johnson D., Perkins C., and Arkko J. (2004). Mobility Support in IPv6. IETF RFC 3775, June.

Ju J. H. and Li V. O. K. (1998). An optimal topology-transparent scheduling method in multihop packet radio networks. *IEEE/ACM Transactions on Networking*, **6**(3):298–306.

Kakas A. C., Kowalski R. A., and Toni F. (1993). Abductive logic programming. *Journal of Logic and Computation*, **2**(6):719–770.

Kant L. (2003). Integrated Fault Performance and Configuration Management System in next generation wireless communications (3G+) Networks. In *Proceedings of the 3G Wireless Conference*, San Francisco, California.

Kant L. and Chen W. (2005). Service Survivability in Wireless Networks Via Multi-Layer Self-Healing. In *Proceedings of the IEEE Wireless Communications and Networking Conference (WCNC)*, Vol. 4, pp. 2446–2452 Baton Rouge, Louisiana.

Kant L., Sethi A. S., and Steinder M. (2002). Fault localization and self-healing mechanisms for FCS networks. In *Proceedings of the 23rd Army Science Conference (ASC)*, Florida, December (Recipient of Best Paper Award).

Katzela I. and Schwarz M. (1995). Schemes for fault identification in communication networks. *IEEE Transactions on Networking*, **3**(6):733–764.

Kaufman C., ed. (2005). *Internet key exchange (IKEv2) protocol*. IETF RFC 4306, December.

Kawadia V. and Kumar P. (2005). A cautionary perspective on cross-layer design. *IEEE Wireless Communications*, February.

Keeney R. L. and Raiffa H. (1976). *Decisions with Multiple Objectives: Preferences and Value Tradeoffs*. Wiley, New York.

Kent S. and Atkinson R. (1998). Security Architecture for the Internet Protocol. IETF RFC 2401, November.

Kent S. and Seo K. (2005). Security Architecture for the Internet Protocol. IETF RFC 4301, December.

Kerravala Z. (2004). As the Value of Enterprise Networks Escalate, So Does the Need for Configuration Management. Enterprise Computing and Networking, Yankee Group Report, January.

Kirkpatrick S., Gelatt, Jr. C. D., and Vecchi M. P. (1983). Optimization by simulated annealing. *Science*, 4598, May.

Kowalski R. and Sergot M. (1986). A logic-based calculus of events. *New Generation Computing*, **4**:67–95.

Kyasanur P., Yang X., and Vaidya N. (2005). Mesh networking protocols to exploit physical layer capabilities. In *Proceedings, IEEE WiMesh 2005*, Santa Clara, CA, September.

Laskey, K. B. (2004). MEBN: First-Order Bayesian Logic for Open World Probabilistic Reasoning. George Mason University, Department of Systems Engineering and Operations Research, http://ite.gmu.edu/ klaskey/papers/Laskey_MEBN_Logic. pdf.

Li Y. and Ephremides A. (2005). Simple rate control for fluctuating channels in ad hoc wireless networks. *IEEE Transactions on Communications*, **53**(7):1200–1209.

Liu J. and Singh S. (2001). ATCP: TCP for mobile ad hoc networks. *IEEE Journal on Selected Areas in Communications*, **19**(7):1300–1315.

Luo J., Ulukus S., and Ephremides A. (2005). Standard and quasi-standard stochastic power control algorithms. *IEEE Transactions on Information Theory*, **51**(7):2612–2624.

Lupu E. and Sloman M. (1997). Conflict Analysis for Management Policies. IFIP/IEEE ISINM.

Lupu E. and Sloman M. (1999). Conflicts in policy-based distributed systems management. *IEEE Transactions on Software Engineering*, **25**(6):852–869.

manet (2007). IETF Mobile Ad Hoc Networks Working Group (manet), http://www. ietf.org/html.charters/manet-charter.html.

Manousakis K., McAuley A., Morera R., and Baras J. (2002). Routing domain autoconfiguration for more efficient and rapidly deployable mobile networks. In *23rd Army Science Conference*, Orlando, FL, December 2–5.

Manousakis K., McAuley A., Morera, R., and Baras, J. (2005). Using multi-objective domain optimization for routing in hierarchical networks. In *International Conference on Wireless Networks, Communications and Mobile Computing 2005*, Vol. 2, June 13–16, 2005, pp. 1460–1465.

Manousakis K. and McAuley A. (2006). Active Maintenance of Hierarchical Structures in Future Army Networks. Army Science Conference.

Maughan D., Schertler M., Schneider M., and Turner J. (1998). Internet Security Association and Key Management Protocol (ISAKMP). IETF RFC 2408, November.

McFaden M., Partain D., Saperia J., and Tackabury W. (2003). Configuring Networks and Devices with Simple Network Management Protocol (SNMP). IETF RFC 3512, April.

Michalski R. S. and Tecuci G., eds. (1994). *Machine Learning: A Multistrategy Approach*, Vol. IV (*Machine Learning*). Morgan Kaufmann, San Francisco.

Mockapetris P. (1987). Domain names—Implementation and Specification. IETF RFC 1035, November.

Moffett J. D. and Sloman M. S. (1994). Policy conflict analysis in distributed system management. *Journal of Organisational Computing*, **4**(1):1–22.

Mohapatra P. and Krishnamurthy S., eds. (2004). *Ad Hoc Networks: Technologies and Protocols*, Springer, Berlin.

More J. J. (1977). *The Levenberg–Marquardt Algorithm: Implementation and Theory*, In G. A. Watson (ed.) Numerical Notes in Mathematics 630. Springer-Verlag, Berlin, pp. 105–116.

Moore B., ed. (2003). Policy Core Information Model (PCIM) Extensions. IETF RFC 3460, January.

Moore B., et al. (2001). Policy Core Information Model—Version 1 Specification. IETF RFC 3060, February.

Moore B., et al. (2004). Information Model for Describing Network Device QoS Datapath Mechanisms. IETF RFC 3670, January.

Moses T. (2005). OASIS eXtensible Access Control Markup Language (XACML) Version 2.0. OASIS Standard, Document Identifier oasis-access_control-xacml-2.0-core-spec-os, February.

Murthy C. S. R. and Manoj B. S. (2004). *Ad Hoc Wireless Networks: Architectures and Protocols*. Prentice Hall Communications Engineering and Emerging Technologies, Prentice Hall, Upper Saddle River, NJ.

Nakamoto G., Higgins L., and Richer J. (2005). Scalable HAIPE Discovery Using a DNS-like Reference Model. IEEE MILCOM 2005, Atlantic City, NJ, October.

National Communications System Technology & Standards Division (1996). Telecommunications: Glossary of Telecommunication Terms. Federal Standard 1037C, General Services Administration Information Technology Service, August.

Natu M. and Sethi A. S. (2005). Adaptive fault localization for mobile, ad-hoc battlefield networks. In *Proceedings of IEEE MILCOM*, Atlantic City, October.

netconf (2007). Network Configuration (netconf) IETF Working Group Charter, http://www.ietf.org/html.charters/netconf-charter.html.

Nichols K., et al. (1998). Definition of the Differentiated Services Field (DS Field) in the IPv4 and IPv6 Headers. IETF RFC 2474, December.

Nichols K., Jacobson V., and Zhang L. (1999). A Two-Bit Differentiated Services Architecture for the Internet. IETF RFC 2638, July.

Nygate Y. A. (1995). Event correlation using rule and object based techniques. In Proceedings of the 4th International Symposium on Integrated Network Management, pp. 278–289.

Paschke A. (2005). ECA-RuleML: An Approach Combining ECA Rules with Temporal Interval-Based KR Event/Action Logics and Transactional Update Logics. IBIS, Technische Universität München, Technical Report 11.

Paskin M. and Guestrin C. (2004). Robust probabilistic inference in distributed systems. In *Uncertainty in Artificial Intelligence (UAI)*.

Paskin M., Guestrin C., and McFadden J. (2005). A robust architecture for distributed inference in sensor networks. In *Fourth International Conference on Information Processing in Sensor Networks (IPSN'05)*, April.

Paton N. W. and Diaz O. (1999). Active database systems. *ACM Computing Surveys (CSUR)*, Vol. **31**, Issue 1, pp. 63–103, March.

Pearl J. (1988). *Probabilistic Reasoning in Intelligent Systems: Networks of Plausible Inference*. Morgan Kaufmann, San Francisco.

Peirce (1958). *Collected Papers of Charles Sanders Peirce*. Vol. 2. Hartshorn et al. (eds). Harvard University Press, Cambridge, MA.

Perkins C., ed. (2002). IP Mobility Support for IPv4. IETF RFC 3344, August.

Perkins C., et al. (2003). Ad hoc On-Demand Distance Vector (AODV) Routing. IETF RFC 3561, July.

Piper D. (1998). The Internet IP Security Domain of Interpretation for ISAKMP. IETF RFC 2407, November.

Poylisher A., Chadha R., Deb B., Littman M., and Sabata B. (2005). Adaptive dynamic server placement in MANETs. In *Proceedings of IEEE MILCOM*, Atlantic City, October.

Poylisher A., Anjum F., Kant L., and Chadha R. (2006). QoS mechanisms for opaque MANETs. In *Proceedings of IEEE MILCOM*.

Qie X. and Narain S. (2003). Using service grammar to diagnose BGP configuration errors. In *Proceedings of the 17th USENIX Conference on System Administration*, San Diego, CA, October 26–31.

Rastogi R., Breitbart Y., Garofalakis M., and Kumar, A. (2003). Optimal configuration for OSPF aggregates. *IEEE/ACM Transactions on Networking*, **11**(2).

Rentel C. and Kunz T. (2005a). On the Average-throughput Performance of Code-Based Scheduling Protocols for Wireless Ad Hoc Networks. Poster presentation, MobiHoc 2005, Urbana-Champaign, IL, May.

Rentel C. and Kunz T. (2005b). Reed–Solomon and Hermitian code-based scheduling protocols for wireless ad hoc networks. In *Proceedings AdHocNow 2005*, Cancun, Mexico, October 6–8.

Rose M. and McCloghrie K. (1990). Structure and Identification of Management Information for TCP/IP-Based Internets. IETF RFC 1157, May.

Rosen E., Viswanathan A., and Callon R. (2001). Multiprotocol Label Switching Architecture. IETF RFC 3031, January.

Sadler C. M., Kant L., and Chen W. (2005). Cross-layer Self-healing Mechanisms in Wireless Networks. In *Proceedings of the 3G Wireless conference*. San Francisco, California.

Sahita R., ed. (2003). Framework Policy Information Base. IETF RFC 3318, March.

Santi P. (2005). *Topology Control in Wireless Ad Hoc and Sensor Networks*. John Wiley & Sons, New York.

Saperia J. (2002). IETF wrangles over policy definitions. *Network Computing*, January, p. 36.

Sethi A. S., Raynaud Y., and Faure-Vincent F., eds. (1995). *Integrated Network Management IV*. Chapman and Hall, New York.

Snir Y., et al. (2003). Policy Quality of Service (QoS) Information Model. IETF RFC 3644, November.

Stallings, W (2003). *Cryptography and Network Security: Principles and Practice*, 3rd edition. Prentice Hall, Upper Saddle River, NJ.

Steinder M. and Sethi A. S. (2001). Non-deterministic diagnosis of end-to-end service failures in a multi-layer communication system. In *Proceedings of ICCCN, Scottsdale, AZ*, pp. 374–379.

Steinder M. and Sethi A. S. (2002a). End-to-end service failure diagnosis using belief networks. In *Proceedings of Network Operations and Management Symposium (NOMS)*, Florence, Italy.

Steinder M. and Sethi A. S. (2002b). Increasing robustness of fault localization through analysis of lost, spurious, and positive symptoms. In *Proceedings of IEEE Infocom*, New York.

Steinder M. and Sethi A. S. (2003). Probabilistic event-driven fault diagnosis through incremental hypothesis updating. In *Proceedings Eighth IFIP/IEEE International Symposium on Integrated Network Management*, Colorado Springs, CO, March.

Steinder M. and Sethi A. S. (2004a). Probabilistic fault diagnosis in communication systems through incremental hypothesis updating. *Computer Networks*, **45**(4): 537–562.

Steinder M. and Sethi A. S. (2004b). Non-deterministic fault localization in communication systems using belief networks. *IEEE/ACM Transactions on Networking*, **12**(5):809–822.

Steinder M. and Sethi A. S. (2004c). A survey of fault localization techniques in computer networks. *Science of Computer Programming, Special Edition on Topics in System Administration*, **53**(2):165–194.

Strassner J. and Schleimer S. (1998). Policy Framework Definition Language. IETF Internet Draft, http://www3.ietf.org/proceedings/99mar/I-D/draft-ietf-policy-framework-pfdl-00.txt, November.

Strassner J. S. (1999). *Directory Enabled Networks*. Macmillan Technical Publishing, Indianapolis, IN.

Strassner, J. (2003). *Policy-Based Network Management: Solutions for the Next Generation*. Morgan Kaufmann Series in Networking, Morgan Kaufmann, San Francisco.

Strassner J., et al. (2004). Policy Core Lightweight Directory Access Protocol (LDAP) Schema. IETF RFC 3703, February.

Takikawa M., D'Ambrosio B., and Wright E. (2002). Real-time inference with large-scale temporal bayes nets. In *Proceedings of the Eighteenth Conference on Uncertainty in Artificial Intelligence*, Edmonton, Alberta, Canada, August.

Thomson S. and Huitema C. (1995). DNS Extensions to Support IP Version 6. IETF RFC 1886, December 1995.

Toh C.-K. (1997). *Wireless ATM and Ad-Hoc Networks*. Springer, New York.

Toh C.-K. (2002). *Ad Hoc Mobile Wireless Networks: Protocols and Systems*. Prentice Hall, Upper Saddle River, NJ.

Valaee S. and Li B. (2002). Distributed call admission control for ad hoc networks. In Proceedings of the VTC'02.

Verma, D. (2000). *Policy-Based Networking: Architecture and Algorithms*. Technology Series, New Riders Publishing. Indianapolis, IN.

Wahl M., et al. (1997). Lightweight Directory Access Protocol (v3). IETF RFC 2251, December.

Waldbusser S., Saperia J., and Hongal T. (2005). Policy Based Management MIB. IETF RFC 4011, March.

Wang C. and Schwartz M. (1993). Identification of faulty links in dynamic-routed networks. *Journal of Selected Areas in Communications*, **11**(3):1449–1460.

Westerinen A., et al. (2001). Terminology for policy-based management. IETF RFC 3198, November.

White D. J. (1993). *Markov Decision Processes*. John Wiley & Sons, New York.

Wu T. (1992). *Fiber Network Service Survivability*. Artech House, Norwood, MA.

Wu T.-H. and Yoshikai N. (1997). *ATM Transport and Network Integrity*. Academic Press, New York.

Yadav S., et al. (2001). Identity Representation for RSVP. IETF RFC 3182, October.

Yankee Group (1998). IVPN Service backbones: The operations cost angle. *Data Communications Report*, **13**(19).

Yates R. D. (1995). A framework for uplink power control in cellular radio systems. *IEEE Journal on Selected Areas in Communication*, **13**(7):1341–1347.

Yavatkar R., Pendarakis D., and Guerin R. (2000). A Framework for Policy Based Admission Control. IETF RFC 2753, January.

Yemini S. A., Kliger S., Mozes E., Yemini Y., and Oshie D. (1996). High speed and robust event correlation. *IEEE Communications Magazine*, **34**(5):82–90.

Yeong W., et al. (1993). X.500 Lightweight Directory Access Protocol. IETF RFC 1487, July.

Yeong W., et al. (1995). Lightweight Directory Access Protocol. RFC 1777, March.

ACRONYMS

AC	Access Control
ACF	Admission Control Function
AD	Administrative Domain
AF	Assured Forwarding
AGC	Automatic Gain Control
AODV	Ad hoc On-demand Distance Vector
API	Application Programming Interface
APS	Automatic Protection Switching
ATCP	Ad hoc Transport Control Protocol
ATM	Asynchronous Transfer Mode
AVA	Attribute Value Assertion
BB	Bandwidth Broker
BE	Best Effort
BGP	Border Gateway Protocol
BML	Business Management Layer
BLSR	Bidirectional Line Switched Ring
CA	Certificate Authority
CCITT	Comité Consultatif International Téléphonique et Télégraphique
CDMA	Code Division Multiple Access
CERDEC	Communications Electronics Research and Development Engineering Center
CIM	Common Information Modeling
CLI	Command-Line Interface
CM	Configuration Management
CN	Common Name
COPS	Common Open Policy Service

Policy-Driven Mobile Ad hoc Network Management, by Ritu Chadha and Latha Kant
Copyright © 2008 John Wiley & Sons, Inc.

COPS-PR	Common Open Policy Service for PRovisioning
CORBA	Common Object Request Broker Architecture
CPU	Central Processing Unit
CRL	Certificate Revocation List
CSMA	Carrier Sense Multiple Access
CTA	Collaborative Technology Alliance
DAC	Discretionary Access Control
DAP	Directory Access Protocol
DARPA	Defense Advanced Research Projects Agency
DEN	Directory-Enabled Networks
DHCP	Dynamic Host Configuration Protocol
DIB	Directory Information Base
DiffServ	Differentiated Services
DISP	Directory Information Shadowing Protocol
DIT	Directory Information Tree
DLL	Data Link Layer
DLG	Dynamic Latency Graph
DMTF	Distributed Management Task Force
DN	Distinguished Name
DNS	Domain Name System
DoD	Department of Defense
DOS	Denial of Service
DPA	Domain Policy Agent
DRAMA	Dynamic Re-Addressing and Management for the Army
DSA	Directory System Agent
DSCP	Differentiated Services Code Point (or DiffServ Code Point)
DSP	Directory Service Protocol
DTG	Dynamic Throughput Graph
ECA	Event Condition Action
ED	Encryption Device
EF	Expedited Forwarding
EML	Element Management Layer
ES	Encrypted Side
FCAPS	Fault, Configuration, Accounting, Performance, Security
FCS	Future Combat System
FDMA	Frequency Division Multiple Access
FIFO	First In, First Out
FM	Fault Management
FTP	File Transfer Protocol
GSL	Gnu Systems Library
GPA	Global Policy Agent
GPS	Global Positioning System
GUI	Graphical User Interface
HAIPE	High Assurance Internet Protocol Encryptor
HIDS	Host-based Intrusion Detection System

HITL	Human in the Loop
HLI	High-speed LAN Interface
HTTP	HyperText Transfer Protocol
IA	Information Assurance
ID	Intrusion Detection
IDS	Intrusion Detection System
IEEE	Institute of Electrical & Electronics Engineers
IER	Information Exchange Requirement
IETF	Internet Engineering Task Force
IHU	Incremental Hypothesis Updating
INFOCON	Information Condition
IntServ	Integrated Services
IOS	Internetwork Operating System
IP	Internet Protocol
IPSec	IP Security
IPSO/CIPSO	Internet Protocol Searity Option/Commercial Internet Protocol Searity Option
IPv4	Internet Protocol version 4
IPv6	Internet Protocol version 6
IS-IS	Intermediate System to Intermediate System
ISO	International Standards Organization
ISP	Internet Service Provider
ITU	International Telecommunications Union
ITU-T	International Telecommunications Union Telecommunication Standardization Sector
Kb/s	Kilobits per second
LAN	Local Area Network
LDAP	Lightweight Directory Access Protocol
LPA	Local Policy Agent
LSP	Label Switched Path
M&S	Modeling and Simulation
MAC	Mandatory Access Control
MAC	Medium Access Control
MANET	Mobile Ad hoc NETwork
Mb/s	Megabits per second
MBAC	Measurement Based Admission Control
MCF	Measurement Collection Function
MD5	Message-Digest algorithm 5
MIB	Management Information Base
MIP	Mobile IP
MLPP	Multi-Level Precedence and Preemption
MLS	Multiple Levels of Security
MPE	Most Probable Explanation
MPLS	Multi-Protocol Label Switching
NE	Network Element

NEDM	NEtwork Dependency Model
NEL	Network Element Layer
NIDS	Network-based Intrusion Detection System
NML	Network Management Layer
NMS	Network Management System
NP-hard	Nondeterministic Polynomial—hard
OASIS	Organization for the Advancement of Structured Information Standards
OID	Object Identifier
OLSR	Optimized Link State Routing
OSI	Open System Interconnection
OSPF	Open Shortest Path First
OSS	Operations Support System
OTA	Over the Air
PBNM	Policy-Based Network Management
PCIM	Policy Core Information Model
PDA	Personal Digital Assistant
PDP	Policy Decision Point
PECAN	Policies using Event Condition Action Notation
PEP	Policy Enforcement Point
PHB	Per Hop Behavior
PHY	PHYsical Layer
PIB	Policy Information Base
PKI	Public Key Infrastructure
PKIX	Public Key Infrastructure X.509
PM	Performance Management
PSTN	Public Switched Telephone Network
PVC	Permanent Virtual Circuit
QoS	Quality of Service
QAF	QoS Adjustment Function
RA	Revocation Authority
RAP	Resource Allocation Protocol
RBAC	Role-Based Access Control
RCA	Root Cause Analysis
RDN	Relative Distinguished Name
RFC	Request for Comments
RIP	Routing Information Protocol
RSVP	Resource ReSerVation Protocol
SAML	Security Assertion Markup Language
SITL	Software in the Loop
SLA	Service Level Agreement
SM	Security Management
SMI	Specification of Management Information
SML	Service Management Layer
SNMP	Simple Network Management Protocol

SNMPv3	Simple Network Management Protocol version 3
SQL	Structured Query Language
SSID	Service Set Identifier
SVC	Semipermanent Virtual Circuit
TCP	Transmission Control Protocol
TDMA	Time Division Multiple Access
TMN	Telecommunications Management Network
TOS	Type of Service
TTA	Trouble Ticket Administration
UAN	Unmanned Aerial Node
UES	Un-Encrypted Side
UDP	User Datagram Protocol
URL	Uniform Resource Locator
UTC	Coordinated Universal Time
VPN	Virtual Private Network
WG	Working Group
Wi-Fi	Wireless Fidelity
WS	Window Size
XACML	eXtensible Access Control Markup Language
XML	eXtensible Markup Language

INDEX